PREFACE

In many ways, this book began when I first stepped into a college classroom in 1972. The university presented an opportunity to learn about how the world around me—our Western approaches to politics, economics, and society, as well as science and technology—had come about. I sampled a variety of courses and would eventually take my senior year departmental exams in a field called the "History of Ideas." In my junior year, at the suggestion of my advisor, Professor William Jordan, I took advantage of my Spanish language skills and wrote an undergraduate thesis on why the eighteenth-century European Enlightenment had failed to take hold in Spain. This thesis convinced me to study in the future not a nation that had declined and needed "apologists," but a society on the rise that needed people to explain what was happening there and why.

Before graduating in 1976, I was lucky to receive a fellowship from the Princeton-in-Asia Foundation to go to Japan to teach English for two years and study the Japanese language. I had never been outside the United States before. While in Japan, I did what I had done regarding Europe. I read as much as I could about the country's history, literature, economy, and society. It then occurred to me to go on to graduate school and compare Japan's own successful "enlightenment" in the nineteenth century and subsequent modernization and Westernization to the experience of Western countries—and try to understand why Japan succeeded whereas a country such as Spain had languished. I was fortunate to gain admission to an interdisciplinary doctoral program in East Asian studies.

In my second year of graduate school, at the prodding of my next advisor, Professor Albert Craig, I decided to specialize in Japanese management and business history. I then spent more than three years based at the University of Tokyo and completed a Ph.D. thesis for Harvard University on the development of the Japanese automobile industry, focusing on technology strategy and management at Nissan and Toyota. Japan in 1980 had just passed the United States to become the world's largest auto-producing nation, but there was little written on how the leading companies had cultivated their world-class manufacturing skills. In what became my first book, *The Japanese Automobile Industry* (1985), I explored this story and included comparisons to General Motors, Ford, and Chrysler. The American giants had already fallen behind the Japanese, especially Toyota, in productivity and quality, as well as the management of people, suppliers, sales, and probably everything else. At this point, I was much more in the field of operations strategy and innovation management than Japanese studies. Consequently, after completing my degree, I was fortunate again to spend two years as a postdoctoral fellow in production and operations management at the Harvard Business School. Then I joined the MIT Sloan School of Management in 1986 as a member of what is today the Technological Innovation, Entrepreneurship, and Strategic Management (TIES) Group.

Back in 1985, I decided to shift the bulk of my new research to the computer software business and the evolution of software development practices. The reason was that I considered software the next great challenge for Japan to master and a technology that, I was sure, would change the world if only we could make programming less of an art and more like science and engineering. I had been exposed to computers in college (where I learned to use an IBM 370 mainframe to do word processing on punched cards) and in graduate school (where I took short courses on computing systems and applications in order to use another IBM mainframe to write my Ph.D. thesis). While at Harvard Business School, I worked mainly with Richard Rosenbloom on VCR (video cassette recorder) development. I also started a detailed study of software factories as my individual project and finished my second book, *Japan's Software Factories* (1991), after joining MIT Sloan.

How I got from Toyota, Nissan, and Japanese software factories to Microsoft, Netscape, Intel, Apple, Google, and the like is what makes life interesting! Moreover, these firms, combined with my understanding of Japan, have shaped my observations on what makes companies successful—and not so successful—over long periods of time, especially in fast-paced and often unpredictable markets. I had carefully analyzed how big Japanese firms had cultivated world-class capabilities for automobile manufacturing and software engineering. But later I realized that the Japanese too often did not create new technologies or global platforms. They trailed the United States and Europe in understanding how to use personal computers and then the Internet, though the Japanese would emerge as leaders in using mobile phones. Back in 1992, however, I decided to "follow the money," so to speak, and visit Microsoft. It was clear that the real power in information technology had shifted from the mainframe computer makers to the PC hardware and software firms, and that Microsoft was at the heart of the new industry, along with Intel. Fortunately, people at Microsoft knew of my work because of the company's growing interest in software engineering best practices and Japanese quality control. They welcomed a visit, which I did with a master's thesis student sent to MIT from IBM, Stan Smith.

I was immediately impressed with Microsoft's approach to product development and proposed to do another book. This became *Microsoft Secrets* (1995), done with Richard Selby, a colleague in computer science, and then at the University of California, Irvine. The World Wide Web exploded on the scene in 1995, along with Netscape and Windows 95. These events prompted me to launch another study on the rise of Netscape, the challenge the Internet posed to Microsoft, and how "Internet time" was affecting strategy, planning, and product development. The result was *Competing on Internet Time* (1998), done with David Yoffie, a colleague at the Harvard Business School. At this time I also published a book on leading-edge practices in product development in the automobile industry, *Thinking Beyond Lean* (1998), with Kentaro Nobeoka. He is one of my former MBA and doctoral students who had worked as a product engineer at Mazda and is now at Hitotsubashi University. The rising importance of Windows and other platforms in information technology, also apparent by the late 1990s, inspired me and another doctoral student, Annabelle Gawer, to write *Platform Leadership* (2002). She is now at Imperial College Business School in London. The collapse of the Internet bubble, high-technology commoditization, and the shift of value from products to services provided the setting for *The Business of Software* (2004) and the themes fleshed out in this book. I also continued the work on products and services with Fernando Suarez, now at Boston University, and Stephen Kahl, now at the University of Chicago. Both are also former doctoral students of mine at MIT Sloan.

In December 2005, another fortunate event occurred: I received an invitation to deliver the 13th Annual Clarendon Lectures in Management Studies at the University of Oxford in 2009 and, as part of the invitation, to write a book based on these lectures. After consulting with members of the awards committee, including Professor Mari Sako from Oxford's Saïd School of Business and David Musson from Oxford University Press, we decided that an overview of my research would be the best book to write. This advance warning gave me ample time (as intended) to think about how to describe what I have learned in my career. But I like to think that this book has become much more than an overview; my focus has become the *principles* or "big ideas" that create staying power and superior performance. It took a couple of years of pondering, and a lot of thinking about how firms such as Toyota and Microsoft, as well as JVC, Sony, Apple, Intel, IBM, Google, Qualcomm, Adobe, and others were similar and different. The focus of the book became clearer, though, once I realized that, in college and as an academic, and while working with dozens of companies, I still have been concerned with "enlightenment" and understanding best practices in management.

How well I have succeeded is a different matter, but my challenge for the Clarendon Lectures has been to describe what I have learned. The resulting book is not a comprehensive review of the most valuable things we know about strategy and innovation, or competitive advantage more generally. The thousands of articles and books written on these subjects would require several lifetimes to synthesize. I have simply organized this book around a handful of ideas that have been central to my research and that of my co-authors. Perhaps most importantly, I have stated these as principles that, in my experience, should have enormous value for managers *and* that academic researchers in a variety of disciplines have subjected to decades of rigorous empirical and theoretical scrutiny.

To gain some perspective for how to go about this task, I also drew on another student of history, Theodore White. He is probably best known for the 1962 Pulitzer-prize winning chronicle of John F. Kennedy's presidential candidacy. But White is also the author of a fascinating account of his own life and work, beginning with college and then his years covering China as a journalist during the 1940s and 1950s. Published in 1978, *In Search of History: A Personal Adventure* is a book I still find inspiring.

Michael A. Cusumano

Cambridge and Groton, Massachusetts
April 2010

CONTENTS

ACKNOWLEDGMENTS

As might be expected for a wide-ranging study such as this, many people and institutions made this book possible. First, I need to acknowledge my co-authors of the various research projects: Annabelle Gawer of Imperial College Business School and Richard Rosenbloom of Harvard Business School, Emeritus (as well as Yiorgos Mylonadis of London Business School), for the Chapter 1 discussion of technology platforms in computing and video recording; Fernando Suarez of Boston University School of Management and Stephen Kahl of the University of Chicago Booth School of Business for the Chapter 2 discussion of products and services; Richard Rosenbloom (again) and Richard Selby of Northrup-Grumman Corporation and the University of Southern California for the Chapter 3 discussion of capabilities at Japanese VCR producers and at Microsoft in PC software; Richard Selby (again) for the Chapter 4 discussion of pull concepts in iterative product development at Microsoft and other companies; Kentaro Nobeoka of Hitotsubashi University for the Chapter 5 discussion of scope economies in automobile product development; and Fernando Suarez (again) and Charles Fine of MIT Sloan, as well as David Yoffie of Harvard Business School, for the Chapter 6 discussion of flexibility in manufacturing and in strategy and entrepreneurship at Netscape. Bill Crandall (formerly of Hewlett-Packard), Chris Kemerer of the University of Pittsburgh Katz Graduate School of Business, and Alan MacCormack of MIT Sloan (formerly of Harvard) joined in the research on software development discussed in Chapters 4 and 6. Bill, with an initial inspiration from Alan, also took the lead in creating the graphic in the Introduction that maps out the six principles.

Second, I need to thank the Clarendon Lectures in Management Studies Selection Committee at the University of Oxford (Professors

Anthony Hopwood, Mari Sako, Colin Mayer, Tim Morris, Richard Whittington, Paul Willman, and Steve Woolgar of the Saïd Business School, and David Musson of Oxford University Press) for granting me this acknowledgment back in 2005. They required that I deliver three lectures sometime in 2009 and then a book within a year afterwards. The timing was fortunate. I had a sabbatical due from the MIT Sloan School of Management in 2008–9, for which I thank our dean, David Schmittlein. I was fortunate as well to receive an MIT–Balliol College Fellowship that enabled me to spend three highly enjoyable and productive months in Oxford during the spring of 2009. We have nothing at MIT that resembles High Table or even the daily lunches at Balliol College! My thanks to Dean Deborah Fitzgerald of the MIT School of Humanities, Arts, and Social Sciences, and to Balliol College, which jointly administer the fellowship. I also thank William (Bill) Dutton, Director of the Oxford Internet Institute, for being my Balliol College host, and Mari Sako, my host at the Saïd Business School. The collegial environment at Balliol and Saïd were perfect for writing a good chunk of the book, as well as for preparing the lectures. Annabelle Gawer hosted me occasionally at the Imperial College Business School in London, where I must thank the director, David Begg, and David Gann for a longer-term visiting professor appointment. My appreciation as well to Naren Patni, chairman of Patni Computer Systems, for his interview on the history of the company and for arranging a sabbatical office in our Cambridge, Massachusetts, headquarters.

Third, I need to thank my readers of the manuscript drafts. Their comments were invaluable, and I tried to address all their major comments and suggestions. My strategy was first to line up academic colleagues and graduate students to do a first pass at the manuscript or specific chapters. I wanted to make sure I got the ideas, literature, and arguments "right" from the perspectives of research and theory. After incorporating the suggestions from academics, I then sent version 2 of the manuscript to several managers I felt I could impose upon to do another careful reading (mostly former MIT Sloan students of mine). As I reacted to their feedback, the manuscript went through three more revisions. These "builds" attempted to "synchronize and stabilize" the many changes, including a new ordering of the chapters and a new title (suggested by Bettina Hein). My editor at Oxford University Press,

David Musson, also provided important suggestions on the general tone of the manuscript, the examples, and the conclusion. Most important, he believed with me that I could write this book for a general management audience. He was also the first to suggest (at the time of the Clarendon Lectures) that I should start the book with the chapters on platforms and services.

As for my academic readers, I owe the greatest debt—in alphabetical order—to Michael Bikard of MIT Sloan (who happened to be reading for his doctoral general exams), Mel Horwitch of the Polytechnic Institute of New York University, and Michael Scott Morton of MIT Sloan, Emeritus. They read the entire manuscript with great care and provided much needed insights for improvement as well as important inspiration to continue. Michael B. offered seventeen pages of detailed comments, most of which I used! I also greatly valued the ongoing conversations with Michael S. M. about the progress of the manuscript. Equally enjoyable was the chance to reconnect with Mel and bring his always engaging point of view into the revisions. Other colleagues and doctoral students provided feedback on specific chapters. David Yoffie critiqued the capabilities chapter and prevented me from going too far astray. Current doctoral student Phil Anderson of MIT Sloan made several useful comments on the capabilities chapter as well. And I thank Akira Takeishi of Kyoto University for encouraging me to stay focused on capabilities, which he argued have been a central theme throughout much of my research. Annabelle Gawer critiqued (and sometimes corrected!) the platforms chapter. Kentaro Nobeoka assured me that I got the multi-project management story right in the economies of scope chapter. He also produced the graphics in this chapter for our 1998 book.

As for my manager readers, I owe the greatest debt—again, in alphabetical order—to Bill Crandall, Andreas Goeldi of Buzzient.com, and Bettina Hein of Pixability.com. They not only read the entire manuscript but provided pages of meticulous as well as "tough" comments. Their help enabled me to turn what started out as a fairly academic exercise into something that I hope is much more useful for business readers. All three are former MIT Sloan students, and I supervised their master's theses. I hoped they would take the opportunity to turn the tables and critique my work, and I was right!

Bill, a former Leaders for Manufacturing Fellow, is also a co-author of several articles on software engineering. So I knew what to expect—nothing gets past him. I also had Andreas and Bettina in two classes as Sloan Fellows in addition to their theses, and knew they would push me to provide deeper, broader, and clearer insights for managers. I should add that Ben Slivka (formerly of Microsoft), Bob Spence (of 8over8), and my nephew Robert Cusumano (of Continuity Control) provided some general comments on the manuscript structure and style, and helped with the title debate.

Fourth, I benefited from many comments at seminars and lectures where I presented early versions of this work, usually under the title "In Search of Best Practice." I first experimented with the enduring ideas framework for MIT Sloan Executive Education sessions during 2006–8 done for Chinese managers visiting MIT through Tsinghua University. I also delivered the first detailed version of the lectures at the University of Auckland Business School in New Zealand in February 2009. My thanks to Tom McLeod of the Foundation for Research on Science and Technology (FRST) for arranging the lecture and Hugh Whittaker for serving as my faculty host. I benefited as well from discussions with Mitch Olsen of Smallworlds.com in New Zealand on the role of management vision in relation to strategy and capabilities. As planned, I delivered the three Clarendon Lectures at the Saïd Business School in May 2009 and need to thank the many people who asked questions and made comments. In particular, I recall Colin Mayer, Dean of Saïd, nudging me after the opening lecture to explain how my six principles differed from other lists of best practices. I also got to spend some valuable time with Paul David of Stanford and Oxford, Emeritus, whose early work on standards and technological diffusion I found inspiring years ago. Nick Oliver, a former colleague from IMVP and now director of the University of Edinburgh School of Business, provided useful suggestions on the Toyota–Microsoft comparison when I delivered a seminar in Edinburgh in May 2009, co-hosted by the Advanced Institute of Management Research (AIM). My thanks to Neil Pollock for being my faculty host. I also benefited from comments and encouragement from my MIT Sloan colleagues at the Technological Innovation, Entrepreneurship, and Strategic Management (TIES) seminar in September 2009 and the MIT Center for Digital Business seminar in October 2009. My thanks in particular (in alphabetical order) to Howard Anderson, Pierre Azoulay, Jason Davis, Aleksandra (Olenka) Kacperczyk, Matt Marx, and Pai Ling Yin, as well as other faculty, students, and visitors who attended the talks. I received useful feedback as well from attendees at the 24th International Forum on COCOMO and Systems/Software Cost Modeling, where I gave a keynote address. Valarie Kniss, my assistant at MIT Sloan, tracked down articles and prepared the bibliography. The MIT Center for Digital Business as well as Hewlett-Packard provided some early seed funds for the products versus services business models research. I must also acknowledge my copy-editor for Oxford University Press, Hilary Walford, for her suggestions and patience with my last-minute revisions.

Finally, my wife, Xiaohua Yang, who has both an academic (Ph.D. in global political economy) and a business background (from financial services and Internet marketing to dealing in Chinese art), read the first version of the manuscript, as well as some late revisions involving Toyota. She encouraged me to make the arguments as clear as possible. She also joined me for the final month at Oxford and some travel in the UK, helped decide on the title, and generously tolerated my devotion to finishing the book.

Introduction: The Principles of "Staying Power"

Anyone who has ever read a book about "excellent" companies and managers, or even invested in the stock market, has probably thought about what makes some firms and managers consistently better than others. Is it simply timing and good luck? The ability to attract talent? Closeness to customers? Or distance from customers—the foresight to look ahead of the current market? Perhaps it is simply the good fortune to avoid mistakes. There seem to be countless books and articles on how to become a superior manager, build a great company, and sustain your competitive advantage. But, whilst some firms do outperform their peers for years and decades, this is a rare accomplishment. Staying power is particularly difficult when managers confront disruptive change in technology and customers' needs—factors largely outside their control, except for how they may respond. Permanent competitive advantage probably does not even exist—that is, across generations of technologies and customers. Firms that continue to do well usually need to continue reinventing themselves.

The greatest threat to managers, then, may well be inside the firm. Years of success have the potential to breed complacency or arrogance. Both can plant the seeds of decline and make it more difficult to foresee and respond to change. In fact, steady growth and profitability are likely to encourage firms to become more bureaucratic and less attentive to detail and innovation as they transition to managing a larger number of people, more complex products and services, and a bigger scale of operations. And the more competent managers become at doing certain things well, the more difficult it is for them to think "outside the box" or recognize when they are losing their edge. This admittedly difficult challenge—to identify fundamental principles of management that may help managers create competitive advantage and staying power for the firm—is the primary purpose of this book.

Surely one obstacle to sustaining any market position is that managers find it difficult to anticipate change or adapt quickly to change. Whatever "secrets of success" or "best practices" that a firm has mastered for one point in time, these advantages are likely to become obsolete or less effective as conditions and technologies evolve and as competitors improve what they are doing. For example, we once thought that economies of scale and automation were most important in volume manufacturing businesses, and we looked to giant companies such as General Motors and Ford to set the standard. Ford pioneered automobile mass production but in the 1920s was unable to adapt to consumer demands for more product variety. GM offered more product variety and took over as the new dominant firm in the global automobile industry, a position it held for more than half a century. GM then began to struggle in the 1970s as competition from Japan and Europe became more intense.

The Japanese became the new world leaders in the automobile industry from 1980, led by Toyota. This company, which dates back to the 1930s, reinvented the process of automobile mass production in the 1940s and 1950s and gradually established new standards for quality and productivity for the mass market. Due in no small part to Toyota and other Japanese competition, GM and Chrysler both went bankrupt in 2009. GM's fall allowed Toyota—widely touted since the 1990s as perhaps the finest manufacturing company the world has ever seen—to become the largest automobile producer. Then, in late 2009 and early 2010, we learn that Toyota, distracted by the quest to become number one, has been experiencing serious quality and safety problems for more than a decade. Toyota executives had trouble comprehending the seriousness of these complaints. They kept most of them quiet until a series of fatal accidents and government inquiries led to the recall of over 9 million vehicles worldwide by April 2010. The defective vehicles ranged from the Toyota Corolla and Camry passenger cars to the iconic Prius hybrid and even some Lexus luxury models. For Toyota executives, as well as long-term observers and customers of Toyota (including this writer), the world in 2010 seemed to turn upside down.

The number of fatalities involving Toyota vehicles was relatively small compared to other recalls in the automobile industry. Nonetheless, for Toyota products to have any major defects, and for the company to fail to address the problems immediately, shocked customers, government officials, industry analysts, and competitors alike. It is important to realize that Toyota did not slip up in its basic approach to manufacturing, commonly known as the "lean" or "just-in-time" production system. I will have more to say about these techniques as well as Toyota's approach to product development later in this book. More fundamentally, the issues that at least temporarily knocked Toyota from its pedestal are more subtle. But they provide important lessons for any manager thinking about competitive advantage and staying power.

On a technical level, Toyota's problems in 2010 had more to do with design and testing flaws than with manufacturing, production management, or process quality control, which Toyota has mastered like no other company. It is more a lapse in design and testing when even a minuscule number of accelerator pedals get stuck on loose floor mats, or pedal materials become sticky when exposed to moisture and friction. It is also a lapse in design and testing when engineers decide to use problematic "drive-by-wire" software and sensor devices to control braking, accelerating, cruise control, and vehicle stability. The conditions that lead these complex systems to malfunction on rare occasions, even momentarily, have been extraordinarily difficult for Toyota and government safety investigators to replicate in a laboratory setting. A few accidents seem due to driver error. But, an early report found that, between 1999 and early 2010, at least 2,262 Toyota vehicles in the United States experienced unintended rapid acceleration and were associated with 815 accidents and 19 deaths.[1] Later reports cited 50 to 100 deaths. Since the mid-1990s, Toyota has also encountered other quality and safety problems that it mostly kept out of the headlines—such as dangerous corrosion in the frames of Tacoma pickup trucks sold in North America between 1995 and 2000, apparently due to improper antirust treatment. Toyota did not recall these trucks, but silently bought them back from consumers.[2] Some Tundra pickups also suffered from frame corrosion. And there are other complaints about the driving mechanisms in some Corolla, Camry and Lexus models. So there is a pattern here, as well as a problem.

In some cases, Toyota may trace the cause to overly rapid expansion of production and parts procurement outside Japan, and the transition to new components and materials that it did not—apparently—test thoroughly enough. Toyota used to manufacture new models in Japan first for a couple of years, using carefully tested Japanese parts, before moving production of only its best highest-volume models to overseas factories. In the last couple years, however, it has ramped up global production of new models with new suppliers more quickly. In other cases, such as the hybrid models, Toyota may trace the cause to potential deficiencies in how the complex new software systems interact or respond to different conditions, such as bumpy or slippery roads. Beyond the surface, however, the root cause of Toyota's recent problems seems much more internal, and fundamental. It has everything to do with senior management and their increasing—though probably temporary—lack of attention to the business and the technology. It is almost as if management believed their philosophy of continuous improvement (called *kaizen* in Japanese) had already brought Toyota products to the point of perfection and there was nothing more to worry about, even as the technology and suppliers changed and the components became more complex.

During the 2000s, Toyota became a leader in introducing electronics to the mass-market automobile. The new technologies replace older mechanical systems, such as for braking and accelerating. They offer new capabilities, such as for dynamic (real-time) stability control or recharging batteries whilst braking in the gas-electric Prius and other hybrids. Ford in 2010 also recalled several thousand of its Ford Fusion and Mercury Milan hybrid vehicles for a software-related braking problem similar to that in the Prius, so Toyota is not unique in struggling with new technology.[3] But no one expected these kinds of quality and safety problems from the world's best manufacturing company. In hindsight, we can say that Toyota is introducing space-age technologies to control its vehicles. But, apparently, the company has yet to introduce space-age design and testing capabilities. We see these in companies building mission-critical aerospace, defense, or nuclear-power-generation systems, although they have had their share of disasters, too—aircraft, spacecraft, and power plants that fall out of the sky or explode because of faulty designs and materials, or inadequate testing under adverse conditions. Perfection is hard to achieve in a human world. But automakers, if they use space-age technologies, also have the obligation to find cost-effective ways to deploy these technologies as well as countermeasures like system redundancy, fault-tolerance, and creative "stress testing." Again, some systems will still fail, but customers expect companies building products for the mass market to eliminate repeated failures and make catastrophes into the rarest of events. The Toyota way used to be that one defect was one too many. That is the kind of thinking that Toyota must regain.

For Toyota to rebuild customer confidence will probably take a few years. Companies in this degree of trouble must face problems head on and get to the root causes as quickly as possible. Instead, in February 2010, the world saw Toyota executives looking like deer stuck in the headlights of an oncoming bus. CEO Akio Toyoda, grandson of Kiichiro Toyoda, the company founder, and other executives, were clearly unprepared to explain quality and safety problems—to themselves or to the global media. Rather than basking in the glory of their market success, Toyota made its difficulties worse by not paying adequate attention to complaints going back to the 1990s. What is worse, the company responded much too slowly. It allowed bad news to leak out in drips and drabs ("death by a thousand cuts," as they used to say in medieval Japan), whilst continuing to deny there was a real problem.

Toyota engineers will solve the technical flaws in their products, however complex, even if they have to go back and design simpler systems. Toyota the company will survive this unfortunate period and thrive again. But the best outcome will be for Toyota managers to reflect deeply on what has happened and re-create an even stronger company for this new age of high technology and software. They should become better able to handle adversity and change because they now know what failure looks like.

The Toyota debacle also reinforces a major argument in this book: we can learn a lot about best practices and enduring principles of management from looking at exemplar firms. But our observations must also rise above any particular firm. Every company and every market will experience ups and downs. Even the best firms are subject to lapses and may encounter disasters due to chance or mistakes of their own making. Moreover, the practices that lead a company to become number one may be vastly different from the skills and mindset needed to stay there. This seems especially true if, indeed, long-term success does breed complacency, arrogance, or inward-thinking, and thus the seeds of eventual decline.

Many companies have survived product disasters as well as deterioration in their businesses because of radical changes in markets and technologies. In terms of technical calamities, despite its later financial problems, GM rebounded from selling the unsafe Chevrolet Corvair in the 1960s. This unstable product (primarily because of a design that required significantly different front and rear tire pressures to accommodate the powerful rear engine and a suspension system that lacked an anti-sway bar) inspired Ralph Nader to write *Unsafe at Any Speed* in 1965.[4] The book also led to the creation of the National Highway Traffic Safety Commission, which has responsibility for investigating the Toyota problem. Ford had a major recall in the 1970s with the Ford Pinto, which had a gas tank directly behind a small bumper and tended to explode when hit in the rear. Audi survived a major problem with unintended acceleration in the 1980s, probably caused by driver error because the gas and brake pedals were located too close to each other. Johnson & Johnson overcame tampering with its Tylenol products, also in the early 1980s, and quickly became the model for crisis management (though it has had several ethical lapses since that time). Ford had a more massive recall in 2000 of some thirteen million faulty tires made by Firestone and fitted on its Explorer SUVs, reportedly resulting in over 250 deaths and 3,000 catastrophic injuries.[5]

In terms of technological change and business decline, IBM has waxed and waned in significance as mainframes fell out of favor with the rise of PCs and distributed computing. It barely survived huge losses and layoffs between 1991 and 1993. But, though IBM gave way to Microsoft and Intel during the 1980s and 1990s in technical leadership and market value, it regained considerable influence. It is once again a highly profitable firm, now focused on software and services. As the industry shifted once more to the Internet and then mobile computing in the 1990s and 2000s, Microsoft and Intel have had to share their markets and customer attention with upstarts such as Google, Nokia, Qualcomm, and Salesforce.com, as well as that persistent innovator, Apple. Microsoft, in particular, despite its enormous ongoing sales and profits, has been slow to adapt to technological change (graphical computing, the Internet, mobile devices, Internet services) as well as slow to fix quality and security problems in its flagship Windows platform. It has also encountered rising bureaucracy and employee dissatisfaction with the sluggish pace of innovation and political turf-fighting associated with the multi-billion-dollar Windows and Office franchises.[6] On the global front, China and India now rival or surpass the United States, Japan, and Germany in critical industries. Japan stands out for struggling with recession and deflation since its real-estate and stockmarket bubbles burst way back in 1989. It remains an enormously wealthy country but has yet to recover the growth levels and vibrancy that once led us to talk about "Japan, Inc." and "Japan as Number One."

With so much unpredictable change, managers must surely wonder which practices, firms, industries, and geographies will dominate two or three decades from now. Who and what have staying power—or not? The simple answer is that the future is uncertain and therefore unknowable, so be prepared for anything. But that is not a very useful response for managers. Instead, I argue in this book that it is possible to identify a few fundamental principles that can stand the tests of time and help managers overcome both internal and external challenges to their business. I believe these principles will prove especially effective in markets subject to rapid and unpredictable change, even if the companies that pioneered the practices I cite have stumbled somewhere along the way or end up being surpassed by their competitors. To construct these principles, I have looked back on thirty years of my research and personal experiences as a teacher as well as a consultant, director, or advisor for some 100 firms across the United States, Japan, Europe, China, India, and elsewhere, small and large, high tech and low tech. Who should read this book? Anyone interested in what it takes for organizations to perform well over long periods of time, especially as the world around us changes in unpredictable ways.

Innovation and Commoditization

Toyota's recall troubles at least in part reflect a broader challenge facing many firms today: *we have entered an age of simultaneous innovation and commoditization.* Customers around the globe continually demand new types of products and services, sometimes of great technical sophistication. But too often competition drives down prices to unprofitable levels. In some cases, customers want to pay no more than what Google charges for searching the Internet: *nothing.* As a result, many companies struggle to offer increasingly complex products and services at low and often declining prices. The trends are there even in conventional manufacturing such as automobiles, where an exemplar firm, Toyota, has reduced prices to stimulate sales, encountered a series of quality and safety problems, and announced its first ever operating loss in 2009.

The impact of innovation and commoditization as a dual phenomenon is especially clear in businesses dominated by digital technology. Consumers now realize that the marginal cost of reproducing a software program or a text, video, and music file, or sending a telephone call, photo, or text message over the Internet, is essentially zero. Yet much of the value in today's high-technology world, as well as in information products ranging from books to newspapers, magazines, music, and movies, comes in the form of these easily reproducible bits of digital content. Often firms distribute this content at minimal cost through computer servers and "online automated services." How can managers in a wide variety of industries adapt to these changes and still make enough money for their companies to survive and thrive? That is the dilemma.

For businesses dependent on electronic components and software, this trend of commoditization and innovation dates back at least to the 1960s. It accelerated with the increasing application of transistors and then programmable microprocessors. With the dawn of the PC era from the mid-1970s, the "price-performance" of computers and other programmable devices fell rapidly, even as reliability, functional sophistication, and ease of use increased. Many software products, such as for large organizations, held their value until recently. But then the rise of free (free as in "free beer") and open-source software distributed through the Internet caused many of these prices to collapse. The bursting of the Internet bubble in 2000 and then the worldwide recession of 2008–9 merely exacerbated long-term trends. These changes threaten the very existence of many firms, whole industries, and entire professions.

Globalization is an increasing part of this story. In the 1960s and 1970s, dozens of Japanese firms entered manufacturing and high-technology markets and drove down prices whilst gradually offering more sophisticated products. Many American and European firms exited from these markets or struggled to reinvent themselves. Korean firms became major players in the 1980s and 1990s in sectors such as steel and shipbuilding and then semiconductors, consumer electronics, and automobiles. More recently, Chinese and Indian companies have emerged to compete in key segments of manufacturing and high technology. The Chinese in 2010 can make nearly everything, including computers, telecommunications equipment, machine tools, and automobiles. China also ranks as the third largest automobile-producing nation (behind Japan and the United States) and is the world's largest motor vehicle-consuming market (passing the United States in 2009). In addition, China Mobile has become the world's largest cell-phone services provider. It is still true that Chinese firms do not invent very much. But, in manufacturing, China's costs, or, more precisely, its prices, have become the world's prices.

At the same time, Indian firms have learned how to deliver many high-technology services, initially starting with back-office operations but now extending to custom software development and complex research and product engineering work. India is also becoming a major automobile producer. The possibility of disruptive competition increased after India's Tata Motors shipped its first $2,500 Nano car in July 2009. It is only a matter of time before these kinds of innovative but inexpensive products from developing countries reach first-world markets. India's costs, or its prices, have become the world's prices for many sophisticated professional services and a growing number of manufactured goods.

It is no surprise that managers in the most advanced economies—the United States, the United Kingdom, Western Europe, Japan, Australia and New Zealand, Korea, and elsewhere—feel threatened. They puzzle over how to compete with these new global players and continue innovating, especially in markets where prices can fall dramatically, sometimes to zero. Moreover, as we have seen in the case of Toyota in 2010, compromises or short cuts in quality can lead to disasters and global attention for all the wrong reasons.

The Six Principles and Plan of the Book

The best way to survive and thrive in such an uncertain, competitive world—particularly in an age of technological disruptions as well as global economic catastrophes—is to understand how best to prepare firms to perform well over years and decades. This means distinguishing short-term fads in management thinking from more lasting principles that can help managers create and recreate value for customers. But, to do so, organizations must be able to respond quickly and effectively to change—such as when dealing with new quality problems, rapidly declining prices, or subtle but steady shifts in value from hardware to software and services.

My research has mainly looked at how managers can balance efficiency and flexibility in strategy, operations, and product development. In reflecting on what I have learned, I concluded that a handful of principles—I have chosen six—appear to have been essential to the effective management of strategy and innovation over long periods of time. Other authors have discussed each of the underlying concepts before me; in fact, as noted in the Preface, I have focused on principles supported by considerable theoretical and empirical research undertaken by a variety of scholars in different disciplines. Nor do I try to cover all or even most areas in management. What I offer is a selective list, with examples mainly from the automobile, computer software, telecommunications, consumer electronics, and Internet service industries. But, because of their generality, I am convinced that these principles provide essential lessons for managers in nearly any industry.

The first two principles deal with platforms and services, especially for product firms. The ideas are relatively new or at least understudied in the management field. But they are increasingly on the minds of managers and gaining attention fast in business schools. Most importantly, platforms and services take us far beyond conventional modes of thinking about product strategy, innovation, and the usual ways of making money. My argument is that many firms today must develop a whole new set of skills and relationships with partners that offer complementary products and services. These complements can make a platform increasingly valuable to customers and other "ecosystem" participants (such as advertisers or content providers). Modern platforms and software-based services usually depend heavily on information technology. But we can also see these two principles in sectors not usually associated with the high-tech economy, such as the automobile industry.

The next four principles deal with capabilities, "pull" concepts, scope economies, and flexibility. These ideas have a longer history in management practice and research. All describe some aspect of "agility"—a term similar to flexibility but with strong connotations of quickness. My argument is that agility is essential for firms to adapt or respond fast to unpredictable and rapid change in technology, markets, or competitive conditions. We also see the term commonly used in product development as a contrast to the slow, sequential, and inflexible "waterfall-ish" way of doing things.[7] These four principles have been around sufficiently long to become part of *standard* best-practice thinking about strategy and innovation management, with different degrees of emphasis depending on the firm and the market context. I also discuss how these four principles take on new meaning when applied to platform strategies and new types of services.

Figure 1 illustrates the six principles, points to some examples, and suggests how they contrast with an older or narrower view of competitive advantage. In the narrow way of thinking, managers focus on strategy and products, and support these through push-style production systems and a preoccupation with scale economies and efficiency. By contrast, the six principles highlight the building blocks for a new type of firm and a broader way to create and re-create competitive advantage. The focus here is on deep capabilities, not just strategy, and on achieving the agility necessary to adapt to change. Moreover, in many markets today, the most successful firms compete on the basis of industry-wide platforms or complements, not just products, at what we can call the ecosystem level. We also see many products turning into services as well as services turning into products. In this new style of competition,

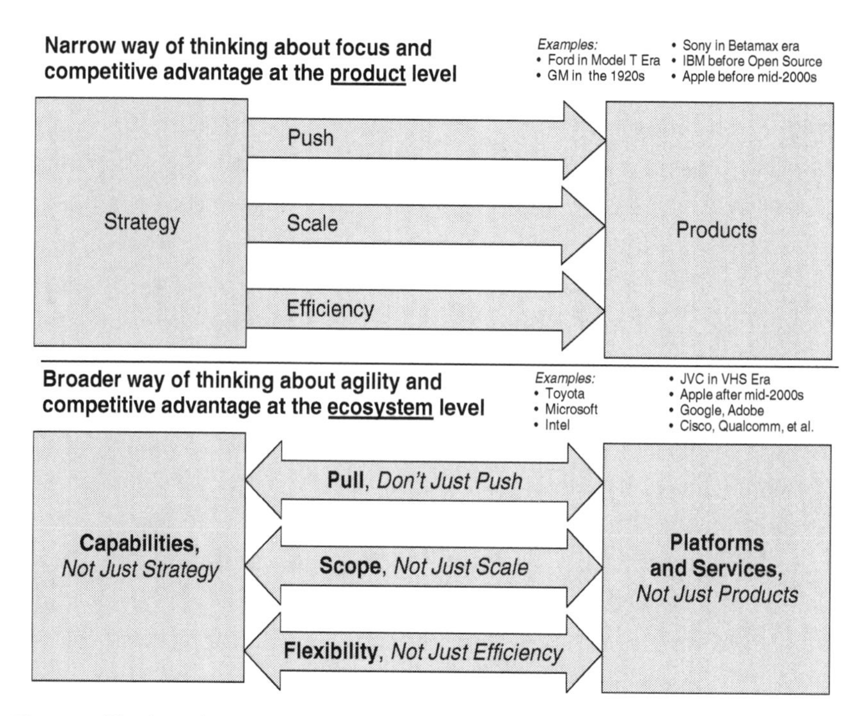

Figure 1. The six enduring principles and competitive advantage

companies support their platforms and services by direct linkages to the market—through pull rather than just push systems. They also take advantage of more complex scope economies and seek flexibility more than efficiency.

Established firms such as Microsoft, Intel, Apple, Toyota, and Nissan illustrate the six principles in different ways, and provide important—positive and negative—examples in this book. I also examine younger companies such as Google, Qualcomm, Adobe, Amazon, and Salesforce.com. Again, it is my intention to transcend particular company examples. The principles that give firms staying power should be timeless. Below is a more precise statement of the six principles and their counterpoints, as well as a brief outline of key examples. The plan of the book follows this structure, with one chapter devoted to each principle.

1. *Platforms, Not Just Products*

Managers (at least in industries affected by digital technologies as well as "network effects" more broadly) should move beyond conventional thinking about strategy and capabilities to compete on the basis of platforms, or complements to another firm's platform. A platform or complement strategy differs from a product strategy in that it requires an external ecosystem to generate complementary product or service innovations and build "positive feedback" between the complements and the platform. The effect is much greater potential for innovation and growth than a single product-oriented-firm can generate alone.

Too many managers in a variety of industries seem unsure what an industry platform is and think primarily at the product level. They find it difficult to formulate a platform or complements strategy that goes beyond individual products and technical standards. In Chapter 1, I describe more of the thinking behind this broader view of industry platforms as well as some of the relevant research. Then I illustrate the principle with comparisons of JVC and Sony in video recorders, and Microsoft and Apple in personal computers. Finally, I analyze the style of platform leadership that has emerged at firms such as Intel, Microsoft, Google, Qualcomm, and Apple since 2004.[8]

2. *Services, Not Just Products (or Platforms)*

Managers (at least in product firms or service firms that offer standardized or automated services treated as products) should use service innovations to sell, enhance, and even "de-commoditize" products or standardized services. Services can also be new sources of revenues and profits, such as an ongoing maintenance or subscription stream. The goals of most firms should be to find the right balance between product and service revenue, and then "servitize" products to create new value-added opportunities and pricing models as well as "productize" services to deliver them more efficiently and flexibly, such as by using information technology and service automation.

Too many managers in product firms treat services as a cost center and a necessary evil to sell or support products, which generally have much higher gross profit margins, especially in industries such as packaged software or digital goods. The reality, though, is that many products or common services, ranging from software and automobiles to banking and overnight delivery, have become commodities. The major source of differentiation for firms in these businesses may have already shifted to customization and the quality or innovativeness of their services. In Chapter 2, I describe more of the thinking behind this view of services as well as some of the relevant research. Then I illustrate the principle with examples drawn mainly from the software products business, including enterprise software as well as software as a service (SaaS) and cloud computing. I also discuss some applications to the automobile industry.[9]

3. *Capabilities, Not Just Strategy*

Managers should focus not simply on formulating strategy or a vision of the future (that is, deciding what to do) but equally on building distinctive organizational capabilities and operational skills (that is, how to do things) that rise above common practice (that is, what most firms do). Distinctive capabilities center on people, processes, and accumulated knowledge that reflect a deep understanding of the business and the technology, and how they are changing. Deep capabilities, combined with strategy, enable the firm to offer superior products and services as well as exploit foreseen and unforeseen opportunities for innovation and business development.

Too many managers rely on strategy to differentiate their firms without making the more difficult, longer-term investments essential to implement strategy successfully and, more importantly, to create an ongoing foundation for product, process, or service innovation. In Chapter 3, I describe more of the thinking behind this view of capabilities as well as some of the relevant research. Then I illustrate the principle using historical comparisons of Toyota and Nissan in automobile technology transfer and manufacturing, Japanese competition in the design and mass production of a home video recorder, and Microsoft in building and marketing PC software.

4. *Pull, Don't Just Push*

Managers should embrace, wherever possible, a "pull-style" of operations that reverses the sequential processes and information flow common in manufacturing as well as product development, service design and delivery, and other activities. The goal should be to link each step in a company's key operations backward from the market in order to respond in real time to changes in demand, customer preferences, competitive conditions, or internal difficulties. The continuous feedback and opportunities for adjustment also facilitate rapid learning, elimination of waste or errors, and at least incremental innovation.

Too many managers rely on sequential processes and planning mechanisms and then have difficulty incorporating feedback and learning from customers, competitors, suppliers, partners, and internal operations, especially when developing new products or managing production and service delivery. Of course, the concept of "market pull" versus "technology push" is an older theme found in marketing, manufacturing, and supply-chain management literature. But here I extend the principle to how managers should think about strategy and innovation, as well as operations, more broadly. In Chapter 4, I describe more of the thinking behind this view of pull as well as some of the relevant research. Then I illustrate the principle with two examples—the evolution of the just-in-time or "lean" system for production management at Toyota, and the evolution of the synch-and-stabilize system for iterative or "agile" product development at Microsoft, with some comparisons to Netscape, Hewlett-Packard, and other companies.

5. *Scope, Not Just Scale*

Managers should seek efficiencies even across activities not suited to conventional economies of scale, such as research, engineering, and product development as well as service design and delivery. Firms usually pursue synergies across different lines of business at the corporate level. But scope economies within the same line of business can be an important source of differentiation in markets requiring efficiency and flexibility, and responsiveness to individual customer requirements. These deeper economies of scope require systematic ways to share product inputs, intermediate components, and other knowledge across separate teams and projects. Firms can also eliminate redundant activities and other forms of waste, and utilize resources more effectively.

Too many managers focus on getting big and chase seemingly easy efficiencies from scale or corporate diversification, without pursuing the more complex but equally powerful efficiencies of scope economies within a particular business. Scope economies of any kind place some constraints on the firm if not designed and managed properly. Reusing good technology in many different products can save the firm huge amounts of money. This practice can also lead to disaster if the technology proves faulty. Overall, however, well-managed scope economies can create competitive advantage precisely because they are difficult to achieve. In Chapter 5, I describe more of the thinking behind this view of scope as well as some of the relevant research. Then I illustrate the principle with software factories in the United States, Japan, and India and compare these to "multi-project" management systems for product development, especially as practiced at Toyota and other Japanese automakers.[10]

6. *Flexibility, Not Just Efficiency*

Managers should place as much emphasis on flexibility as on efficiency in manufacturing, product development, and other operations as well as in strategic decision making and organizational evolution. Their objectives should be to pursue their own company goals whilst quickly adapting to changes in market demand, competition, and technology. Firms also need to be ready to exploit opportunities for product or process innovation and new business development whenever they appear. Moreover, rather than always requiring tradeoffs, flexible systems and processes can reinforce efficiency and quality, or overall effectiveness, as well as facilitate innovation.

Too many managers see flexibility in narrow, technical terms, such as in flexible manufacturing, and neglect the potential benefits of investing in flexibility more broadly throughout the organization and in decision-making processes. In some instances, this may be a short-term versus long-term problem: flexibility has some immediate costs. But, in general, managers should strive to eliminate tradeoffs as they design their technical systems or hire and train their people. They should avoid practices that inhibit the ability of the firm to respond quickly to new information from the market or their own operations. In Chapter 6, I describe more of the thinking behind this view of flexibility as well as some of the relevant research. Then I illustrate the principle with examples from electronics component manufacturing, software development at Microsoft and other companies, and strategy evolution at Netscape and Microsoft as both companies learned how to compete on "Internet time."[11]

Pause for Reflection: Some Comments on the Research

Before continuing to the main chapters, I need to explain more about the thinking behind these principles and examples. First, whilst I have tried to describe big, enduring ideas, they are relatively high-level abstractions. Managers must still figure out how to apply these concepts and lessons from the different cases to their particular situations. The contexts can vary widely, such as by timing (early or late mover, before or after imitation); stage of the industry or technology life cycle (early, mature, end of life); nature of the technology or innovation (software versus hardware, product versus process, incremental versus disruptive); "clock speed" of the industry and other differences (fast-moving, R&D-intensive, or capital-intensive); and the environmental or institutional setting (regulated or not, "Japan, Inc.," or "China, Inc."). Academics and business writers who try to identify what works best for managers and organizations also need to make these adjustments in their research and add the appropriate "controls." Too often they do not. This oversight is one reason why supposedly "excellent" or "great" companies cited in the past quickly falter or even go bankrupt (see Appendix I). The problem is not necessarily in the principles or specific concepts, or in the company examples, but in the research methods and claims of the researchers.

Managers also need to be aware that each of the six principles comes with different potential tradeoffs, and they need to manage these as well in order for the principles to work for them. For example, designing a product and competitive strategy to enhance your chances of becoming an industry-wide platform will probably involve sharing revenues and profits with ecosystem partners. On the one hand, this strategy is probably costly in the short term. On the other hand, it can be far more profitable over time, even if not completely successful (winner-take-most as opposed to loser-gets-nothing). Investing in services can hurt profit margins for a product firm, especially those with potential 99 percent gross margins (companies in the business of packaged software or digital media). But, if product sales or prices collapse, as has occurred in software products, newspapers, magazines, books, and music publishing, then those magical gross margins are worthless. The real value of these businesses has probably shifted already to complementary services or service-like versions of the products. Moreover, the services generate not only revenue but also intimate customer knowledge important for staying in business and continuing to innovate.

We also know (and Toyota's experience in 2009–10 reinforces this observation) that implementing a pull system in manufacturing does not overcome design or architectural flaws in a product or inadequate management attention to the business and customer complaints. Just-in-time (JIT) approaches can also complicate relationships with suppliers and may reduce buffer inventories to a dangerously low level. But pull systems, within certain limitations, do give companies an immediate "pulse" on at least short-term fluctuations in market demand—in addition to saving operating costs and forcing ongoing process and quality improvements. Similarly, pursuing scope economies in production, product development, and services may be costly today and impede flexibility at some level, or prove disastrous if the company broadly reuses a faulty component—which Toyota, our example again, seems to have done with gas pedals and antilock braking control software. But strategic reuse also can be highly effective as a source of competitive differentiation. Finally, flexible firms are probably more complex to manage because their people must be able to respond to change or share knowledge and responsibilities on the fly, outside formal structures and processes. But, over long periods of time, the more flexible firms should be able to withstand unpredictable change or adversity better than their competitors. This is really what staying power is all about.

Academics and business writers who have researched the sources of superior performance in its various forms usually encounter another dilemma: case studies provide the detail to uncover the subtle realities of strategy, structure, process, and decision making—and many more elements critical to understanding success as well as failure. But we cannot generalize, or not very confidently. Large-sample studies, with and without sophisticated statistical analysis, provide the power to argue with greater levels of confidence. Nonetheless, the bird's-eye view from "50,000 feet" and statistically significant results usually do not provide enough detail to help managers move beyond generalizations. Intermediate research approaches that combine case studies with large datasets can be much more informative. But these studies require more varied research skills and research time, and are subject to their own methodological constraints.

Most academics probably believe that managers need to think in terms of "contingency frameworks" to understand when to apply particular principles or specific practices. Building such frameworks is a valuable exercise for managers and academics alike. I will refer to several in this book. But academics also tend to put in too many "if this, then that" kinds of relationships. As the number of contingencies rise, the frameworks can become too cumbersome to be useful to managers who need to make quick decisions with limited information. For this reason, I have created a relatively simple framework: six principles, with counterpoints and detailed examples to help illustrate the underlying ideas. In the thirty years of research that preceded these principles, my colleagues and I have spent countless hours thinking about the implications of these ideas, as well as control variables and alternative explanations for the phenomena. I will explain only a little bit of the research process as I go through the literature and different examples. But, at this juncture, I am as confident as I can be about what we know regarding strategy and innovation and their relationship to competitive advantage. And I am reasonably certain about what we do not know, and have tried to make this clear as well.

The invitation to deliver the 2009 Clarendon Lectures in Management Studies at the University of Oxford provided me with the opportunity to reflect on my research and the strategy and innovation fields more broadly. My research style has varied with the questions and the data available. But usually I have tried to combine qualitative and quantitative case studies that probe a phenomenon deeply with a few larger-sample empirical analyses and "longitudinal" (historical) data that help us understand the bigger picture. I am convinced that the six principles are of enduring importance and transcend specific company examples precisely because they derive from relatively high-level abstractions and detailed cases that cut across very different industries and settings. Moreover, I really have spent the better part of three decades thinking about the underlying ideas, as well as collecting data, testing hypotheses, and discussing the concepts with university colleagues, students, managers, and consultants. Many researchers in different fields have provided important validation of these principles as well.

The six principles and examples in this book also reflect my conviction that people who write about strategy and innovation, or the sources of competitive advantage more broadly, should first study—in great depth—what leading firms are doing. But they should not stop there. Academics need to bring in their research skills and powers of generalization in order to go beyond specific examples and help managers identify the deeper elements driving superior performance and staying power. Managers or academics usually cannot do this by themselves, and that is why business research is important—or at least why it is important to do this well.

Unfortunately, business schools have experienced pressure to move beyond qualitative case studies to become more "rigorous" and "quantitative." They have followed the lead of established disciplines such as economics, finance, quantitative sociology, and operations research, where many of the new business school faculty come from. This is not necessarily bad. But too many of the current editors of academic journals are also trained in these disciplines and seem to care much more about methodology than ideas or relevance to managers. As a result, the practical value of business school research is declining even whilst the rigor of the research improves. Some academics are lobbying to reverse this trend and others refuse, most of the time, to publish any more in the top academic management journals. A few practitioner-oriented journals survive, and they are much better at translating research into concepts useful for managers. But very little of the most rigorous work makes it into these outlets. The challenge for business schools and firms is to encourage more relevant research and evolve management practice together.

It would be good to know more about when the principles I propose are likely to be more or less effective, as well as how they can help a firm respond to adversity as well as disruptive change. We need to know more about how the different concepts complement one other and potentially work together as a "system of practice." Nonetheless, I am convinced that managers who can turn their products into industry-wide platforms or complements whilst cultivating global ecosystems of innovation, and develop services to make their product businesses more distinctive and resilient, will find themselves better able to survive and thrive in many different conditions. And managers who cultivate distinctive organizational capabilities tied to a deep knowledge of the business and the technology, rely on pull as much as or more than on push concepts, exploit economies of scope as well as scale, and pursue flexibility as much as efficiency will create agile organizations able to learn from their mistakes and withstand the dual challenge of innovation and commoditization.

Notes

1. There are numerous reports on the Toyota problem in the media and information available from Toyota directly. But a particularly important early document is Kane et al. (2010). Later reports cited 50 to 100 deaths.
2. Toyota's buyback program covered Tacoma pickup trucks made between 1995 and 2000. See Walters (2008).
3. Bailey (2010).
4. See Nader (1965).
5. Ackman (2001) and Wikipedia (2010).
6. Brass (2010).
7. See, e.g., Highsmith (2009). Some members of the software community have even issued a set of precepts they call "The Agile Manifesto." See http://agilemanifesto.org (accessed Apr. 16, 2009).
8. The initial study of platform leadership at Intel was the thesis subject of my former doctoral student, Annabelle Gawer, of the Imperial College Business School. See Gawer (2000).
9. All the database and recent theoretical and empirical work on product and service business models I have done jointly with Fernando Suarez of Boston University School of Management, and Steven Kahl of the University of Chicago Booth School of Business. Both are former doctoral students of mine at MIT Sloan.
10. The initial work on multi-project management was done by my former master's (MBA) and doctoral student Kentaro Nobeoka, who was a Mazda product engineer and is now a professor at Hitotsubashi University in Tokyo. See Nobeoka (1993, 1996).
11. The printed-circuit board (PCB) study was the thesis subject of Fernando Suarez. See Suarez (1992).

1

Platforms, Not Just Products

The Principle

Managers (at least in industries affected by digital technologies as well as "network effects" more broadly) should move beyond conventional thinking about strategy and capabilities to compete on the basis of platforms, or complements to another firm's platform. A platform or complement strategy differs from a product strategy in that it requires an external ecosystem to generate complementary product or service innovations and build "positive feedback" between the complements and the platform. The effect is much greater potential for innovation and growth than a single-product-oriented firm can generate alone.

Introductory

A powerful new idea has appeared in strategy and innovation practice as well as research over the past several decades, with important implications for staying power. The new challenge is to compete in *platform* markets within an industry and to innovate through a broader "ecosystem" of partners and users not under any one firm's direct control. Platform leaders are difficult to dislodge. They can retain dominant market shares for decades, and not only when by chance they design a hit product. But to compete on the basis of platforms, and not only on products, requires a different approach to strategy and business models. It also requires a broader application of the principles discussed in later chapters—services, capabilities, pull mechanisms, scope economies, and flexibility.

For example, a successful platform strategy benefits from particular skills in product architecture and interface design. It also requires negotiations with other firms to build products and services that complement the platform and make it more useful. We have strong pull effects in platform markets as well, but the most important come from "network effects" between the platform technology (such as the VHS (Video Home System) video recorder, the Windows-Intel PC, or the Amazon cloud) and complements (such as tapes recorded on the VHS standard, or applications written only for Windows or Amazon Web services). Similarly, economies of scope and flexibility play critical roles for platform companies, but in somewhat different ways from products that companies encounter in markets not subject to network effects. Platform leaders or "wannabes" must decide what complementary products or services to create themselves and which ones they will help partners or users—the ecosystem—to provide.

The term "platform" first came into wide usage in the management field as a word meaning foundation of components around which an organization creates a related but different set of products (or services). Toyota's Corolla sedan, Celica sports car, Matrix hatchback, and Rav-4 sports utility vehicle are different products built in separate projects. But they share the same underbody as well as other essential components such as the engine. Microsoft builds the Office suite (mainly the Word, Excel, and PowerPoint products) around shared components, such as the text-processing, file-management, and graphics modules.[1] In the 1990s, many researchers in operations and technology management as well as in strategy and economics popularized this concept of an in-house product platform used to create a family of related products, particularly when discussing modular architectures and component reuse.[2]

This chapter uses the word differently—following my 2002 book with Annabelle Gawer, *Platform Leadership*.[3] In that study and in subsequent articles, we distinguished between an in-house "product platform" and an "industry platform." The latter has two essential differences. The first is that an industry platform is a foundation or core technology (it could also be a service) in a "system-like" product that has relatively little value to users without complementary products or services. The platform producer often (but not always, as seen in the case of Microsoft) depends on outside firms to produce the essential complements. The Windows-Intel personal computer and a "smartphone" (a Web-enabled cell phone that can handle digital media files as well as run applications) are just boxes with relatively little value without software development tools and applications or wireless telephony and Internet services. Cisco (founded in 1984) has a platform that has evolved from a specialized computer system called a router that connected corporate networks with the Internet to a software layer, the Internetworking Operating System (IOS). IOS has little value by itself but becomes much more useful when customers deploy this software with a variety of networking equipment, such as different types of routers, computer servers, telecommunications switches, and wireless devices, from Cisco and other vendors. For these reasons, a potential industry platform should have relatively open interfaces in the sense of being easily accessible technically and with inexpensive or free licensing terms. The goal is to encourage other firms and user communities (such as for Linux) to adopt the platform technology as their own and contribute complementary innovations. These external innovators form the platform ecosystem.

The second essential difference between a product and an industry platform, as various authors have described, is the creation of network effects (see Figure 1.1). These are positive feedback loops that can grow at geometrically increasing rates as adoption or usage of a platform grows. The network effects can be very powerful, especially when they are "direct," such as in the form of a technical compatibility or interface standard. This exists between the Windows-Intel PC and Windows-based applications or between VHS, DVD, or Blu-Ray players and media recorded according to those formats. The network effects can also be "indirect." Sometimes these are very powerful as well—such as when an overwhelming number of application developers, advertisers, content producers, and buyers or sellers adopt a platform with specific technical interfaces or connection standards. Examples include not only the Windows-Intel PC and the VHS versus Betamax video cassette recorders, but also the eBay marketplace, Google search bar and cloud-computing platform, or the Facebook,

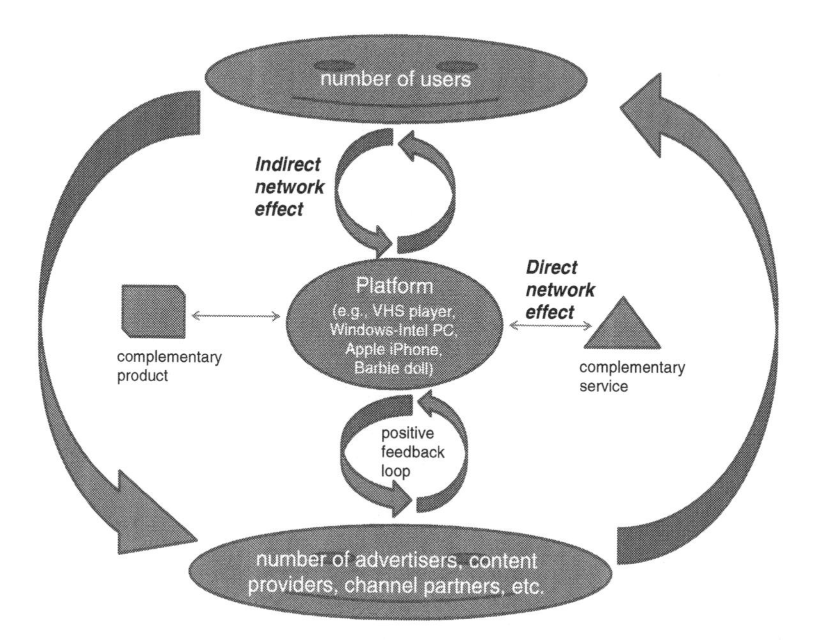

Figure 1.1. The ecosystem of platforms, complements, and network effects

MySpace, LinkedIn, or Twitter social networking portals, among many others.

Perhaps most important, a network effect means that, the more external adopters in the ecosystem that create or use the complementary innovations, the more valuable the platform (and the complements) become. This dynamic, driven by direct or indirect network effects or both, should encourage more users to adopt the platform, more complementors to enter the ecosystem, and more users to adopt the platform and the complements, almost ad infinitum.[4]

We have seen many platform-like battles and network effects in the history of technology, mainly in cases where competitions emerge because of incompatible standards and when a product by itself has limited value. Standards are not platforms either; they are rules or protocols specifying how to connect different products or modules and use them together. Prominent past examples of platforms incorporating specific standards include the telegraph (what format or language to use for coding and decoding messages and sending the electrical signals), the telephone (how to do the same thing as the telegraph but with voice signals), electricity (the battle between alternating versus direct current), radio (struggles over the establishment of

AM and FM standards, broadcasting technology, and content), television (what standards to adopt initially, and then the movement from black-and-white to color), magnetic-tape video recording (VHS versus Beta formats and content), and computer operating systems (from IBM mainframes to PCs, the Macintosh, and Linux).

Other recent hardware and software platform battles have emerged over Internet portals, search, and content delivery; online marketplaces; smartphone operating systems and transmission technologies; video-game consoles and games; electronic payment systems; foreign-exchange trading systems; electronic stockbrokerage systems; electronic display technologies; advanced battery technologies; alternative automotive power systems; and social networking sites. Even the human genome database has become a platform of data and knowledge for researchers and pharmaceutical companies as they compete (and sometimes cooperate) to analyze how genes function and discover new drug products. In fact, the more you look inside modern society and its technological artifacts—the computer, cell phone, media player, home entertainment systems, office equipment, or even the automobile—the more you will see platforms, and platforms within platforms, as well as direct and indirect network effects.

We also can see platform competition and network effects surrounding non-technology products and services—reinforcing the idea that this principle is not simply for high-tech managers. Wal-Mart, for example, has created a global supply-chain platform to feed its retail stores. Marks & Spencer has done the same thing on a smaller scale. Best Buy is doing the same thing in electronics-goods and home-appliances retailing. Suppliers make particular investments to become part of these networks and cannot so easily switch.

Other examples include CVS and Walgreens. They are starting to use their networks of pharmacy retail stores as platforms to offer an increasing variety of customer services from new internal divisions and acquisitions as well as partners. They started with filling prescriptions but now offer photography, flu shots, and basic healthcare in their retail locations, at people's homes, or in their workplaces. There are some network effects and switching costs to the extent that customers register specific medical, insurance, and financial information with these providers. The information may not be so easy to transfer. Moreover, CVS and Walgreens can build very detailed customer profiles with the data and increasingly improve their ability to add or refine these various services. There is a reputational network effect as well in that, the more customers who use these services and are satisfied with them, the more likely it is that other customers will come to CVS or Walgreens rather than to their doctor or a hospital emergency room for basic healthcare.

Another non-technology example is the Barbie doll. This toy, owned and trademarked by Mattel, Inc., and first introduced in 1959, has become a multi-billion-dollar platform business. It serves as a foundation for many variations of the doll itself as well as a growing variety of complementary products (clothes, fashion accessories, toy cars, toy houses, companion dolls) and services (online videos, games, music, shopping).[5] Mattel makes some of these complements itself as well as licenses to hundreds of partners the right to make these products or to offer new services. Just like Microsoft, Cisco, Intel, Google, Qualcomm, and other high-tech firms, Mattel too has been engaged in a fierce battle over intellectual property rights. MGA Entertainment launched the competing Bratz family of dolls and accessories in 2001, before being stopped by a Mattel lawsuit. MGA-E then introduced the Moxie Girlz family of dolls in 2009.[6]

Not surprisingly, we see a growing amount of research on industry platforms, initially by economists but increasingly by scholars of strategy and innovation.[7] Competition in the consumer electronics and computer industries spurred a great deal of thinking on this topic in the early 1980s, just as the arrival of the World Wide Web did so again in the mid-1990s. Influential early work mostly focused on theory, with few detailed examples and no large-sample studies. But the key concepts are all there and are now familiar to researchers and managers alike: how technical standards and compatibility or user adoption affect the course of platform industries and product designs, the phenomenon of network effects and positive feedback, and the role of switching costs, pricing, and bundling.[8] More recent economics work has focused on models that improve our understanding of how "multi-sided" platform markets function.[9]

In strategy and innovation, recent studies also analyze multi-sided platform competition as well as how to manage complementors, use the ecosystem for innovation, and compete as a complementor.[10] For example, the battle between Netscape and Microsoft in the browser wars illustrated the use of one-sided subsidies. By this term I mean the strategy of "free, but not free"—give one part of the platform away, such as the Internet browser, but charge for another part, such as the Web application server or the Windows operating system. Adobe has done the same thing by giving away the Acrobat Reader and charging for its servers and editing tools, technical support, and online services. Some firms give one part of the platform system away to some users (students or the general consumer) but charge others (corporate users). Intellectual property too can be "open, but not open" or "closed, but not closed." By these terms I mean that firms can make access to the interfaces easily available but keep critical parts of the technology proprietary or very distinctive. Netscape did this with the Navigator browser and an array of servers, special versions of scripting and programming languages, and intranet and extranet combinations. Microsoft has done this with the entire set of Windows client and server technologies, as well as Office and other Windows applications.[11] At the broader ecosystem level, we see the emergence of "keystone" firms—industry leaders ranging from Wal-Mart to Microsoft and automobile companies that encourage innovation by cultivating networks of firms to make modularized components.[12] We also have important work by Eisenmann, Parker, and Van Alstyne—which I will return to later—on the conditions that make for, and prevent, "winner-take-all" markets.[13]

Given the breadth and growing popularity of this topic, it is important to be clear about what an industry platform is not. Although it is not a technological standard, technology-based platforms usually incorporate existing industry standards and help establish new ones. Microsoft and Intel, by promoting certain standards within Windows and the "x86" line of compatible microprocessors, did this with applications programming and connectivity standards for the personal computer, beginning with the first IBM PC. Cisco, by bundling certain protocols within its operating software for routers and other equipment, did this with networking.

Nor is an industry platform the same as a "dominant design," though a successful platform is, by definition, widely adopted. My MIT Sloan colleague James Utterback, and the late William Abernathy of the Harvard Business School, defined a dominant design as a particular configuration of a product that wins allegiance from the majority of users and thus influences what subsequent designs look like. The QWERTY keyboard, the Ford Model T, and the IBM PC have all played this role in their industries.[14] But, just as different product designs may compete to become the dominant form, an industry may generate multiple platform candidates that compete for dominance. Some industries never experience a dominant design or a dominant platform. In any case, though, industry platforms differ from dominant designs in that they are part of a system—the platform and the complements—and are not stand-alone products. They also require network effects for the platform to grow in value to users. In addition, the dominant designs of Utterback and Abernathy appear in the latter stage of an industry's evolution as part of the maturation process and managerial shift of attention from product design to the production process. It may happen that platforms emerge later in an industry's development. But they can appear early as part of a competition to establish a dominant platform.[15] And some competing platforms may persist for long periods of time without any one leader emerging.

Even if a company fails to establish the dominant platform or if the market never adopts one platform, platform strategy can still be a valuable tool for *strategic marketing*. Simply thinking hard about whether a firm is in a platform market or a winner-take-all environment provides deep insights into competition and a product's broadest possible potential. Any effort a firm makes to promote adoption of its technology or service by other firms, and to create even a small ecosystem of complementors and users, should enhance its reputation and sales. Moreover, these kinds of strategic insights are as useful for would-be complementors as for potential platform leaders.

As noted in the Introduction, I illustrate the various dimensions of a platform strategy and the capabilities required to become a platform leader with several examples. First, I discuss how a platform strategy differs from a product strategy by reviewing the cases of Apple versus Microsoft in personal computers and Sony versus Japan Victor Corporation (JVC) in video recorders. Next, I describe the platform-leadership model refined at Intel and other established platform leaders. Finally, I look at how relatively new firms can turn a product strategy into a platform strategy as well as help a market "tip" in their direction. Most of my examples deal with information technology because these are the industries I follow most closely. But, as I have indicated, platform dynamics is a much broader phenomenon and will certainly become more important for managers and policy-makers in a variety of industries going forward.

Product versus Platform: Apple and Sony, et al.

There is no doubt that a product strategy can turn into a platform strategy, and a best-selling product is an excellent start to a successful industry platform. But what managers need to know first is how, in the early stages of competition, a *product* strategy differs from a *platform* strategy, and what are the potential consequences for both the innovator and the users or adopters, such as advertisers and content providers in the case of digital businesses. We can learn a lot about this dilemma from observing the behavior of Apple and Sony—two great product companies where managers have not always thought "platform first."[16]

Apple versus Microsoft

To begin, we must acknowledge that Apple, founded by Steve Jobs and Steve Wozniak in 1976, ranks as one of the most innovative product companies in history. The list of "insanely great" Apple products—Jobs's promotional mantra for the Macintosh personal computer, introduced in 1984—is truly impressive. But, in the past, Apple often chose not to adopt an explicit *industry* rather than a *product* platform strategy, at least initially. Consequently, Apple has missed out on some enormous business opportunities as well as the chance to make our lives much easier than they have been. We all *should have been* users of the Macintosh personal computer and, more recently, the iPod and the iPhone products as well as the iTunes digital media service. Instead, the vast majority of us became users of cheap and powerful but clumsy DOS and then Windows PCs. Apple also trails in the global smartphone market by a large margin, except for the United States. And, though the iPhone is gradually doing better overseas, Google has now entered the smartphone market with its "open" Android software platform and its own line of phones.

My lament is because Apple, with the Macintosh, pioneered graphical user interface technology (albeit inspired by Xerox—another great product company with missed platform opportunities) for the mass market. Other landmark Apple products include the first mass-market PC, the Apple II, introduced in 1977; the PowerBook, which in 1991 set the design standard for laptops; the unsuccessful though still pioneering Newton PDA, first sold in 1993; and the iMac all-in-one "designer PC," released in 1998. More recently, we have seen the iPod digital media player (2001), the iTunes digital media service (2003), the iPhone (2007), and the iPad (2010). Jobs did not himself design these products. He was absent from the company during 1985–97 and returned only when Apple acquired one of his less successful ventures, NeXT Computer. But, even then, NeXT technology and the UNIX operating system provided the basis for another hit Apple product released in 2001, the Mac OS X operating system. Most importantly, Jobs created the design culture and hired or supervised the people (such as Jonathan Ive, chief designer of the iMac, the iPod, and the iPhone) most responsible for the company's historical legacy and recent revival.

The world truly would have been a different place if Steve Jobs earlier in his career had thought a bit more like his arch-rival, Bill Gates. Microsoft, founded a year before Apple in 1975, generally has not tried to develop "insanely great" products. Occasionally, some have been very good—such as BASIC for the Altair PC kit, the 1990 version of Excel, Internet Explorer version 4 (1997), and Windows 7 (which, in 2009, finally caught up to the Macintosh OS, after twenty-five years). Mostly, Microsoft has tried to produce "good-enough" products that can also serve as industry platforms and bring cheap and powerful computing to the masses (and mega-profits to Microsoft). DOS, Windows, and Office have done this since 1981.[17] And Microsoft continues to try with new platform candidates, such as a version of Windows technologies (.NET) for enterprise computing, where it has been relatively successful. In other markets it has made less progress, such as Windows for smartphones and handheld devices or the tablet computer, the Xbox video-game console as a new hardware–software platform, and recent online versions of Windows and Office that come under the rubric of "software as a service" or "cloud computing."

Sony versus JVC

Not only has my experience studying Microsoft and Intel influenced my view of Apple, but so did my first platform-related research. In the mid-1980s Richard Rosenbloom and I examined the race between Sony—another great product company in its heyday—and JVC to introduce a home video cassette recorder (VCR). As the years unfolded, I realized that the development story needed an ending and started a follow-on project to understand why VHS so convincingly dominated Beta in the marketplace.[18]

To explain the outcome, we need to go back to 1969–71. During this period, Sony engineers had compromised their technology goals when designing an earlier device using ¾-inch-wide tape, called the U-Matic. They compromised in order to get the support of other firms in Japan and elsewhere. As a result, the large, bulking, and expensive U-Matic failed to attract home users. But institutions such as schools and police stations did purchase the machines. These customers provided Sony as well as JVC and other vendors with the inspiration to continue and enough feedback to design a more successful home product. When Sony introduced their smaller ½-inch tape machine in 1975, dubbed the Betamax, company executives again tried to persuade other firms to adopt their technology as the new industry standard. Sony's goal was to replace the ¾-inch format as well as competing formats under development at several firms. But this time Sony engineers refused to alter the Betamax design to accommodate other firms in Japan or in the United States. General Electric, for example, wanted a much longer recording time for American consumers. The original Betamax recorded for only one hour.

JVC, backed by its giant parent Matsushita Electronics (recently renamed Panasonic after its US brand name), in fall 1976 came out with its own product, VHS. This offered two hours of recording. Within five months, Sony matched the two-hour time by, for example, using thinner tape. Some observers also thought VHS was technically inferior to the Beta machines. This reputation, along with improvements in the recording time, should have provided Sony with more staying power in this market. But JVC and Matsushita continued to match Sony reasonably quickly with new features and longer recording times, and comparable prices. Sony eventually came out with an unmatched eight hours of recording time in 1982 (see Table 1.1).

Table 1.1. *VHS and Beta recording—playing time comparison*

Year/Month	Beta	VHS
1975 May	1 hour (Sony)	
1976 October		2 hours (JVC)
1977 March	2 hours (Sony)	
1977 October		4 hours (Matsushita)
1978 October	3 hours (Sony)	
1979 March	4.5 hours (Sony)	
1979 August		6 hours (Matsushita)
		4 hours (JVC)
1979 December		6 hours (JVC)
1982 March	8 hours (Sony)	
1982 September	5 hours (Sony)	

Source: Cusumano, Mylonadis, and Rosenbloom (1992: 77, table 7).

Features and prices ultimately mattered little because the VHS and Betamax machines were very comparable technically and hard for users to differentiate. Simultaneously, however, there were powerful network effects. VHS and Betamax, though both based on the U-Matic, utilized different cassette sizes and incompatible signal-encoding formats. At the time, the machines were sufficiently expensive for consumers to be unlikely to own more than one format. We know from research by Eisenmann, Parker, and Van Alstyne that three factors in combination—(1) little room for platform differentiation, (2) strong network effects between the platform and the complements, and (3) the unlikelihood of users buying more than one platform, which they call "multi-homing"—should lead to a winner-take-all or winner-take-most market. Indeed, this is what happened.

Of equal importance, we can see that the market dynamics here did not simply unfold through some natural or random process. JVC and Matsushita deliberately tried to position VHS as a new *industry* standard and worked very hard to make this happen. The JVC executives and development team humbly visited competitors and potential partners, asked for feature suggestions, and did their best to accommodate them. JVC and Matsushita also broadly licensed the new technology on inexpensive terms to some forty firms. They provided essential components (like the helical scanner, which was very difficult to mass produce) until licensees were able to do the manufacturing themselves. In contrast, the Beta group totaled merely twelve firms at its peak, with Sony doing the bulk of the manufacturing (see Appendix II, Table II.1).

JVC and Matsushita, with great foresight (which was lacking in Sony at the time), aggressively cultivated a complementary market in prerecorded tapes and retail distribution. Matsushita even used its engineering resources to build machines that replicated tapes at very high speeds for the prerecorded market. All of these very deliberate moves—which we called "strategic maneuvering"—helped establish the VHS technology as a new platform for the consumer electronics industry and "tip" the market toward VHS. The network effects increased in strength as the much larger number of firms licensing VHS brought more production capacity to their standard, which encouraged more tape producers and distributors to make many more prerecorded VHS tapes. Retailers increasingly used their limited shelf space for VHS machines and prerecorded tapes. Users responded and bought more VHS machines, which encouraged more firms to license the VHS standard and then more tape producers, distributors, and consumers to adopt VHS. Betamax went from a 100 percent share in 1975, the beginning of the market, to zero by the later 1980s (Appendix II, Table II.2).

Apple's Evolution

The Macintosh story resembles the Betamax story, with a critical difference. Apple's product survived, even though it remained for many years only on the periphery of the PC industry in terms of market share—stuck at a fraction until newer product designs and exploding sales of the iPod and then the iPhone spilled over into higher computer sales, at least in the United States. Poor responses to Microsoft's Windows Vista operating system, introduced in 2006 and then replaced by the much improved Windows 7 in 2009, also persuaded many users to switch over to Apple. Still, the US market share for the Mac peaked at around 10 percent during 2008–9, and seems to have leveled off or dropped. The main point is that Apple's strategy never got the Macintosh beyond 2 or 3 percent of the global personal computer market, compared to 90–5 percent for Windows-Intel PCs.[19] Of course, the Mac's innovative software and hardware designs have attained great "mind share" or attention in the industry, and forced responses from Microsoft and PC hardware manufacturers. This competition remains vitally important to stimulating innovation and is the reason we now have Windows 7. Nonetheless, there are unfortunate similarities between Sony and Apple.

Like Sony, Apple chose to optimize the Mac's hardware–software system and complete the design on its own as well as control the flow of revenues (and profits) from the product. By contrast, a *platform* strategy would have meant licensing the Macintosh operating system widely and working much more openly and actively with other companies to evolve the platform *together* and create complementary applications. Microsoft and its ecosystem partners have done this for the Windows-Intel PC. Apple did not do very much of this platform evangelism and has remained (with a brief exception many years ago) the only producer of the Mac. This product-centric strategy has kept prices high (historically, about twice the cost of a Windows-Intel PC with comparable levels of power and memory) and diffusion low. Moreover, the relatively closed and expensive Macintosh did not stimulate the enormous mass market in applications that Microsoft and Intel have done for the PC. The Macintosh lived on initially as a minor second standard mainly because it found two niches—desktop publishing and consumers (including institutions such as primary schools) willing to pay more for an easier-to-use and more elegant product.

This brings me to more recent "insanely great" products from Apple that have done much better in the market. They also have enormous industry platform potential—some of which Apple has finally tapped! The iPod, with its unique "click wheel" interface and new touch screen, is the best-selling music player in history, with its own near monopoly—about a 70 percent market share. It has attracted complementary hardware innovations that have made it more valuable, such as connectors for a car or home stereo system, or add-ons that turn the iPod into an FM radio, digital recorder, or camera. Initially, however, Apple introduced the iPod as another "closed" product system that worked only with the proprietary Macintosh computer and the relatively open iTunes music warehouse. It did not support non-Apple music formats or software applications, though any content provider could join iTunes. Eventually, it seems that consumer and market pressure persuaded Apple to open up the interfaces to the iPod software (but not the hardware) so that it could play some other music formats (but not those championed by Microsoft or Real). Apple also started out with proprietary digital rights management (DRM) technology on the iPod and its iTunes store, creating problems with potential ecosystem partners as well as customers, although the service and the Apple devices have been more open since 2009.

The iPod, and not the Macintosh, seems to have taught Apple how to behave more like an *industry* platform leader. In 2002, it introduced an iPod compatible with Windows and then opened a Windows version of the iTunes online store in 2003. By mid-2008, the iTunes store had become a near monopoly in its own right, with about a 70 percent share of the worldwide market for digital music.[20]

Then, in 2007, Apple introduced the iPhone—what I called "the most exciting electronics product to hit the market since the Macintosh."[21] But quickly the debate ignited again over whether this was a product or a platform. The iPhone was distinctive first because of another remarkable user interface (there is a pattern here!) driven both by touch and virtual keyboard technology. But the original iPhone would not run applications not built by Apple, and it would not operate on cell-phone networks not approved by Apple (initially only AT&T in the USA, but later Deutche Telekom/T-Mobile in Germany, Telefonica/O2 in the UK, and Orange in France). Fortunately for consumers, hackers around the world found ways to "unlock" the phone and add applications. A black market also developed for "hacked" devices. This market pressure again seemed to persuade Apple management that its latest great product was also becoming a great new platform, at least in the United States, and so the interfaces needed to be more open to outside application developers and other complement producers.

It is possible that Apple executives all along planned to open up the interfaces gradually, if the product won broad market acceptance. The facts are that the opening did happen, but slowly and painfully for many users. In March 2008, Jobs announced that Apple would license Microsoft's email technology to enable the iPhone to connect to corporate email servers. By April 2010, there were nearly 200,000 applications available for the iPhone through the official App Store. Some applications were free, and many vendors continued to sell unauthorized "illegal" applications over which Apple had no control—something to which Apple, unlike Microsoft, is unaccustomed.[22] Apple also had yet to allow consumers to use the iPhone on any service network they chose. Apple's repeated attempts to control applications that work on its iPhone platform led to several very public confrontations with Google, banning of some very useful technology (such as Google Voice), and the resignation of Google CEO Eric Schmidt from Apple's board of directors. Google's expansion into mobile operating system software and applications has transformed it from being Apple's partner in the competition with Microsoft over Internet search and desktop software applications ("the enemy of my enemy is my friend") into Apple's rival in the cell-phone business.[23]

Despite some gaps in its historical strategy, Apple finally seems to have figured out how to play on both sides of the industry platform game and to create platform-like synergies and network affects across several of its product lines as well as complements. The iPod, iPhone, and iTunes service all work particularly well with the Macintosh computer, and have some interoperability with Windows—a kind of "closed, but not closed" strategy. And providing its own essential complements—like Microsoft has always done for DOS and Windows—has become critical to Apple's success. The iPod is not very valuable without external digital content such as music and video files. These complementary innovations also make the versatile iPhone and other smartphones much more valuable than ordinary cell phones. Here, Apple cleverly found a way to provide the key complements—the iTunes Store and the iPhone App Store. Moreover, these are *automated services*, with low costs and high potential profit margins. Apple is being smart and sharing most (about 70 percent) of these revenues with the content owners. Since 2000, Apple has also been creating more software applications for the Macintosh to reduce its dependence on Microsoft, Adobe, and other independent software vendors.[24]

We can see the results of these product and platform efforts in Apple's much-improved financial performance and market value (Table 1.2). Few people probably know that, in 1995, Apple was nearly *twice* the size of Microsoft in annual revenues (about $11 billion to $6 billion). However, Apple's market valuation was only about *40 percent of revenues*, whereas Microsoft's value was nearly *six times revenues*. Not surprisingly, Microsoft's

Table 1.2. *Microsoft and Apple financial comparison*

Year	Microsoft			Apple		
	Revenues (% mn.)	Operating profits (%)	Year-end market value (% mn.)	Revenues (% mn.)	Operating profits (%)	Year-end market value (% mn.)
2011	69,943	39.0	218,380	108,249	31.0	376,410
2010	62,484	38.0	238,784	65,225	28.0	295,886
2009	58,437	34.8	246,630	36,537	21.0	180,150
2008	60,420	37.2	149,769	32,479	19.3	118,441
2007	51,122	36.2	287,617	24,006	18.4	74,499
2006	44,282	37.2	251,464	19,315	12.7	45,717
2005	39,788	36.6	233,927	13,931	11.8	29,435
2004	36,835	24.5	256,094	8,279	3.9	8,336
2003	32,187	29.7	252,132	6,207	(loss)	4,480
2002	28,365	29.2	215,553*	5,742	0.3	4,926
2001	25,296	46.3	258,033*	5,363	(loss)	7,924
2000	22,956	47.9	302,326*	7,983	6.5	5,384
1995	5,937	35.3	34,330*	11,062	6.2	4,481

Note: Fiscal year data. Market value is for calendar year, unless marked with an asterisk, which indicates fiscal year.
Sources: Company *Form 10-K* annual reports and Financial Times Global 500 for 2010–2011.

operating profit margin was also about six times Apple's (35 to 6 percent). Apple shrank in subsequent years whereas Microsoft's sales exploded, with Windows 95 becoming the basis for a new generation of Internet-enabled consumer and enterprise products, including Office. Not until iPod sales began to catch on in 2005 did Apple's revenues, profits, and valuation turn around. Since 2003, Apple's revenues have risen several times more quickly than the overall PC industry. They jumped from $6.2 billion in 2003, with an operating loss, to over $36 billion in 2009, with a 21 percent operating profit margin. In addition, Macintosh computers in 2009 made up only 38 percent of Apple's revenues, down from 72 percent in 2003. The iPod (including the iPod Touch—in essence, an iPhone without the telephony function) accounted for 22 percent of 2009 revenues, music products 11 percent, and the iPhone about 18 percent. Software and services as well as hardware peripherals (the rest of the complete user experience) generated the other 12 percent of sales. It is particularly striking how Apple's market value remained less than its annual revenues for so many years, whilst Microsoft's market value was 8–13 times revenues. But here too, by 2005, the tide had turned. Apple's value has risen, reaching nearly five times revenues in early 2010—now in Microsoft territory, since Microsoft's valuation has been on the decline, owing, at least in part, to commoditization in PC hardware and software markets.

Most important for our purposes in this chapter is to recognize that Apple's resurgence reflects, at least in part, the value of a *platform company* compared to a product company. The remarkable financial turnaround since 2005 began with some new "hit" products, and this demonstrates the importance of having a strong product strategy to go along with a platform strategy. But Apple also now has a portfolio of products that have become or are becoming industry platforms, including essential complementary services platforms (iTunes, App Store and iBooks). They all work together and reinforce each other, through strong direct and indirect network effects. Moreover, Apple now benefits from a vibrant ecosystem around the iPod and iPhone, which means it no longer has to do the lion's share of innovation itself! It is finally allowing ecosystems to form that can rival the Windows world, even though Apple at times is clashing with Google, Palm, and other partners and users with regard to how open to make the iPhone and iTunes. In 2010, Apple also introduced the iPad. This is a more elegant tablet computer than Microsoft's earlier design, and uses the same remarkable touch-screen technology as the iPod Touch and the iPhone. The iPad has some technical limitations—such as the inability to run more than one application at a time, and the lack of support for Adobe's rival Flash video technology (which the iPhone does not support either, even though Flash is used for the vast majority of videos and advertisements on the Web). But Apple was also reaching agreements with major book and newspaper publishers as well as encouraging iPhone developers to build applications that will make the iPad a new platform for surfing the Internet and handling digital content (music, photos, books, magazines, newspapers, videos, and documents).

Apple's recent successes illustrate my general point: if Steve Jobs and Apple had tried to make "insanely great platforms" first and "insanely great products" second, then most personal-computer as well as smartphone users today would probably be Apple customers. We would have lived much more in an Apple, rather than a Microsoft world. Apple has grown from being merely a fifth of Microsoft's size in terms of sales as late as 2003 to just over half in 2009. Apple's rate of growth suggests that it may once again surpass Microsoft in revenues, though this may not be so important. It is sobering to realize that General Motors in 2008 had revenues of about $150 billion—two and a half times that of Microsoft—along with billions of dollars in losses and then a US taxpayer bailout. Revenues are only part of the story for a firm; the real bottom line for investors is market value, which is driven by elements other than sheer scale.

On these dimensions, Apple has improved markedly in just a few short years. But it still is much less profitable than Microsoft and is not likely to reverse this situation any time soon. Apple will always struggle to maintain the distinctiveness of its products and to convince new customers beyond the first wave of early users to pay those premium prices. Customers will spend more for a product when it is new and path-breaking. The difficulty arises when the novelty wears off and cheaper copy-cat products appear that are "good enough." Bill Gates learned this lesson early on in his career and ruthlessly (effectively?) exploited this characteristic of the market. We can see this not only in the way Windows mimicked the look and feel of the Macintosh, but also in how Word and Excel in the 1980s and 1990s mimicked the functionality of WordPerfect and Lotus 1-2-3. The Windows NT and Windows 2000 server also took billions of dollars in revenues from Novell and UNIX vendors.

Apple probably has the world's happiest and most loyal customers, but that is not enough to keep its growth rates high. It needs more new customers, especially outside the United States. Apple probably cannot charge higher prices than it has done already in the past few years; in fact, it dropped prices on the iPhone significantly in 2008–9. Prices on this and other products such as the iPad will probably fall as well whenever there is a recession and as competition intensifies.

The Microsoft-Intel ecosystem has at least one advantage: its customers do not have to love their product to buy it and do not have to pay premium prices. Most users do not even choose Microsoft or Intel products directly. For example, in fiscal 2009, only about 20 percent of Microsoft's Windows desktop (client) revenues were direct sales to consumers, and this amounted to a little more than 5 percent of total revenues.[25] Overall, only 30 percent of Microsoft's sales were directly to consumers (20 percent of Windows desktop and 20 percent of the Office division, and all of Online Services and Entertainment and

Devices sales). Most of the Windows desktop and server as well as Office sales were either to OEMs (the PC makers) or to enterprises and other large organizations (Appendix II, Table II.3). This remains true in 2010, despite open-source and "free" software. In addition, Apple has still not created the enormous recurring revenues that Microsoft's ecosystem and enterprise customers generate, with those continuing sales of Windows and Office to PC manufacturers and corporations, as well as individuals—who will mostly upgrade their PCs if not their software products, eventually.

More importantly, Microsoft has those wonderful profit rates generated from the software product business.[26] The cost of reproducing a software product is essentially zero. Since 2000, Microsoft has typically had gross margins of 65–80 percent and operating margins (profit before taxes and investment income) of around 35 percent. This compares to gross margins for Apple of 34 percent and operating profit margins of 18–19 percent in 2007–8, after years of much lower profit (and revenue) levels. In addition, though Apple won the battle for digital media players with the iPod, that product, like personal digital assistants (PDAs), is likely to disappear in favor of smartphones. Apple may yet win the global smartphone battle, but the iPhone still trails RIM's Blackberry and Symbian/Nokia smartphones by a wide margin, especially outside the United States, where the Macintosh has a tiny following. Nokia, Samsung, Palm, and other firms using Google's Android software platform are also introducing products that look and feel similar to the iPhone. Google has even designed its own phone, called Nexus One and introduced for marketing in 2010, to mimic the iPhone features and take special advantage of the Android software.[27] In addition, RIM, Nokia, and Palm have growing online stores for their smartphones. And Windows 7 is an important step forward for Microsoft in reducing the usability gap between PCs and the Macintosh.

In the long run, if hardware and software products both continue to experience commoditization and declining prices, then the most valuable part of the Apple franchise might end up being iTunes. The hardware products may simply become platforms to sell high-margin automated digital services, including music and video content. The acquisition in December 2009 of Lala, the streaming Web music service, also gives Apple the technology to allow users to store their music and listen to songs from different devices, anywhere and anytime.[28]

Platform or Product—or Both?

Perhaps the most challenging question for managers gets into the heart of strategy and innovation: is it possible for a firm with Apple's creativity, foresight, and independence to think "insanely great platform" first and still produce such great products? Based on Sony's experience with VCRs, or Microsoft's with DOS and Windows, it appears that platform companies do need to make technical and design compromises in order to work effectively with other industry players and encourage them to be partners and complementors rather than competitors. Nokia has done this reasonably well by convincing some competitors to join its Symbian consortium to develop an alternative mobile operating system to Microsoft and then making this an independent non-profit as well as open-source entity. I hear from Apple people that Steve Jobs and other executives have been acutely aware of the product versus platform distinction and deliberately chose not to follow an open platform strategy until recently. They have preferred to control "the user experience" and take most of the revenues and profits for Apple, though more recently with a "closed, but not closed" approach. It appears that a more open industry-platform strategy is only a secondary consideration. But the fact that Apple did open up its platforms eventually without losing their distinctiveness as products suggests the company could have pursued product and platform leadership simultaneously. The challenge here is to be open, but not so open that the platform leader makes it too easy for competitors to imitate the essential characteristics that make the original product so appealing.

Of course, despite the many examples, not every market is or will become a platform industry (though most related to information or digital technology are) and not every product can become an industry platform. Annabelle Gawer and I considered this issue in a recent article and concluded that, for a product or component technology to have platform potential, it should satisfy two conditions.[29] First, the product or technology should perform at least one essential function as part of a "system," like the scanning mechanism and playback format in a home video recorder, or the operating software and microprocessor hardware in a personal computer. The function is essential if it solves a critical system-related problem for the industry, such as how to encode video signals or control the operations of a personal computer or a smartphone. Second, the product or technology should be relatively easy for other companies to connect to with their own products, components, or services in order to improve or expand the functionality of the overall platform system, for both intended and unexpected uses.

Some complementors also become platform leaders within a platform. Adobe, founded in 1982 to make laser printer software for Apple computers, falls into this category. It has become one of the most profitable software companies in the world—with 2009 revenues of $2.8 billion, a gross margin of 90 percent, and an operating profit rate of 23 percent. It rivals Microsoft in sales productivity and profitability. Adobe gives away or sells platform technologies and tools (Acrobat readers and servers, Photoshop, Illustrator, Flash and Dreamweaver, Cold Fusion, Air, etc.) for printing and editing digital files, including text, photos, and videos, as well as for creating Web content. Other firms build complementary hardware and software products such as laser printers, special font sets and editing tools, or applications with Flash video clips that use Adobe technology. Still more firms use Adobe products to offer their own digital content and online services. But Adobe's main products (though not those using technologies that directly threaten alternatives from Microsoft and Apple) are also wonderful complements for the most common platforms in the software business—Windows personal computers and smartphones from Apple, RIM (Blackberry), Microsoft, and Google.[30]

It is important to realize as well that a company does not have to be the first to market or to have the best technology to become the platform leader and achieve the dominant market share in its industry. But platform leaders and wannabes do need to encourage innovation around their platforms at the broad industry level. The reason is that platform leaders usually do not themselves have the capabilities or resources to create all possible complementary innovations or even complete systems in-house. Yet the value of their platforms depends on the availability and innovativeness of these complementary products and services. In addition, based on the history of other platform technologies, where wars over incompatible standards often led to market confusion and wasted innovation, we can say that platform industries generally need architects. This is where platform leadership becomes important.

The Concept of Platform Leadership

In 2002, Annabelle Gawer and I described the concept of "platform leadership" as motivated, at least in part, by "a vision that says the whole of the ecosystem can be greater than the sum of its parts, if firms work together and follow a leader."[31] We identified four "levers" or strategic mechanisms that companies such as Intel, as well as Microsoft and Cisco, used to influence producers of complements.[32] The lever terminology and the four categories are from us. But the dimensions of platform leadership came from our observations over several years. We also believed that firms who wanted to become platform leaders (wannabes) needed to figure out a coherent strategy along these four dimensions, though it was equally clear from our research that there were different paths to this "holy grail" of platform leadership.

The Four Levers

The first lever we called the *scope of the firm.* By "scope" in this context, we meant a kind of corporate diversification, or the breadth of what the platform leader does itself: specifically, what complements does the platform leader or wannabe make in-house versus what it encourages outside firms or partners (or users) to make. This dilemma resembles the "make versus buy" debate in vertical integration strategy. But, rather than buying complements, platform leaders generally try to influence other firms to decide on their own to produce products or services that make the platforms more valuable. The key idea is that platform leaders or wannabes need to determine whether they can or should develop an in-house capability to create their own complements or whether they are better off letting "the market" produce complements. They can also take an intermediate approach, such as to cultivate a small in-house capability.

For example, since 1980, Microsoft has encouraged many third parties to develop applications using the interfaces embedded in its part of the PC platform—DOS and then the Windows operating systems. At the same time, Microsoft has developed the capabilities to make many complements—including the most important complements—itself. Thus we see Microsoft introducing a DOS version of Word and the predecessor to the Excel spreadsheet in 1983–4. Microsoft also introduced a mouse in 1983 for the DOS version of Word, which was especially useful later on for graphical versions of these applications. Microsoft started designing these products in 1981, at Apple's request, and then began selling them for the Macintosh in 1984.[33]

With its forays into the other "side" of the PC platform—from the operating system to the complementary applications—Microsoft was able to ensure that new generations of DOS and then Windows (which was good enough to start selling well from version 3, released in 1990) had the most critical complements available and were optimized for the next generation of the platform. It is very different for other platform companies and wannabes, such as Apple, Nokia/Symbian, IBM (WebSphere), Palm, and Red Hat/Linux. They rely much more heavily on third parties to provide complementary products. Intel also has been Microsoft's partner as the PC platform leader. Intel and Microsoft took over this position in the 1980s from IBM, which designed the original PC but did not control rights to the operating system or microprocessor design. But Intel lacks the in-house software capabilities (though it employs thousands of programmers) to develop mass-market consumer applications and systems software. So even a firm as powerful and wealthy as Intel has to rely on Microsoft and other firms to produce new generations of operating systems, hardware peripherals, and software applications that take advantage of new generations of its microprocessors—a dilemma that David Johnson, a senior Intel manager, described as "a desperate situation."[34]

The second lever is *product technology (modularity of the architecture, and openness or accessibility of the interfaces and intellectual property)*. Platform leaders need to decide on the degree of modularity for their product architectures and the degree of openness of the interfaces to the platform. In particular, they must balance openness with how much information about the platform and its interfaces to disclose to potential complementors, who may use this information to become or assist competitors. We know from various studies that an architecture that is "modular" and "open"—rather than "integral" and "closed"—is essential to enable outside firms to utilize features or services in the platform and innovate around it. The original Macintosh, as well as the early versions of the iPod and the iPhone, are all good examples of closed integral architectures, in both their hardware and software. For the PC, application developers use programming interfaces that are essential parts of Windows, which Microsoft owns. But detailed information and examples for how to use these interfaces to develop applications are open to anyone and free with the Windows Software Development Kit (SDK).

The third lever is *relationships with external complementors*: platform leaders need to determine how collaborative versus competitive they want the relationship to be between themselves and their complementors, who may also be or become competitors (such as the relationship between Microsoft and IBM/Lotus, Apple, SAP, Oracle, Adobe, Intuit, and many other software product firms). Platform leaders need to worry about creating consensus among their complementors and partners. The biggest concern is that they may have to resolve conflicts of interest, such as when the platform company decides to enter complementary markets directly and turn former complementors into competitors. Microsoft generally limited the scope of its business, but it has always maintained it would compete with complementors if the market seemed sufficiently attractive. Accordingly, from programming languages and operating systems, beginning in the early 1980s, Microsoft has moved into desktop applications (to compete with WordPerfect and Lotus) and personal finance software (to compete with Intuit, though not very effectively), in addition to networking software (Novell), databases (Oracle and IBM), browsers (Netscape), media players (Real and Apple), online content (Yahoo!), search engines (Google and Yahoo!), video games (Electronic Arts and many others), mobile operating systems (Nokia/Symbian, Palm, and the Linux community), and business applications (SAP and Oracle)—to name only a few examples. Microsoft's strategy is generally to enter any "horizontal" (as opposed to industry-specific or "vertical") business, because anyone with a computer potentially becomes a customer. Its strategy for Windows has also been to ward off potential competition by enhancing the operating system with numerous features that complementors often sell as separate products—sometimes bringing

Microsoft into conflict with the antitrust authorities as well as its complementors.

The fourth lever is *internal organization.* More specifically, platform leaders can reorganize to deal with external and internal conflicts of interest. They may decide to keep groups with similar goals under one executive, or separate groups into distinct departments if they have potentially conflicting goals or outside constituencies. For example, Intel established a virtual "Chinese wall" to separate internal product or R&D groups that might have conflicting interests among themselves or clash with third-party complementors, such as chipset and motherboard producers. The latter relied on Intel's advance cooperation to make sure their products were compatible. When Intel decided that these chipset and motherboard producers were not making new versions of their products fast enough to help sell new versions of microprocessors, Intel started making some of these intermediate products itself—to stimulate the end-user market. But it still kept its laboratories in a neutral position to work with ecosystem partners.

By contrast, Microsoft claimed not to have such a wall between its operating systems and applications groups—despite the potential conflicts. Microsoft also insisted that "integration" of different applications, systems, and networking technologies (such as embedding its own Internet browser, media player, and instant messaging technology into Windows) was good for customers because it improved performance of the overall system. There is some truth to this. It is one reason why the user experience with the far more integral Macintosh system is better than the Windows-Intel PC experience, which has always mixed and matched hardware and software from many different vendors. But Microsoft leveraged the market power of Windows and its other platform, Office—which by the latter 1990s had evolved into another set of services and tools used by various companies to build their own desktop application products—to influence the direction of the software business.

It is not illegal under US or most other antitrust regulations to have a monopoly or any particular share of a market. Microsoft has controlled as much as 95 percent of the desktop operating systems market. Intel has produced 80 percent or more of PC microprocessors. Cisco has sold perhaps 70 percent of basic Internet routers in its peak years. ARM PLC licenses the microprocessor designs in some three-quarters of all smartphones. Qualcomm has had a similar dominant share in cell-phone wireless chips using the Code Division Multiple Access (CDMA) technology. In recent years, Apple has gained a dominant market share with the iPod and iTunes. But it is illegal to utilize a monopoly position to harm consumers and competitors, such as through predatory pricing or contracts that impede competition and supply of the product. It is also illegal to use a monopoly in one market to enter an adjacent market by tying products together and thereby limiting consumer choice and restraining competition. Microsoft, as we know, committed these kinds of violations when it bundled Internet Explorer with Windows and did not charge extra for it. Microsoft also pressured PC makers not to load Netscape Navigator on their machines—essentially destroying Netscape's browser business and reducing competition in this market. Microsoft argued that the browser was an integral part of Windows. But Microsoft also sold or distributed the browser as a separate product, as did Netscape and several other companies, so this argument made little sense. Again, antitrust enforcement in the United States, Europe, and Asia has frequently forced Microsoft to adjust its behavior, though usually too late to make much difference in the current market. In browsers, for example, in December 2009 Microsoft reached a settlement with the European Commission to update versions of Windows used and sold in Europe through to 2015. The software update offers users the ability to select from several alternatives, including browsers from Apple, Google, and Mozilla Firefox (the open source successor to Netscape Navigator).[35]

A major focus of the *Platform Leadership* book was to dissect the case of Intel and then compare it to other established platform leaders that had followed somewhat different paths, such as Microsoft and Cisco. Gawer and I also analyzed several leader wannabes—Red Hat (which was pushing Linux), NTT Docomo (which was trying to export its dominant i-mode cell-phone platform overseas), and Palm (which was pushing both the Palm operating system as an industry platform and selling its own PDAs as early handheld computers). Based on these examples, we came to several conclusions with regard to the four levers.

How a platform leader or wannabe should position itself on Levers 3 and 4 seems relatively clear. Although they have many organizational and strategic options to choose from, firms in a potential or actual leadership position with a platform technology need to rely on cooperation (as JVC did) to encourage outside innovation around their platforms. They also need to deal internally with potential conflicts of interest if they make their own complements that compete with partners such as OEM licensees or complementors.

How to deal with the choices inherent in Levers 1 and 2 is more complex (see Figure 1.2). Whether or not to make complements yourself, and how open (or how closed) to make your platform—and thereby subject your technology to the scrutiny of potential competitors as well as complementors—has continued to vex platform leaders and wannabes. We have seen managers struggle with this issue not only at Apple but also in recent years at SAP (with NetWeaver, a "middleware" software program that integrates externally built enterprise applications with SAP's internally built applications and development tools) and EMC (with WideSky, another middleware software program designed to control different data storage systems).

Various cases, especially that of Microsoft, suggest that the "best place" to be, first, is to have a strong capability to make your own complements, whilst still offering incentives to encourage outside firms to do the same. And, second, to have a platform open enough for

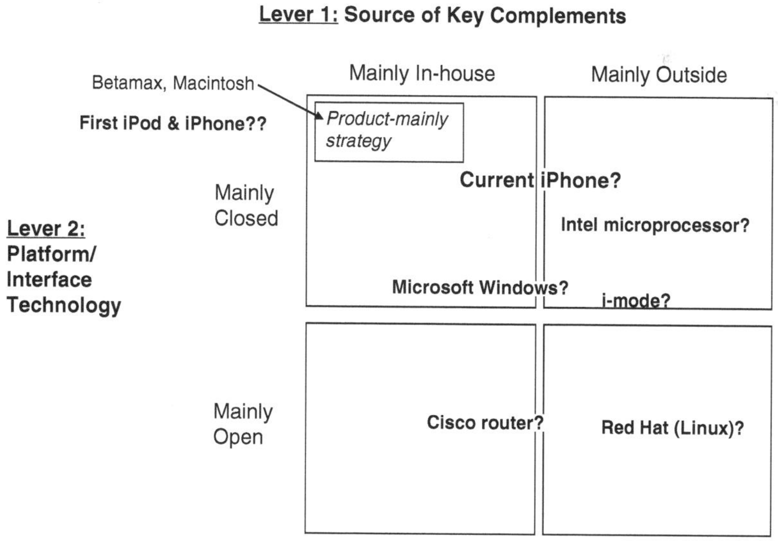

Lever 1: Source of Key Complements
Mainly In-house
Mainly Outside
Betamax, Macintosh
First iPod & iPhone??
Product-mainly strategy
Current iPhone?
Lever 2: Platform/ Interface Technology
Mainly Closed
Intel microprocessor?
Microsoft Windows?
i-mode?
Mainly Open
Cisco router?
Red Hat (Linux)?

Figure 1.2. The strategy spectrum for Levers 1 and 2.

complementors to thrive but closed enough to protect the core technology from easy imitation, such as through patents or proprietary ownership with special licensing agreements. Cisco is vulnerable, because its platform evolved from the router to the IOS operating system. It contains proprietary technology but relies heavily on open networking standards and technologies—primarily the Internet. Red Hat, as the primary distributor of Linux, is probably in the *worst* place. The platform technology is completely open, not to mention free (if you download Linux from the Web); and most of the complementary innovations that make Linux valuable as a platform (such as Apache or Web server hardware) come from outside firms or the open source community, over which Red Hat has limited influence. Red Hat does have options, though. It built a service capability to make money, and used its own programmers to enhance Linux and build special utilities (such as for installing and updating the system) as well as to create some open-source applications.[36]

Somewhere in the middle is probably the best place strategically, because then a firm can benefit from the best of both worlds. For example, Microsoft makes its own key complements and it has cultivated an enormous ecosystem of hardware and software manufacturers that has kept it ahead of Apple. As Apple has moved closer to a similar position of multiple platforms, and complements, and more balance between being open versus closed, it has greatly improved its financial performance.

With regard to how to *behave* as a platform leader, Intel has generally been a good role model for other firms.[37] It did not flaunt antitrust regulations as openly as Microsoft did before losing the antitrust trial, though Intel has recently attracted a lot of antitrust scrutiny, particularly because of clashes with rival microprocessor manufacturer AMD. Be this as it may, Intel has provided an excellent model for the *process* of platform leadership. Job 1 should always be to sell your basic products (in this case, microprocessors) and protect the core platform technology from imitation. But Job 2 has been to encourage complements. In so doing, Intel has taken risks to open up its microprocessor interfaces, assist complementors, and give away a lot of important technologies. It has also tried to help key complementors and partners make money—necessary to keep the ecosystem vibrant and to reduce the potential of complementors becoming competitors. In retrospect, we can see that Intel, through the 1990s and early 2000s, followed a specific set of measures to encourage complementors to adopt and continue supporting its microprocessor platform:

- Create and communicate a *vision* of platform evolution.
- Build a *consensus* among a small group of influential firms for the vision and new initiatives.
- Identify and target *system bottlenecks.*
- Distribute *tools and enabling technologies* to help outside firms develop complements fitting the vision.
- Highlight business opportunities and help leading firms to stimulate the market in different areas (Intel called these firms "rabbits").
- Facilitate *multi-firm initiatives* to reduce system bottlenecks and promote new standards, interfaces, and applications.

Challenges for Platform Leaders or Wannabes

Intel as well as other established platform leaders such as Microsoft and Cisco have maintained their market positions for decades. This kind of staying power is impressive, because new companies continually emerge that want to become the next generation's platform leader. These wannabes encounter special challenges when they tackle incumbents. For example, they may need to turn a product market into a platform market, such as by gradually becoming at least an architectural leader for the next-generation technology or new, broader applications. Hence, the four levers themselves may not be enough for wannabes to develop specific action plans, such as in two particular areas: becoming "core" or essential in an emerging platform market, and helping a market tip in their direction when there is more than one competing platform.

How to Become "Core" to a New Industry Platform

The first challenge, which Gawer and I called "coring," requires a leader-wannabe to resolve a major technical problem affecting a system-like product with industry platform potential. Most companies choose to protect proprietary knowledge if this approach is likely to help them get a high return on their investment. But platform-leader wannabes must also encourage other firms to adopt their solution, join the emerging ecosystem, and alter their R&D plans to develop complementary applications rather than a competing platform or incompatible complements. Netscape and Microsoft faced this problem when they tried to convince Web masters and developers to optimize their websites and Web-based applications for Navigator versus Internet Explorer. It is therefore useful for the leader-wannabe to introduce a platform that is "open, but not open" or "closed, but not closed," as well as "free, but not free." That is, the platform should appear open enough in the sense of adopting as many publicly available standards as possible and be easy or cheap enough for outside firms to connect to, but still contain enough protected or proprietary technology to facilitate some way for the leader to make money. It is also essential for the new platform technology to generate direct or indirect network effects with any complementary products or services, such as through technical standards and compatibility-dependent formats or interfaces that make it difficult for complementors and customers to switch to another technology.

Managing the *technology side* of the platform is one problem; the leader-wannabe also has to manage the *business side* by creating the appropriate economic incentives for companies to join the ecosystem of another company and a potential rival. To understand these issues more fully, our follow-up research looked at numerous cases of successful, failed, and inconclusive coring initiatives. We hoped these would lend some insight into how best to implement a platform strategy for an industry that did not yet have a core technology or a platform leader.

Google provides an excellent and commonly understood example of successful coring. It started off as a simple search engine company in 1998, founded by former Stanford graduate students in computer science Larry Page and Sergey Brin.[38] They went on to establish their proprietary technology as a foundation for navigating the Internet. The company's algorithms solved an essential technical problem—how to find anything on the World Wide Web, which, even in the late 1990s, was adding millions of websites, documents, and other content each year. Most cleverly, Google distributed its technology to website developers and users as an "automatically embedded" toolbar that was easy to connect to and use, and free. Then Google made its search content available to various outside parties for their own products and services, such as an application combining search information with local maps and restaurants or other location-specific information. It was essential that Google found a way for itself and partners—mainly advertisers in the beginning—to make money on the Internet by linking focused context-specific advertising to user searches in a way and on a scale that earlier search engines could not. Some 70 percent of Internet shopping begins with search, and Google gets paid whenever users go to websites its posted ads recommend.

Moreover, everything that Google does is reinforced by network externalities—the "increasing returns" or positive feedback generated by advertisers, users, and affiliated websites that embed the Google search bar—even without the benefit of strong direct network effects. It is now obvious that the more users and advertisers who use Google, the better the searches and the advertising information, and the more money that flows to Google and its partners. And Google can continue to expand its automated services, which have evolved into a broad platform. The strongest network effects are indirect, and the benefits are mostly around advertising. Google refines its search algorithms and results every time somebody does a search, so there are network effects here, though probably with some diminishing returns. But advertisers want to advertise where there is the most search traffic, and that by far is Google. So the more search traffic Google acquires, the more advertising it acquires, and the higher the prices it can charge. In fact, advertisers bid for priority listing for their sponsored ads!

There are different ways to measure market share (search volume, number of hits), but they all show that Google continues to grow, mainly from its base in the United States. It started 2007 with a US share under 55 percent and ended 2009 with around 65 percent. Meanwhile, Yahoo! dropped from 28 percent in 2007 to under 20 percent in 2009. Microsoft has remained with about 10 percent, though it was also actively trying to increase its share. In 2009, Microsoft introduced a new search engine, called Bing. This did not seem superior in general features but supported more refined searches than Google. Microsoft also reached a ten-year agreement with Yahoo! to provide search technology in return for 12 percent of Yahoo!'s add revenue. In addition,

Microsoft was negotiating with content sources such as News Corp., publisher of the *Wall Street Journal*, to give exclusivity to Bing.[39] And Google's overseas market share, comparable to that in the United States, was not growing as fast. It may even contract as specialized and language-specific search engines gain momentum, especially after Google moved its Chinese search operations to Hong Kong in March 2010.[40]

It is worth re-emphasizing that Google did not start out as a *platform* company; it was not even first to this market. The founders simply wanted to produce a better search engine *product*, which they delivered as an automated service over the Web. Page and Brin discovered how to rank website pages in terms of popularity measured by linkages to other websites. Nonetheless, the Google founders and senior executives, led since 2001 by CEO Eric Schmidt (formerly of Sun Microsystems and Novell), realized that search technology was not "sticky" enough by itself. Anyone can switch search engines with a simple click of a mouse, even though they get attached to their list of favorites (which users can export to other search engines). Google itself exploited the lack of stickiness in Internet search when users switched over from earlier search engines—Altavista, Inktomi, Yahoo!, and others.

But, to counter the absence of strong direct network effects such as benefited Microsoft, Intel, and JVC, Google gradually adopted a platform strategy to attract and keep users. It leveraged the search technology to create a broad portal for various products and services, as well as applications from third parties. Google now offers everything from email to basic desktop applications (competing with Microsoft Office) and cloud-computing services. In addition, clearly heading toward a direct confrontation with Microsoft, in 2008–9 Google entered the infrastructure software business. It released the Chrome Internet browser, then the Android operating system and development platform for smartphones, and finally the Chrome operating system for small "Netbook" computers connected to the Internet.[41] All these software products are free and open source, and supported by Google's search advertising revenue.

Yet, despite various initiatives to draw users to its platform and to grow its product and service offerings, Google is unlikely to turn Internet search into a global "winner-take-all" market. This occurred with PCs and VCRs, with the dominant firms ending up with

90–100 percent of the market. Search more likely will remain a case of "winner-take-most," more like Internet routers and microprocessors. Going back to the Eisenmann, Parker, and Van Alstyne framework, we can see why. First, there is room for differentiation and niche strategies; in China, Brazil, and a few other countries, local search engines were gaining market share in 2009, whilst Microsoft's Bing was gaining attention. There are also many specialized search engines, such as for video content, that should become more important in the future. Second, the network effects for search are more indirect than direct. However, the more Google adds products and services, the "stickier" the platform becomes, and the more search users it acquires, the greater its share of Internet advertising revenue. And, third, users still can almost effortlessly switch or use more than one search engine (multi-homing).

Qualcomm is another prominent case of successful coring. This company, founded in 1985 by former MIT engineering professor Irwin Jacobs, quickly became a leader in wireless communications technology for the cellular phone industry and then gradually diversified into PCs and other devices.[42] Like Google, Qualcomm solved a basic technical problem for the industry—the incompatible and inefficient wireless technologies of the late 1980s and early 1990s. Irwin's company invented the CDMA technology, which lets many users operate on the same channel by assigning specific codes, breaking the signals into small bits, and then reassembling them later, much like the Internet does with data packets. Qualcomm also sold chipsets that were easy to adopt and customize. Similar to Google though not as fast or as much, Qualcomm became a multi-billion-dollar firm with enormous profits and some astounding years—2003–5 in particular (see Table 1.3). On the business side, unlike Google, Qualcomm has not allowed its ecosystem partners to earn very much money. It has charged very high license fees (this technology is not free!) and vigorously enforced a large number of patents. In response, competitors such as Nokia and Broadcom, and overseas governments such as China, have sought technical alternatives and challenged Qualcomm's patents and fees in court.

Of course, there have been many failed attempts to disseminate a core technology or service and create a new industry platform. One

Table 1.3. *Qualcomm and Google financial comparison*

	Qualcomm			Google		
	Revenues ($ mn.)	Operating profits (%)	Year-end market value ($ mn.)	Revenues ($ mn.)	Operating profits (%)	Year-end market value ($ mn.)
2011	14,957	33.6	86,665	37,905	31.0	210,000*
2010	10,982	33.9	67,115	29,321	35.4	96,501
2009	10,416	31.2	77,744	23,651	35.3	97,782
2008	11,142	33.5	62,724	21,795	30.4	116,684
2007	8,871	32.5	68,728	16,594	30.6	104,596
2006	7,526	35.7	79,774	10,605	33.5	98,268
2005	5,673	42.1	56,519	6,139	32.9	53,030
2004	4,880	43.6	48,251	3,189	20.1	27,286*
2003	3,847	40.9	28,304	1,466	23.3	n.a.
2002	2,915	28.8	27,785	440	42.3	n.a.
2001	2,680	1.5	38,831	86	12.8	n.a.
2000	3,197	22.6	45,529	19	(loss)	n.a.
1999	3,938	10.6	56,212	2	(loss)	n.a.

Notes: Fiscal year data, except when marked by asterisk, which indicates calendar year.
n.a. = not available (Google went public in 2004).
Sources: Company *Form 10-K* annual reports.

such example involves General Motors, which launched OnStar in 1995 with the goal of giving wireless capabilities to the automobile for navigation systems, directions, accident notification, remote diagnostics, maintenance reminders, Internet connectivity, remote opening of locked vehicles, and other services. GM established OnStar as a wholly owned subsidiary in collaboration with its EDS and Hughes Electronics subsidiaries before selling them off. The technology platform consists of hardware, software, and service agreements with a wireless provider. Initially, GM convinced several automakers (Toyota/Lexus, Honda, Audi/Volkswagen, and Subaru) to adopt the OnStar platform. Fairly quickly, however, these firms concluded that the OnStar capabilities and, in particular, the information on the customer that the system generated about driving habits, was too valuable to let a competitor control. Consequently, they decided to build or buy other systems and stopped licensing OnStar. In retrospect, GM created impressive technology but failed to create proper economic incentives for its service to become a neutral industry platform. It might have spun off OnStar as an independent company. Or GM might have followed Intel and created the equivalent of a "Chinese wall" around OnStar.[43]

Consulting firms create these kinds of walls all the time, to protect client confidentiality. GM continues to sell OnStar as an in-house service, but the lost opportunity to create a new industry platform is enormous.

Another problematic example involves EMC, a market leader in data storage technology founded in 1979. It launched an effort in the early 2000s to establish its WideSky technology as a new industry-wide platform.[44] EMC invented this middleware software layer to integrate and manage third-party storage hardware. In theory, WideSky solved an important technical industry problem—how to connect a growing assortment of storage systems from different vendors. In practice, EMC was unable to convince competitors—principally IBM, Hewlett-Packard, Hitachi, and Sun Microsystems—to adopt this technology as their own. Instead, these firms collaborated by establishing a new open-standards platform managed by their own "neutral" organization, the SNIA (Storage Networking Industry Association). EMC eventually joined SNIA as well. Like GM, EMC succeeded on the technology side by creating a potential core for a new platform but failed at the business side.

A potentially enormous market that lacks both a platform leader and a core technology is the digital home. Many technology vendors are vying to become a force here, but the market is still in a very early stage. The goal of several firms since the mid-1990s has been to connect entertainment devices (for example, television, stereos, and music players) and appliances (for example, heating or air conditioning systems, refrigerators) with a home computer network to enable centralized or remote control as well as billing. To further this vision, several companies in 1999 formed a group called the Internet Home Alliance, including Sears, Panasonic (Matsushita), General Motors, Intel, and Cisco.[45] Many other firms have joined this and a successor organization in later years. At the moment, Microsoft and Intel are once again trying to become the leaders, though it is not clear that either will succeed. Apple, Sony, Hewlett-Packard, Samsung, and several other firms already produce key software and hardware components, some of which could become core elements in a digital home platform. But the market is so diverse that it may never converge around one hardware or software technology. Or the industry might require a different type of platform leader, such as a government agency or industry organization with the ability to influence regulation. We can already see signs of this happening. A large non-profit industry coalition for home builders, the Continental Automated Buildings Association, has taken over the Internet Home Alliance and continues working on the long-term platform goals. Key directors of this organization include executives not only from builders such as Tridel and Leviton but also from technology companies, including Bell Canada, Honeywell, Hewlett-Packard, Microsoft, AT&T, Invensys, Cisco, Siemens, Panasonic, Whirlpool, and Trane.[46]

How to "Tip" a Market

The second challenge for a leader-wannabe is to help a market "tip" or move to a strong market share for its platform—when at least two platform candidates compete. As with coring, successful tipping requires managing both the technology and the business sides of the platform. Most commonly, firms have used price to attract users, but this is rarely sufficient. Our view of the tipping problem is broader.

At the simplest levels, leader-wannabes can use their R&D skills not only to create a core technology that solves an important industry-wide problem, but also to create a high-demand feature or product (a "killer app") that is compelling to users. Apple is superb at doing this—a major reason why the company has several times put itself in an excellent position to tip a market toward its product as the industry platform. Another common tipping strategy is bundling—leveraging a strong position in one market in order to move to an adjacent market. Microsoft has often done this, such as to enter the browser, Internet server, media player, and enterprise computing markets on the backs of Windows and Office. It has also run into antitrust violations. Other companies that have used bundling with fewer legal controversies include Cisco, Intel, Qualcomm, and Nokia, among others. Each has expanded the capabilities of one platform to move into adjacent or similar platform markets. (Eisenmann, Parker, and Val Alstyne call this strategy "platform envelopment."[47]) Other tipping strategies include economic incentives for adopters, such as inexpensive or free licensing terms; subsidies that help one side of the market in order to attract the other, such as providing money or technology assistance to application developers in order to get them to build applications that attract users to the platform, or subsidies that make the platform inexpensive for users. There are also coalitions, such as the group that opposed EMC, or those promoting non-Microsoft technology. Well-known examples of the last technique include Symbian, a consortium Nokia formed to develop smartphone software, as well as the open-source communities behind Linux and Eclipse (a Java development platform and common user interface, originally developed by IBM).

There have been a number of successful tipping efforts. JVC and Matsushita used broad licensing terms to OEM producers and aggressive promotion of content (prerecorded tapes) to push the market toward VHS. IBM, Microsoft, and Intel all worked hard (and together) in the 1980s to recruit application developers such as WordPerfect and Lotus as well as makers of printers to the PC. In the 1990s, Microsoft created the Office bundle and eventually convinced users to switch from WordPerfect, Lotus 1-2-3, and Harvard Graphics over to the Word, Excel, and PowerPoint applications. Microsoft also beat Netscape in the browser market, even though it lost the antitrust case, by bundling Internet Explorer with Windows 95 and later versions. More recent examples include Linux in the back office as a Web server operating system (but not as a desktop operating system, where the overwhelming availability of complementary innovations in the form of application software and cheap PCs based on the Windows standards has limited its diffusion).

Palm is another case of failure to tip, at least in part because management could not decide early enough whether to be a product company or a platform company. It was founded in 1992 by Jeff Hawkins and Donna Dubinsky and in 2000 spun out of 3Com as a separate public company. The firm is most known for pioneering the PDA market with the Palm Pilot, which it sold to great acclaim during 1996–9. But Palm tried to do two things at once, and did neither well enough: (1) establish its Palm device as the pre-eminent PDA product, and (2) promote the Palm OS as an industry software platform available for license to PDA competitors and later on to smartphone producers. Palm has suffered as well from digital convergence—the PDA market is being absorbed by the smartphone market.

Complement producers seem to prefer neutral platforms so that they do not have to compete so directly with the platform leader. Not many firms can get away with what Microsoft does by competing aggressively on both sides of the platform; more common fates would seem to be what GM experienced with OnStar and EMC with WideSky. Palm also did poorly and recognized the conflict. In 2003, the board of directors split the company into two pieces: palmOne for the PDA devices and PalmSource for the OS. But this was too late to overcome momentum behind competing systems and PalmSource became increasingly dependent on palmOne as its main customer. In 2005, a Japanese firm bought the software company, though a newly unified Palm (which had merged with Handspring, maker of the popular Treo smartphone) repurchased rights to the operating system a year later. Then a private investor came in with a major cash infusion in 2007 to keep the company afloat. Today there is less confusion between Palm as product and Palm as a platform, but other platforms still have much more market share in this space. At the end of 2008, Palm announced it would no longer make PDAs and would concentrate instead on smartphones and a new operating system. In mid-2009, Palm finally introduced a competitor to the iPhone.

We can see tipping in action in another arena as well: "Web 2.0" social networking websites, which are characterized by user contributions and support from advertising. These sites make it possible for individuals and for-profit companies to post content (text, video, audio, blogs, advertisements for products and services, and even some application programs) on a main site, using tools such as simple create-and-upload menus. One of the early leading sites was MySpace, founded in 2003 by Tom Anderson and others and then purchased by Rubert Murdoch's News Corp. (Fox Interactive Media) in 2005. This site was overtaken in membership during 2008 by Facebook, a similar site co-founded in 2004 by Harvard student Mark Zuckerberg. This remains an independent company, though Microsoft became a minority investor in late 2007.[48] We also have YouTube, which Google purchased in 2006, as the most prominent site for video-posting. These Web 2.0 platforms compete but differ in their approach to openness.

For example, MySpace in the past has strictly controlled the features embedded in its site and loosened this hold only gradually. By contrast, Facebook learned quickly from Microsoft and, since 2007, has been functioning very much like a software development company—hosting programmer conferences and sharing its mark-up technology (a special version of HTML) as well as its application programming interfaces (APIs) so that outsiders can develop and post applications. Independent developers can also sell advertisements or incorporate tools for conducting online transactions and keep all the resulting revenue.[49] Another new entrant, and the fastest-growing player in this space, is Twitter, the short-messaging and blogging site created by Jack Dorsey, Evan Williams, and Biz Stone in 2006 and funded primarily by venture capital. By the end of 2009, there were some 2,000 third-party applications available for the Twitter platform.[50]

Then we have a highly competitive market that has consolidated over time but may never tip toward one permanent leader—video-game consoles. Sony (PlayStation), Microsoft (Xbox), and Nintendo (the Wii) are the three remaining contenders. Every several years new generations of these consoles appear with different features, triggering a new series of investments and competition. The key complements are software games, some of which work on all three platforms. More importantly, the three leader-wannabes have followed very different platform strategies, reflecting their varied histories as consumer electronics, PC software, and game companies.

Sony won the round prior to the Wii, with a 70 percent market share for PlayStation 2. This company has focused on the high end of the market and "hard-core" players. PlayStation 3 (PS3), not surprisingly, is the most sophisticated and expensive system. In the past, Sony was slow to adopt a platform strategy and did not work very hard at encouraging outside game developers. In 2008–9, this changed. But PS3 was late to market because it incorporated too many state-of-the-art technologies, including the new Blu-Ray DVD format, which was expensive and slow to catch on as the new standard.[51] Sony has also frustrated software companies such as at FixStars in Japan, which used the open characteristics of PlayStation 3 to load its special parallel-processing (multi-core) version of Linux and then build supercomputer-class applications for a relatively inexpensive array of hundreds or even thousands of the consoles. (Full disclosure: I have been an advisor to FixStars since 2008.) Inexplicably, the latest version of PlayStation 3 (called "Slim") no longer permits the loading of a second operating system.[52]

Microsoft, the newest player in consoles, has approached games much like the PC market. It has developed a highly modular software architecture based on Windows and early on disseminated Windows-like programming tools to facilitate game development. It has tried to rally the largest possible number of developers. Microsoft is also strong in online gaming and designed the Xbox console to work well with PCs and the Internet. So far, Microsoft has subsidized one side of the platform (the consoles). It appears willing to lose money on each box but someday recoup its losses from the software complements side of the platform—the license fees that game developers pay. Microsoft's Entertainment and Devices Division, which makes the Xbox, lost some $3.3 billion in 2006–7, but finally made a small profit in 2008 and 2009 on annual revenues of around $8 billion (see Appendix II, Table II.3).[53]

Nintendo, the loser in the prior round of console wars, started out as a manufacturer of playing cards and has always focused on the gaming business.[54] It generally sells the cheapest platform whilst developing in-house or through a tightly controlled network of developers a smaller number of games but potentially bigger hits. Its consoles share a lot of technology with previous generations, making new games cheap to develop. In the most recent round, Nintendo thrilled consumers with a clever system-level innovation combining hardware and software that changed the player's experience: a wireless remote control for its Wii console. This new technology has attracted first-time users ranging from children to the elderly who are interested in exercising and "virtual" versions of golf, boxing, and baseball. Since mid-2007, the Wii has been outselling competitors by a large margin.[55]

If we again apply the Eisenmann, Parker, and Van Alstyne framework, we can see why the video-game console market is unlikely to tip permanently in favor of one platform. There are strong network effects for content exclusive to the individual platforms. But the other factors seem even more important. The consoles have not yet become commodities, and the vendors have quite different backgrounds, capabilities, and market strategies; no one seems vastly superior in this market. As a result, there is differentiation in performance, features, and available complements (the games). Furthermore, the consoles are inexpensive enough for many users to buy more than one (multi-homing) and take advantage of the different features and content (games).

Lessons for Managers

First, implementing a platform strategy or a complements strategy requires a very different mentality and set of actions and investments compared to a product strategy. There are different risks and higher short-term costs. But the long-term economic rewards from a successful platform or complements strategy—especially when one firm creates both, like Microsoft and now Apple have done—can be enormous. Second, managers must still invest in their own innovation and have a strong, multi-generation product strategy. But it is no longer necessary to have the "best" product all the time to win a platform contest. Platform leaders win their battles by having the best platform, and that requires attracting the most or at least the most compelling complements, which will then attract the most users.

On this first point, we can say that platform leaders and wannabes have clear tasks ahead of them. The four levers define the basic game plan: they must design relatively open product architectures and correctly manage intellectual property rights. They must decide what components and complements to build in-house versus allowing the ecosystem to provide. They must work closely with external partners and share the financial pie with them. They must figure out how to organize internally to minimize potential conflicts when stimulating and competing with partners. At the same time, platform wannabes need to solve system-level problems for users and competitors that draw them to the platform, and they must do whatever they can to help the market tip in their direction. Lots to do!

Complementors have a similar agenda, with equal or even greater risks: if there are competing platforms, they must decide which ones to support and how fully to give their support.[56] They have to select which complements to produce and which to let other ecosystem partners or the platform leader make. They need to work closely (but not too closely) with the platform leader or multiple leader-wannabes, and always have something compelling and proprietary to offer. Otherwise, if the business opportunity is large enough, or if the complementor is too independent, the platform leader will probably try to absorb their product or service into its platform. This can be a delicate balancing act. But complementors also have their own power. It can be their product or service that causes a market to tip and stay tipped.

The benefits of success are clear. Platform leaders can have significant leverage over an entire industry for decades—like Microsoft, Intel, Cisco, Google, Qualcomm, and Adobe, or Wal-Mart and Mattel. They benefit from innovation across an entire network of firms, not just within their own boundaries. Moreover, even if one firm does not take a dominant share, platform initiatives can be invaluable for cultivating broad strategic partnerships to improve sales, profits, and innovation capabilities. If the market is growing and becomes very large, platform leaders will also grow and become large. Initial scale by itself, though, is not essential to establish a platform. It is obvious when you think about it, but all the leaders cited in this chapter began as small firms. In fact, Microsoft, Intel, Apple, Cisco, Google, Qualcomm, and Adobe *became large and enormously valuable* precisely because—and when—their platform or complement strategies became so successful.

But, whilst we can distinguish a product from a platform strategy, my second point emphasizes the need to connect the two. It seems hard to succeed with an industry platform strategy if you do not first have a very good (though not necessarily the "best") product. No amount of strategic maneuvering can make up for a product that customers do not want to buy or use. At the same time, platform-leader wannabes do not always have to produce the industry's best product generation after generation to get a market to tip and stay tipped. To strive for "insanely great" products like Steve Jobs has done at Apple is a wonderful way to compete for a firm with unique design capabilities. But, for most firms, it is probably smarter to adopt a strategy that does not depend on always having the most elegant or sophisticated product in the market. Again, the best product does not necessarily win a platform competition; rather, the winner is more likely to be the platform that ends up with the most support from complementors. Complementary products and services create value for the platform and draw in users. In a platform market, garnering support from a broad ecosystem of innovative partners and users is far more important than winning a features contest or a product design award. And it generally requires a specific set of actions that come under the heading of a platform strategy.

Then there is the practical question: if focusing on platforms (or complements to an existing platform) rather than on stand-alone products has such obvious benefits, why do not *all* managers and firms embrace this principle? One reason surely is that managers in the past have not always fully understood the opportunity. Sometimes they preferred not to exploit the opportunity—which I think describes the behavior of Apple before the mid-2000s. Some managers may believe that platform dynamics apply only to high-tech companies. But examples such as Mattel's Barbie doll, Wal-Mart, Marks & Spencer, Best Buy, and CVS and Walgreens demonstrate this is not true.

Some managers may hesitate because they realize how difficult it is to become a platform leader and maintain this position; they may think they are better off selling a product, keeping it relatively closed, and reaping the revenues and profits for themselves and not risking opening themselves up more to possible imitation. Most firms also find it safer to complement some other firm's platform. Platform leaders have to invest heavily in R&D as well as perform a delicate balancing act: whenever a company creates dependence on its technology—as Microsoft, Intel, Qualcomm, Cisco, Nokia, ARM, and others have done—then they also create resentment among their customer partners and, frequently, invite antitrust scrutiny. Nonetheless, when we look deeper, we see platform battlegrounds emerging almost everywhere, in products and in services. What is more, each platform seems to contain other more specialized platforms, which means that many platform companies are also complementors of some larger platform. This is true of all the platform companies mentioned in this chapter. Adobe, for example, has its own platforms for handling text and graphics. But its technologies are also wonderful complements for Macintosh and Windows computers, and a variety of smartphones as well. Even Google search is really a complementary application for any type of Internet-enabled computing and communications device. Google is also perhaps the most compelling complement for Internet service providers and makers of netbooks.

When thinking about platforms and complements, it is important today to think not only about software and hardware products, but also about services. These include both a wide range of value-added services as well as automated or semi-automated services delivered over the Internet. I take up this topic in the next chapter.

Notes

1. See Cusumano and Selby (1995: 384–97) and Cusumano and Nobeoka (1998).
2. There is an enormous literature on product platforms. I usually begin with Sanderson and Uzumeri (1996) and Meyer and Lehnerd (1997). For more academic treatments, see Meyer and Utterback (1993), Ulrich (1995), Sanchez and Mahoney (1996), Baldwin and Clark (1999), Meyer and DeTore (2001), and Meyer and Dalal (2002).
3. See Gawer and Cusumano (2002).
4. I say "almost," because Kevin Boudreau of the London Business School has presented some evidence that suggests too many complementors can reduce the incentives of new complementors to invest. See Boudreau (2006).
5. See http://barbie.everythinggirl.com (accessed Nov. 12, 2009).
6. See the Wikipedia entry for "Barbie," www.wikipedia.com (accessed Nov. 12, 2009).
7. Annabelle Gawer recently published a book containing the latest platform-related research from multiple disciplines. See Gawer (2009).
8. See, e.g., David (1985), Farrell and Saloner (1986), Farrell and Shapiro (1988) Arthur (1989), Katz and Shapiro (1992), Langlois (1992), Shapiro and Varian (1998), Bakos and Brynjolfsson (1999), and Nalebuff (2004).
9. See Rochet and Tirole (2003, 2006). Also Bresnahan and Greenstein (1999) and Schmalensee, Evans, and Hagiu (2006).
10. See Yoffie and Kwak (2006) and Adner (2006).
11. Cusumano and Yoffie (1998).
12. Iansiti and Levien (2004).
13. See Parker and Van Alstyne (2005), Eisenmann (2006), and Eisenmann, Parker, and Van Alstyne (2006, 2007).
14. Utterback (1996).
15. My thanks to Michael Bikard for this observation.
16. This section elaborates on an earlier discussion in Cusumano (2008b).
17. See the discussion of Microsoft's strategy in Cusumano and Selby (1995: 127–85).
18. See Rosenbloom and Cusumano (1987) and Cusumano, Mylonadis, and Rosenbloom (1992).
19. Yoffie and Slind (2008), and other public sources.
20. Yoffie and Slind (2008: 11).
21. Cusumano (2008b: 24).
22. Kane (2009) and Wortham (2009).
23. See Vascellaro and Kane (2009). Also, e.g., Schroeder (2009).
24. Yoffie and Slind (2008: 6).
25. According to its 2009 10-K report, Microsoft estimated that about 20% of Windows client sales of $14.7 billion or about $2.94 billion were directly to consumers. This is 5.0% of total revenues in fiscal 2009, which were $58.4 billion.

26. Cusumano (2004: 43–6).
27. Vascellaro and Sheth (2009).
28. Stone and Cain Miller (2009).
29. Gawer and Cusumano (2008).
30. See all the various Adobe products, activities, and partnerships at www.adobe.com (accessed Nov. 20, 2009). For a detailed description of Adobe's platform strategy, see Adobe Systems Inc. (2009), *Form 10-K*.
31. Gawer and Cusumano (2002: 269).
32. Gawer and Cusumano (2002: 8–9). In addition, the four levers are also summarized in Cusumano and Gawer (2002). This section closely follows another summary in Cusumano (2004: 75–7).
33. For the specifics of this chronology, see the Appendix to Cusumano and Selby (1995: 451–9), as well as p. 140.
34. Gawer and Cusumano (2002: 1). Also quoted in Cusumano and Gawer (2002: 51).
35. O'Brien (2009: B2).
36. Cusumano (2004: 124–5).
37. See Gawer and Cusumano (2002) for a detailed discussion of Intel.
38. The general ideas in this discussion of Google follow Cusumano (2005a).
39. See Arango and Vance (2009), Lohr (2009), Needleman (2009), and Singel (2009). Also Sullivan (2009).
40. CBC News (2008).
41. Vascellaro and Morrison (2009).
42. This discussion of Qualcomm is based primarily on Yoffie, Yin, and Kind (2005).
43. This description of OnStar benefited from public information as well as my informal discussion with the president of OnStar, Chet Huber, at the MIT Sloan School on Apr. 4, 2007.
44. Saghbini (2005).
45. See Thurrott (2001).
46. See www.caba.org (accessed Apr. 19, 2009).
47. See Eisenmann, Parker, and Van Alstyne (2007, 2008).
48. Perez (2007); Shankland (2006).
49. See Allison (2007) and Auchard (2007).
50. See http://oneforty.com.
51. *The Economist* (2007).
52. See Carnoy (2009). For more information on FixStars, see www.fixstars.com/en/index.html.
53. Microsoft Corporation, *Form 10-K* (annual).
54. Sheff (1993).
55. *Boston Globe* (2007).
56. In software, which platform to support was an issue with Netscape, and it decided to support Microsoft Windows as well as the Macintosh and UNIX. See Cusumano and Yoffie (1998, 1999).

2

Services, Not Just Products (or Platforms)

The Principle

Managers (at least in product firms or service firms that offer standardized or automated services treated as products) should use service innovations to sell, enhance, and even "de-commoditize" products or standardized services. Services can also be new sources of revenues and profits, such as an ongoing maintenance or subscription stream. The goals of most firms should be to find the right balance between product and service revenue, and then "servitize" products to create new value-added opportunities and pricing models as well as "productize" services to deliver them more efficiently and flexibly, such as by using information technology and service automation.

Introductory

Services account for 70 percent or more of gross national product for industrialized nations and are of increasing importance to developing countries such as India and China. This fact alone makes them worth studying. But several changes in various markets since 1990 have led researchers to probe the increasing role of services in what we normally think of as product firms. Within product firms, we can also include banks that primarily sell standardized services such as checking accounts, mortgages, certificates of deposit, or term insurance, in contrast to customized value-added services such as investment advice or portfolio management that vary for each individual (which banks also now offer). We can include as well telecommunications service providers and the cable companies, which offer standardized telephone or cable TV service, but also customized bundles serving residential and commercial needs for voice, video, and data. Hence, as the second of my enduring principles, I propose that managers in a growing number of product firms should view services as a much more integral part of their competitive strategies.

All the principles discussed in this book can play a role in making services more central to product firms. Industry platforms, ranging from Windows to SAP's applications to Apple's iPod, iPhone, and iPad can be great sources of service demand. These platforms have little value without services provided either by the platform leader or by its ecosystem partners. In the case of SAP, the critical services are for customization, implementation, training, and maintenance. For Apple, the critical services are for online content and applications (iTunes and App Store, or electronic books and magazines). There are also critical differences between the capabilities of product firms compared to service firms, which service departments of product firms must overcome. In particular, service departments need to leverage their deep knowledge of how customers use products to have more impact on product (or platform) strategy and innovation. Firms can best manage services using pull concepts because they need to deliver most services in real-time directly to the customer and often with the direct participation of the customer. Generally there are no conventional economies of scale in labor-intensive services. Consequently, managers need to think creatively about economies of scope—how to reuse knowledge, technology, and project artifacts such as design frameworks or modules across different customer engagements, as in Japanese or Indian software factories. Services also require great flexibility in operations to deliver custom solutions and assistance to customers when they need it.

One change that has brought the importance of services more to the forefront has occurred in the software products business, which I have been studying since the early 1990s. This industry has been "servitizing"—that is, turning into a hybrid product-plus-services business. The evolution has occurred gradually. It has proceeded despite the fact that most managers in this industry dislike services that detract from the potential 99 percent gross margins of the "pure" product business. I have pointed out this dilemma often whilst reminding managers that their software product prices can and do fall to zero—resulting in the "99 percent of zero equals zero" problem.[1]

A second change is that many companies in the computer and telecommunications industries have come to rely more on services for revenues and profits (Figure 2.1). Occasionally these include financial services as well as applications hosting. But, mainly, these kinds of firms offer professional services (product customization, strategic consulting, training, assistance with implementation and integration with other products) and maintenance (annual payments that come with the right to receive product updates and quality "patches" as well as basic technical support). The services supplement (and de-commoditize) products or replace revenues when product prices or product sales decline. This is part of the simultaneous "innovation and commoditization" trend discussed in the Introduction to this book. IBM has been the leader in this transformation, which made sense, since it always has had a large services capability. IBM added more services manpower in 2002, when

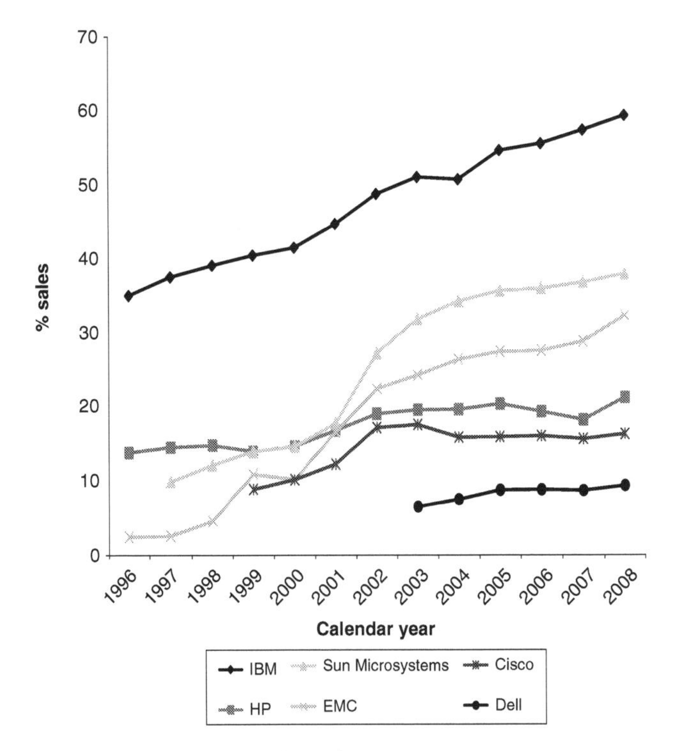

Figure 2.1. Service revenue, selected hardware firms, 1996–2008

Source: Annual company *Form 10-K* reports.

it acquired the consulting arm of PriceWaterhouseCoopers. Indeed, for more than a decade, IBM has been divesting commodity hardware businesses and instead focusing on value-added services and software products, especially software products that generate professional services and maintenance.

Other recent examples include Hewlett-Packard, which in 2008 acquired EDS and its 140,000 employees for $13.9 billion. This acquisition adds greatly to HP's services capability and has enabled it to compete better with IBM in the enterprise computing market. Even Dell, once renowned for the efficiency of its supply-chain operations and Web-based product sales and distribution capability, has been turning to services for additional revenue. In 2009, it purchased Perot Systems and its 23,000 employees for $3.9 billion. We might also conclude from the Dell case that even the cost-conscious customers Dell traditionally targeted now insist on some basic services to complement their product purchases, such as technical advice or help setting up home or small-business networks.

A third development has occurred in other manufacturing industries, such as automobiles. Since 2000, General Motors and Ford have generated much and sometimes all of their operating income, not from their product business (selling cars) directly. Instead, they have increasingly relied on financial services (loans and leases, and insurance). We can see this especially clearly at GM in the period 2001–4, before it started to encounter massive operating losses and sold a majority interest in its financial services unit, GMAC, to raise cash. During these four years, services accounted for 16 percent of total revenues but contributed over $10 billion in operating income against total net income of about $9 billion (Appendix II, Table II.4).

In other words, GM still lost a billion dollars on its non-services business, even during its profitable years. GM's financial services business remained profitable on an operating basis through 2007, even whilst the company lost some $50 billion during 2005–7. Overall, during 2001–8, GM lost nearly $73 billion—an astounding sum—at the same time as its financial services unit generated nearly $19 billion in positive income. Similarly, at Ford during 2001–8, financial services accounted for anywhere from 11 percent to 19 percent of annual revenues, and averaged 14 percent over these eight years. Ford also lost some $31 billion during this period, whilst financial services earned over $17 billion. Unfortunately, GM no longer has this cushion. Other sectors of the automobile industry, such as insurance and repair services from independent firms, also seem to have done much better financially than the largest product companies, at least in the United States during the late 1990s.[2] This is all besides the fact that automobile dealers have long relied primarily on services such as maintenance, repair, financing, and extended warranties to generate their profits—much more so than their small profit margins on vehicle sales.

Not all automobile companies rely on services to make money. At Toyota, for example, financial services accounted for merely 5 percent of revenues during 2001–8 and $7.7 billion in income. The global leader in the automobile industry reported a $4.4 billion loss for fiscal 2009 as sales around the world plummeted, and probably could have used more service revenues then. But Toyota has also been reluctant to lay off workers, since it has ample cash to wait out the recession and no significant debt. Most instructive are the comparisons: GM, Ford, and Toyota had roughly comparable aggregate revenues during 2001–8—ranging from $1.3 trillion (Ford and Toyota) to just under $1.5 trillion (GM). GM and Ford also generated about $185 billion each in service revenues and $17 to $18 billion in service income. Yet GM and Ford combined lost a total of $104 billion during these years, whilst Toyota earned net income of $87 billion. Again, the Toyota story is not merely one of scale; it is also about efficiency and flexibility, and other intangibles, such as brand and reputation—admittedly under threat at Toyota in 2010 as management deals with design quality and safety lapses that have been building for a decade. The GM and Ford stories are also not about lack of scale, nor about insufficient service revenue to complement their product revenue. To the contrary, GM and Ford would have gone bankrupt years ago without financial services income!

A few years ago, I was looking at the poor performance of GM's product business and spoke with several GM managers who attended a 2004 International Motor Vehicle Program meeting in Cambridge, England. I suggested that they "jettison" their capital-intensive automobile business and focus solely on selling financial services as well as the OnStar telematics service. Then GM's operating profitability and market value, I thought, would skyrocket. Even at this point, it was probably too late to save the automaker, but I believe the idea was right. For product companies whose primary business has declined or collapsed because their products are no longer distinctive, services can be a life-saving source of revenues and profits as well as help de-commoditize their products. In the short and long runs, of course, product companies need to build products that customers want. Service revenues may do no more than buy them time and provide funds and customer knowledge to help them revitalize their products. Nonetheless, the combination of services and products together can create a more enduring and profitable revenue stream than products alone.

My GM example leads to another point that exposes the special nature of many services in a product company: the tying or coupling between the two. After they recovered from my suggestion that GM stop making cars, one of the managers responded that GM sold financial services—mostly car loans, leases, and insurance, though home mortgage sales were also on the rise—and most of the OnStar services (though not all, initially (see Chapter 1))—to GM product customers. So it would be difficult to jettison the car business and still sell the same amount of financial services. This is no doubt true. But we can still see the strength of a hybrid business model and how the *product* can be a *platform* to sell services. This holds, even though the complementarity makes it difficult to separate the service from the product as well as to tease apart which side is really driving sales and profits.

Another instructive case is General Electric. Especially under former CEO Jack Welch, GE aggressively pushed services of different types. At some level, GE tried to de-commoditize product businesses such as aircraft engines, locomotives, home appliances, and heavy equipment by selling maintenance and repair as well as financial services (loans, leases, insurance, warranties, etc.). Then managers realized they could sell these same services to customers of other companies' products as well—GE's maintenance staff or leasing agents, for example, were similar to underutilized physical assets. We can see the results in 2009. Product services represented about 37 percent of GE's non-financial product revenue (about $105 billion) and nearly 25 percent of total revenues ($157 billion). Product services also generated $13.4 billion in operating profits, compared to total operating profits of

$10.3 billion (earnings before corporate eliminations and charges, and taxes). These service numbers exclude GE's enormous financial services business, which in 2009 accounted for 33 percent of revenues ($52 billion). These two categories—product services and financial services—in 2009 accounted for no less than 58 percent of GE's total revenues and most of the company's operating profits.[3]

Finally, another development related to the Internet and digital technology more generally has caught the attention of many observers. Beginning in the mid-1990s, we saw a new generation of service companies emerge that take advantage of the network effects inherent in the World Wide Web. Companies with names like Yahoo!, Google, eBay, Expedia.com, and LendingTree.com are in fact *automated services* with potential or actual profit margins comparable to the best software product companies. In fact, they seem no different from software product companies in their potential gross margins because they use standardized features and electronic delivery. They have nearly *productized* a service, using digital technology, computers, and the Internet as a delivery mechanism. Software product companies have started to do the same thing—deliver their standardized products as a simple service, accessed over the Internet with a browser. For several years, we have been calling this "Software as a Service" (SaaS) or "software on demand." We also now hear the term "cloud computing," which refers to the more general infrastructure that makes software as a service possible. Several companies are now competing to offer these infrastructure services as a new type of technology platform.

The key point is that, for more than a decade, we have been witnessing a transformation: the gradual "servitization of products" as well as the "productization of services," often enabled by digital technology. We also see both innovation (lots of new services and product features) and commoditization (prices falling, sometimes to zero) in various types of businesses. Moreover, many product companies today in industries ranging from software to automobiles have become hybrid companies and can no longer ignore services or treat these as a necessary evil. Services have become strategic necessities to sell products and often generate the revenues and profits necessary to keep the product business alive.

There is some but not enough research on how services fit into the strategy, business models, and operations of a product firm. Most of the work has come from academics in operations management, but we do see some studies in strategy and marketing. There has been much more research on defining services as well as explaining how they differ from products, and some studies of what is innovation in the context of services.

One problem has been to define services in a more satisfactory way than the usual definition, which is anything you cannot drop on your foot—that is, anything that is not a tangible good.[4] There are other commonly accepted characteristics of services as opposed to tangible products. Vikram Mansharamani, a former MIT Sloan doctoral student who worked with me on services innovation and business models, summarized these in a literature review. The usual list of service characteristics includes being perishable, simultaneously produced whilst consumed, and "heterogeneous" in the sense that the service differs (at least slightly) with each customer. Researchers disagree on whether customers need to be involved in creating the service, or whether services need to be labor-intensive.[5] We know, for example, that some services, such as airline transportation or overnight package delivery, have a high labor component and can be very capital-intensive (planes and trucks are expensive). By contrast, a fully automated digital service like Google search, or a mainly automated service like Expedia.com, is neither labor-intensive nor very capital intensive (though the size of server farms is continually rising).

Nor are the traditional definitions useful for modern digital products, such as downloaded software programs, music files, blog RSS feeds, and online magazines and books. We cannot readily drop electronic bits on our feet unless they come packaged in boxes or embedded in hardware appliances, but they have other characteristics of tangible products. They are not readily perishable, consumed when used, or intrinsically heterogeneous.

In addition to the problem of how to treat digital goods, we have other types of hybrid services, such as product repair or customization. Whilst the services are intangible at some level, they are hard to separate from the physical products. Sometimes product vendors also bundle these kinds of services with their products and do not charge for them separately, such as by providing "free" (but not free?) technical support or maintenance, even for a limited period. Perhaps a more useful approach (at least from the point of view of a product firm) is to define services in terms of the unique kinds of knowledge they require and generate, based primarily on how customers use the products.

We know a little more about what is *innovation* in the context of services, primarily from research in operations and technology management. Most scholars tackling this question have tried to apply product theories of innovation and product development to services or talked about innovation processes within services firms.[6] There is some sense that service innovation can be very different from tangible product innovation. For example, some services exhibit a "reverse product cycle" to the extent that incremental innovation precedes the "era of ferment" or rapid growth rather than the other way around, as is more common in manufacturing industry life cycles. Moreover, much of service innovation, especially in financial services, aims at improving delivery efficiency, and this in turn can lead to quality improvements and then new service offerings.[7] Other work has focused on the importance of real-time experimentation with live customers as part of the process for service innovation, since it is usually impossible to build and test physical prototypes.[8]

Service firms and departments also experience the fundamental tension expressed in my six enduring principles and their counterpoints: They generally want to deliver something different to each customer, which we can easily associate with differentiated capabilities, pull, scope, and flexibility. But they also need to standardize operations—which we can also associate with more rigid strategies as well as push, scale, and efficiency concepts. "Pure" service offerings are completely different (heterogeneous) for each customer and not very scalable (though modules might be). "Pure" products are by definition standardized and therefore much more subject to economies of scale (and automation through software, computer hardware, and the Internet).[9] Service organizations or hybrids usually try to find a balance somewhere in between these two extremes. For example, there is the case of Harrah's Entertainment, owner of various casinos and restaurants. This company, led by CEO Gary Loveman, a former professor of service operations management at the Harvard Business School and an MIT Economics Ph.D., implemented unique customer relationship management (CRM) software. The technology enabled Harrah's to track each customer's activities and predict individual needs, whilst also categorizing customers and deploying various strategies to maximize revenues per customer and activity—simultaneous management of differentiation and scale.[10]

We are touching on the related principles of economies of scope as well as flexibility, rather than conventional economies of scale and notions of efficiency. But innovation for many services firms is probably oriented toward achieving this balance. For product firms, service innovation is more complex and usually tied to the product offerings themselves, such as to de-commoditize them.

As noted in the Introduction, I illustrate how services can be of critical importance for product companies with some general ideas and a few examples of product firms using services to enhance, extend, or substitute for products. Then I discuss more specifically how platform leaders and wannabes can use services to encourage adoption of their platforms. Finally, I look at the software products business in some detail, as well as software as a service and cloud computing, to show how manual and automated services have become an essential part of the business. Again, my examples rely heavily on information technology, which is at the leading-edge of combining services with products and turning products into services. But the potential implications of this principle, like that of platforms, are very broad. Any firm encountering commoditization can benefit from services that make those standard offerings more personalized and valuable to customers. They may also be able to increase or create new demand by delivering and pricing products more like services.

Services and Product Firms

Managers in many different product firms are finally recognizing the strategic role of services in the business models and life cycles of their industries. IBM, arguably the best-known example of this shifting emphasis toward services, began the change under CEO Lou Gerstner in the early 1990s. More recently, IBM has launched a campaign to promote "services science."[11] As mentioned earlier, automobile companies and other industrial firms generate a large portion of their income from financing and insurance as well as other product-related services such as maintenance and repairs.[12] But few researchers have analyzed the value of services for the product firm in any depth. This gap prompted me and Fernando Suarez and Steven Kahl to explore both the theoretical side of the topic as well as the practical implications of this transition for firm performance and market value.[13]

Researchers have usually seen services as complements that help product companies sell their goods. Some writers have theorized that services become more important as products mature and become harder to differentiate. Others have noted that services often provide a more stable source of revenue than products and that service revenue streams such as from maintenance and repair can outlast the life of the products themselves.[14] In addition, under some conditions, services can actually substitute for product purchases. This happened in the computer and copier industries during the 1950s and 1960s, when most customers leased rather than bought the machines. Early computer users also engaged vendors or consulting firms to build custom software applications when products were unavailable. More recently, we see users renting software products and hardware services in a SaaS or cloud-computing arrangement. In the automobile industry, Zipcar as well as leasing and car rentals are actually services substituting for product purchases.

After thinking about this problem, we developed a three-part taxonomy (Table 2.1). First, some services are indeed complementary and *enhance* the product by making it easier to purchase and use. Second, some services are complementary but mainly *extend* the product by introducing new uses or adapting the product to changing environmental conditions. Third, in some situations services actually *substitute* for product purchases. Services also can play an enormous role in generating knowledge about how customers use products and want to use products, which helps firms develop new products and features as well as new services.

But several reasons explain why, at least in past years, many executives at product firms seem to have underinvested in services. Scott McNealy, when he was CEO of Sun Microsystems in 2004, was quoted in the *New York Times* as saying, "Services will be the graveyard for old tech companies that can't compete."[15] I found this comment typical of technology industry executives at the time. They disliked the lower

Table 2.1. *Taxonomy of services offered by the product firm*

Complementary		**Substitution**
Enhance	Extend	Substitute
• Financing • Warranty/Insurance • Implementation • Maintenance/Repair • Technical support • Training in basic uses • Customization that makes existing product features easier to use	• Customization that creates new features specific to a customer • Training or consulting that introduces new uses • Integrating the core product with new products	• Before product diffusion (e.g., early mainframe computing services) • After product diffusion (e.g., software application hosting, automobile leasing, and temporary rentals)

Source: Adapted from Cusumano, Kahl, and Suarez (2008).

margins of services and saw declining product sales relative to services as signaling the end of their growth phases (and high market value). As Sun's hardware product business continued to deteriorate, in 2006 McNealy moved out of the CEO position, though he remained chairman of the board. Another executive, Jonathan Schwartz, took over as CEO. Schwartz had the opposite view of services, and also had a background in software. As seen earlier in Figure 2.1, in 2008 approximately 38 percent of Sun's revenues came from services, compared to 10 percent in 1997. It is fair to say that Sun itself would be in the graveyard were it not for services to complement its products business, though there was still not enough profit for Sun to remain independent. The board of directors decided to sell the company in 2009, first negotiating with IBM and finally settling on Oracle as a suitor.

Developing innovative service capabilities and efficient service operations often requires product firms to make investments that seem (but may not really be) beyond what they need to design, produce, and sell their products. In addition, mass-market product firms tend to focus on building products for many "average" customers and are generally not so effective at cultivating or managing relationships with individual customers, which is necessary for a services business.[16] Different types of services also require various levels of investment or specific capabilities. Services that complement products (such as basic financing, insurance, or warranties) should be easier to create for product firms than services that enhance or even substitute for products. But when should product firms invest in services, and what kind of services should they invest in?

Market uncertainty and technological complexity seem to play a major role in determining the nature and characteristics of the knowledge boundary between the product firm and its customers. We think these factors largely shape the *type* of services the product firm is likely to offer. The state of the industry, such as whether it is growing or contracting, is another factor. But this relates primarily to the level or *relative importance* of services to the product firm in terms of their business models—that is, how they expect to generate revenues and profits.

In some cases, the environment is relatively stable and the product technology is not very complex to use. In these instances, services should be simple and most likely will be to "enhance" the product. If the environment and the technology are more complex and uncertain, at least from the customer's point of view, such as in the early stage of a new industry or technology, then we may see more sophisticated services such as those that "extend" the product. At extremely high levels of environmental or technological uncertainty, potential customers may find existing products inadequate or too risky or expensive to purchase. This context makes "substitution" (such as product leasing or rental) more appropriate.

As for when a product firm should place more emphasis on services as part of its strategy and business model, we believe that (1) the economic state of the industry, (2) environmental or technology uncertainty, and (3) complexity of the product or its usage are the three key factors in this decision. For example, product firms should focus more on selling products rather than services when product demand is high, such as during periods of rapid industry growth. We also expect the relative importance of services to increase for product firms with complexity of the product or uncertainty in the industry, such as when a disruptive platform transition seems to be occurring. In addition to basic support services, complex products, especially in transitional periods, are likely to require more elaborate, higher value-added services such as consulting, extensive customization, training, or integration and installation.

An increase in either technical uncertainty or market uncertainty is likely to increase the importance of services for product users. For instance, rapidly changing technology may require more frequent servicing or upgrades of the product to make it work with newer components or complementary products (technical uncertainty) or to adapt it for new uses (use uncertainty). Increased uncertainty may also imply the need for more consulting services to help customers make sense of the changes in technology and the marketplace. This occurred during the latter 1990s, when many firms needed help to move parts of their business to the Internet. However, very high levels of uncertainty may lead to a situation where customers find it too risky to buy an unproven technology that is likely to fail or change substantially. In such situations, product firms may resort to selling services as substitutes for products until they can convince customers of the product's benefits or improve the performance of the product and reduce the cost (that is, reduce uncertainty and risk to the customer).

Services and Platform Competition

Another aspect of how services can play a role in the strategy and capabilities of a product firm relates to the previous chapter—platforms and complements. Prior research on platforms has looked at ways firms try to become platform leaders or tip markets in their direction, such as through pricing, complementary products and services, network effects, or specific technical elements or features that differentiate one product or platform from another. But researchers have paid relatively little attention to the potentially broad strategic role of services in platform competition. Fernando Suarez and I recently addressed this topic in a separate paper.[17] I will summarize some of our key arguments here.

Whether the services come from the platform-leader wannabe directly or from a network of complementary service providers, such as independent IT services firms or divisions, services can help tip a platform competition in several ways. Services can reduce the risk to customers to adopt a new platform, provide feedback for further product or platform innovation, and enhance the value of the platform through integration with complementary innovations from other firms or even other platforms. In addition, services can be a tool for platform leaders and wannabes to influence market dynamics. They can act as a type of subsidy for the platform in multi-sided markets (if the companies give the services away or provide them below cost) or increase at least indirect network effects between the platform and complements. Furthermore, in mature markets or product markets particularly vulnerable to price competition and commoditization, services can provide an important source of revenues and profits. These additional revenues can be especially important in commoditizing markets and help the platform leader remain financially viable, as we saw with General Motors and Ford, or Sun Microsystems.

Until a platform winner emerges in an industry, customers may be reluctant to buy if they are unsure who the winner will be.[18] No one wants to get stuck with a product or a platform that quickly becomes obsolete or does not have much support from producers of complementary products and services. In these situations, platform-leader wannabes may offer services to try to minimize the customer's risk. For instance, independent consulting services can help customers better understand the differences between competing platforms and assess the potential impact of choosing a platform that may not become dominant. Likewise, platform firms and their complementors can provide on-the-job training by keeping staff on the customer sites, thus lowering the difficulty of transitioning to a new platform. In very risky situations for the customer, platform firms may offer services as temporary substitutes for the platform products. Again, this was the case when IBM and Xerox, in the early days of mainframe computers and large-scale copier machines, provided leasing (as well as maintenance and other services). It is also the case today with some SaaS products or SaaS versions of packaged software—customers prefer to rent rather than buy.

By providing integration services between the platform and complementary systems, a platform vendor may be able to increase the appeal of its platform and encourage broader adoption. Platform firms may also increase their broader ecosystem-level innovation capabilities from those service interactions with complementors. For example, during the 1990s and 2000s, Intel and Microsoft both worked with complementors such as makers of multimedia and telecommunications software, peripheral devices, and microprocessor chip sets to add various capabilities to the PC platform, ranging from plug-and-play compatibility to video conferencing.[19]

More in line with the discussion in Chapter 1 is the potential use of services as a subsidy for one side of a platform. Most of the literature on platform pricing has focused on subsidizing the product. For example, video-game console makers generally sell the console hardware below cost and try to make money from charging royalties on the complements—that is, the video-game software. In the commercial software products business, similarly, Adobe gives away the Acrobat Reader and hopes to make money by selling editing tools and servers. At least in the case of complex platforms that require extensive technical support, product customization, and other professional services such as consulting and training, platform companies can potentially subsidize the services around their platform rather than give away or subsidize the product itself.

A platform leader or wannabe should also be able to increase at least indirect network efforts and then the value of its platform by cultivating a network of service providers. In the computer software and hardware industries, companies ranging from Microsoft and SAP to IBM, Hewlett-Packard, Cisco, and Sun Microsystems have all followed this strategy of building up a global network of third-party service providers to supplement their in-house service departments. Automobile manufacturers and other makers of complex products do the same thing by creating networks of certified service technicians and franchised dealers. Other things being equal, we should expect customers to prefer products or platforms with a large service network.

Several researchers have argued that services should also rise in importance for product firms as an industry matures. Again, service revenues such as maintenance often outlast the life of the products themselves.[20] Or, if margins or prices are under severe pressure, product firms may turn to services in order to increase revenue and protect margins, particularly during economic downturns. Data on the software products business reflect this trend toward services, both professional services and maintenance, as the industry has aged. Other research confirms that manufacturing companies in general have shown a growing interest in services.[21]

If commoditization dynamics play a key role in the rise of services in product industries, then the platform leader should enjoy significant market power where one platform dominates. Moreover, we should expect less commoditization pressures—and less pressure for platform leaders to retreat into services. In other words, platform dominance should help protect product revenues and margins. It should therefore lessen the incentives to move aggressively toward services. Indeed, this seems to be the case in operating systems and multimedia software. The leading firms (Microsoft and Adobe) have relatively tiny (though growing) levels of service revenue.

Another important question is who should dominate the provision of services—platform leaders or independent service companies? In many industries, firms other than the product or platform producers provide services. Moreover, these networks of service providers usually help the product and platform producers sell their products and expand their market coverage as well as promote product loyalty—encouraging an indirect network effect. In the automobile industry, we can see independent service organizations ranging from repair and maintenance shops to driver education and insurance. In the computer industry, it is common for major platform vendors and product companies to have hundreds of third-party firms assist them with consulting, training, customization, installation, and maintenance work.

To what extent the platform or product company dominates the provision of services around its offerings probably depends on how complex and product-specific is the knowledge required to provide those services. How dominant a platform is likely to become should also influence how many independent service providers enter the market, either to complement the platform producer or to compete with it. If uncertainty over which platform will win lessens with time, and if the technology becomes more familiar, then more independent service firms should appear to support the platform. This seems true as long as the market is large enough to create sufficient incentives for entrepreneurs or managers at established firms to invest in services for a particular platform. If the market is too small, the platform company may have to provide all the services itself.

The degree of modularity in the platform should affect this dynamic as well. As discussed in Chapter 1, highly integral architectures, particularly with partially closed platform interfaces, will probably discourage third-party service providers from entering the market. Apart from this factor, however, modularity can result in simpler designs and reduce the importance of product-specific knowledge possessed by the platform leaders. It can encourage more independent firms to enter the service market. This suggests that platform leaders may have less of an advantage in providing services for highly modular platforms. In fact, this seems to be the case in the computer industry, where many independent firms provide services for Windows PCs as well as platform-like products such as SAP applications and Oracle databases and applications.

The Software Products Business

Software products, particularly for enterprise customers, present an extreme case of how a product business can quickly turn into a hybrid or nearly pure services business. This is because the value of the products can and do fall to zero—leaving some product firms with little apart from services to generate revenues. But services have long been the dominant portion of sales for firms in the software business more generally. In the latter 2000s, the software products market and related service businesses had global revenues of at least $1 trillion dollars (not including business process outsourcing or the value of software that organizations built for their own internal use). Software services—IT consulting, product customization, custom software development, integration work, technical support, and maintenance by independent firms—accounted for about two-thirds of these sales. Packaged or standardized software products, like Microsoft Office and Windows, or SAP and Oracle applications, as well as services provided by the software product companies, accounted for the remaining one-third or so of these global revenues. The United States consumed about half of global sales and Europe at least a quarter. Japan, with 10–15 percent of global product and service sales, remained the second largest country market for software products and services. Domestic sales in China and India probably constituted no more than 5 percent of the global software market combined in terms of domestic demand. But these markets are growing fast and for years have been major sites for outsourcing services and offshore development centers.[22]

The Importance of Services

Given this context, managers should easily recognize how important services are in the software business. In fact, the first software companies created in the 1950s were pure services firms, because the technology was so new and complex to use. System Development Corporation, which built the first software factory in the United States (see Chapter 5), was originally part of the Rand Corporation and worked on the SAGE air defense system in the 1950s. Two other early software services firms include Computer Sciences Corporation, founded in 1959, and EDS (Electronic Data Systems), founded in 1962 and now part of Hewlett-Packard. These and many other firms, as well as hardware vendors such as IBM, built complex systems for governments and large enterprises from scratch. An independent software products business came into being only gradually, as IBM machines and some other mainframe computers became more common in the 1960s and 1970s. It was also important that IBM de-bundled applications software from its basic software and hardware sales around 1970. As user needs became more standardized, entrepreneurs created software product companies, such as the database vendors Cullinane (founded in 1968) and Software AG (founded in 1969).

Most product firms originated in the United States, with the notable exceptions of Software AG and SAP, founded in Germany in 1972. A key reason must be that the US market has seen the largest number of commercial computer users since the 1950s. This also helps explain why, even today, most global software product companies and the largest firms, led by Microsoft (established 1975) and Oracle (established 1977), remain American. Elsewhere around the world, custom-built software continues to dominate most markets. Not until the rising popularity of the personal computer in the 1980s did the software products business take off and become a mass market. However, even in the United States today, most software products sold to enterprises are still customized.[23]

In sum, we can say that the software business was primarily a services business in the 1950s and 1960s. It underwent a transition toward more products in the 1970s and 1980s, but now has shifted again toward services, of both old and new varieties. The software products business may also be an extreme example of the dual trends of commoditization and innovation, as product values seem to have fallen whilst innovation continues in different forms. In addition, the software business provides many useful examples of how product firms can utilize services to de-commoditize and help sell their products as well as generate new types of revenues.[24]

The basic change is that traditional software product sales (which generally consist of up-front "license fees") have declined, whilst software product company revenues have shifted to professional services as well as annual maintenance payments. The latter are usually set at 15–25 percent of the initial product price. Maintenance entitles customers to receive quality fixes or "patches," minor upgrades, and some technical support—often in "perpetuity," as long as they continue paying their annual fees. Also underway is the SaaS trend to deliver and price products more like a hosted or managed service. This shift is especially pronounced among small enterprise software vendors, led by Salesforce.com, with its customer relationship management (CRM) product, or early stage firms such as Hubspot.com, with Web marketing tools that work with the Salesforce CRM platform.[25]

We can clearly see these trends in the case of Siebel, whose product sales fell dramatically before Oracle acquired the company in 2005. In the mid-1990s, even Oracle and SAP experienced what I called the *crisscross*—service and maintenance revenues rising to where they exceed product revenues. It is not possible from these data to determine whether product sales were dropping or product prices were falling (Figure 2.2). But the effect is the same: services (including maintenance, which we estimate typically accounts for 55–60 percent of service revenues for a relatively mature enterprise software firm) became more important than product revenues. Again, though, as in the case of General Motors, the product revenues and the service revenues can be closely tied to each other.

There are some exceptions to this crisscross phenomenon. Product sales continue to account for nearly all game-software revenues, although online-gaming service revenues are growing. Industry platform companies like Microsoft—which has a large ecosystem of PC manufacturers as well as enterprise and individual users driving sales of Windows and Office—continue to generate nearly all their revenues from products. Adobe, the leading seller of multimedia tools, also sells primarily packaged software products. Its services include technical

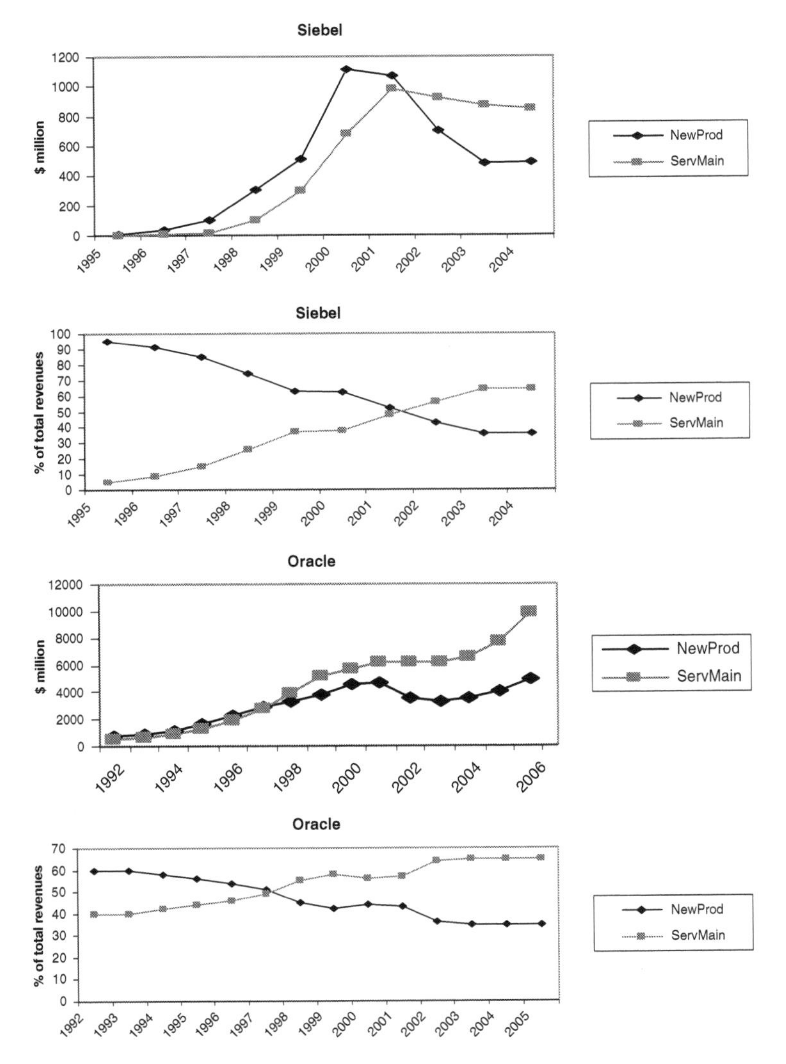

Figure 2.2. Crisscross graphs of Siebel and Oracle

Source: Annual company *Form 10-K* reports.

consulting, training, maintenance, and support. They were 6 percent of 2009 revenues, up from merely 1 percent in 2002.[26] Even Microsoft is encountering change. As noted in Chapter 1 (see Appendix II, Table II.3), services accounted for 5 percent of fiscal year 2009 revenues, up from 4 percent the prior year (20 percent of the $14 billion in Server and Tools segment revenues).[27] Just a few years ago, Microsoft, as well as Adobe, derived 100 percent of revenues from product sales.

Microsoft's Online services division accounted for another 5 percent of 2009 revenues (as well as $2.3 billion in losses).

Services have been growing in importance at software product firms for at least a decade, and we can measure this trend. The advent of free and open source software (which seems to have driven down software prices), as well as Y2K and the Internet boom and bust, appears to have accelerated the move away from either large product unit sales or high product prices. In either case, the result is relatively more revenue from services. In general, since the collapse of the Internet bubble in 2000, many enterprise and individual customers have rebelled against paying a lot of money for standardized or commodity-type software products, especially if there are comparable free products available. This is now true for enterprise software products such as operating systems (Linux), Web servers (Apache), databases (MySQL), CRM products (SugarCRM), and many others.

Another issue is how to treat software delivered as a service, either from a product vendor directly or coming through a neutral cloud-computing vendor such as Amazon or Google. Microsoft is also contemplating the delivery of much more software as online versions of its platform products—Windows Live and Office Live. I believe we should count fully automated digital products delivered online as digital products, with some service costs broken out, if possible. Suarez, Kahl, and I have so far counted SaaS revenues as product revenues. We did this even though vendors deliver and price SaaS products differently from conventional software products, and there are some service elements (such as hosting and basic technical support, and maintenance) either bundled into the SaaS price or eliminated. In fact, some firms, such as Oracle, now list their on-demand SaaS product offerings as service revenues. But these amounted to 3 percent or less of Oracle's total revenues during 2006–9, and were not listed as service revenues before 2006.[28]

"Free but not free" software from companies such as Google, Yahoo!, and Microsoft is also complex to categorize. These firms took what used to be for-fee software products ranging from search and email to basic desktop applications and now deliver them as a nominally free service. The user does not pay directly for the software (unless you count the time spent watching advertisements), but advertisers pay the software service vendor. Vendors such as Salesforce.com still count SaaS as product subscription revenues, and keep these separate from professional services. However, the SaaS pricing model eliminates maintenance payments—a major source of service revenues for software companies—and often includes some bundled technical support—a source of costs and some revenues. Hosting also involves more costs than simply selling copies of a software product.

Therefore, the SaaS model has confused the traditional separation of product and service revenues as well as costs. This should result in a decline in service revenues for software product companies at least because of the elimination of maintenance payments.[29] Nonetheless, for the time being, we can consider SaaS as a new type of digital product (or a new type of service) in that the providers generally deliver a completely standardized offering and do so automatically, using computers and the Web. The amount of software delivered in this way was relatively tiny in 2010. In the future, though, we may want to create a separate category for all automated digital services and treat them as a new type of product.

Inevitable Life Cycle or Strategic Choice?

SaaS and the broader trend of cloud computing will become much more important in the future and may well hasten the transition toward services in the software product business. But, even if we count them as service rather than product revenue, SaaS and cloud computing are not the main reasons why software product companies have seen their product sales decline dramatically over the past decade. In Cusumano (2004), I pointed out (with some alarm) the steep drop in product sales at Siebel, Oracle, SAP, and several other major product firms, and suggested that perhaps at least non-platform companies go through a life-cycle change (Figure 2.3). They may start with most of their revenues coming from product sales. But, if finding new customers becomes harder, or if prices drop because of competition or free alternatives, then these companies will find the bulk of their revenues gradually shifting to a combination of products and services, including professional services and maintenance payments. If companies get no new customers, they shift from hybrid status to service-oriented product firms. The "installed base" of customers paying maintenance can

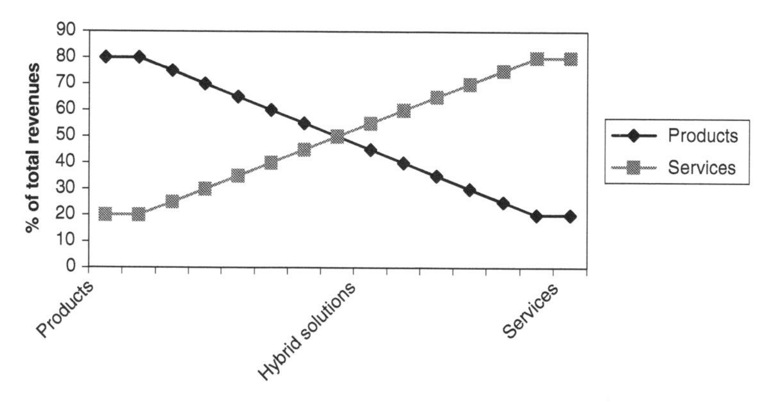

Figure 2.3. Business or life-cycle model for software product firms

Source: Cusumano (2004: 26).

become an enormous and highly profitable revenue stream for product companies. Gross margins can approach those of pure product sales, except for some technical support and quality releases. My question, however, was mainly whether this transition was "inevitable" for most software product companies or whether it was a "deliberate" strategic choice that managers were making.

Managers should want to focus on software products because they can generate very high gross margins, given that the marginal cost is zero to copy a piece of software or any other digital good. By contrast, margins for labor-intensive services in the software business can be 30 percent or lower. We can see this in the case of Business Objects, the leading producer of business intelligence applications now owned by SAP (and a firm with which I worked closely from 1995 to 2000). Gross margins on software product sales rose from 92 percent in 1993 to 99 percent by 2000; meanwhile, gross margins on services declined from 72 percent to under 60 percent during the same time period. Revenues also rose during this time, from $14 million in 1993 (with 82 percent of sales coming from products) to $455 million in 2002 (with 54 percent of sales coming from products).[30]

As competitors appear, software product companies may have trouble getting new customers. They often lower prices because of competition from similar firms or free software. Then the great profit opportunity from selling "pure" software products becomes theoretical and hard to maintain in reality. And, whether they like it or not, the revenues of software product firms gradually shift to professional services and recurring maintenance fees.

There may be more going on here than either an inevitable life-cycle effect or, in some cases, explicit managerial decisions to emphasize services more than products. James Utterback and William Abernathy pointed out decades ago that at least manufacturing firms often encounter a specific "product-process" life cycle. In the beginning of their histories, these firms tend to pay more attention to product innovation and compete on the basis of innovative designs. If and when a "dominant design" emerges, then companies tend to shift their emphasis to the process side and focus more on efficiency in making this design. That is how mass-production technology emerged at Ford and many other companies in different industries.[31] In the software business, there has been such a shift in emphasis from product design in the 1960s to software engineering in the 1970s and 1980s, culminating in "software factories" in Japan as well as the Capability Maturity Model (CMM) in the United States. This shift from software services to products to software engineering has many parallels with the life-cycle stories of manufacturing industries.

Service innovation is another aspect of the life cycle that might affect software and some other industries. For example, if the product has become a commodity—widely available with little differentiation, and priced low around the world—and after a company has wrung maximum efficiency out of process improvement, then management might turn its attention to services. On the other hand, what we are seeing in the software business might be related more to "S-curves" and "disruptive technologies," such as discussed by Clayton Christensen, James Utterback, and others.[32] In software, not only do we have maturity setting in for different product segments and companies shifting their emphasis to services. Some new technologies now support different kinds of business models, including different ways of pricing and delivering software, and reaching different kinds of customers (see Figure 2.4).

Obviously, the Internet and wireless technologies enable all sorts of on-demand or transaction-based pricing models or Google-type advertising-based business models. In addition, a platform transition seems to generate demand not only for buying new products but

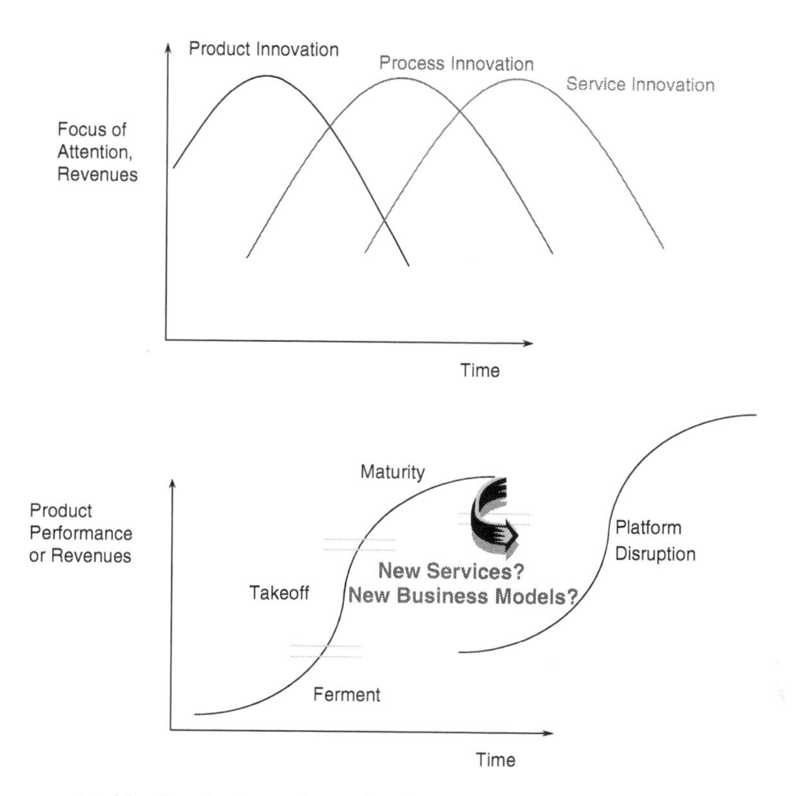

Figure 2.4. Models of service innovation and industry life cycles
Source: Adapted from Utterback and Abernathy (1975), as well as Christensen (1997).

also for consuming transition-related services. For example, we have seen millions of firms switch from mainframe platforms to personal computers and work stations for their employees or from desktop PCs to the Internet or from stationery to mobile computing. These corporate customers need lots of services, such as strategic guidance on what to buy, how to rewrite applications and move data from one system to the other, or how to retrain employees. In other words, platform transitions such as we have experienced since the birth of the computer could also generate as much or more revenue from services as from products, especially since many products are now free or low-priced.

To get some empirical data, I launched a research project in 2003 with the help of Suarez and Kahl to examine this shift from products to services for companies in software and other industries. The first database we created, covering 1990 through 2006, is a comprehensive list of firms that consider themselves software product companies selling "prepackaged software," listed under US Standard Industrial Classification (SIC) code 7372 (Figure 2.5). These data include foreign firms such as SAP and Business Objects listed on US stock exchanges, as well as game-software firms that sell products almost exclusively. The cumulative data contain about 500 distinct firms and peaked in 1997 at about 400 firms listed for that one year. By 2006, the number was down to fewer than 150 firms—indicating a dramatic consolidation of the software products business, at least as measured by the number of surviving public companies.

The second database consists of firms that compete in IT services under several different SIC codes. These data, illustrated in Figure 2.6, also show listed companies peaking in 1999 at approximately 500, and declining to less than 200 in 2006. The strong rise in IT services companies in the 1990s suggests that the transition from desktop PC to Internet platforms provided as many or more opportunities for service firms as it did for software product firms. Both the services and products sides of the business, nonetheless, have experienced significant consolidation since that time.

The fact that the number of public software product and IT services companies is shrinking suggests that the software business is maturing, even though total sales may be growing. Other data collected by a former MIT student, Andreas Goeldi, suggest a strong rise in start-up

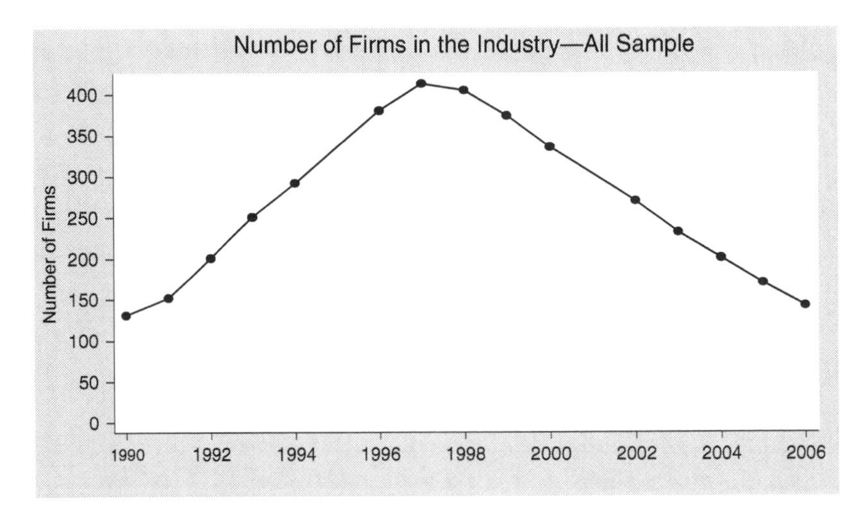

Figure 2.5. Publicly listed software product firms, 1990–2006

Source: Suarez, Cusumano, and Kahl (2008).

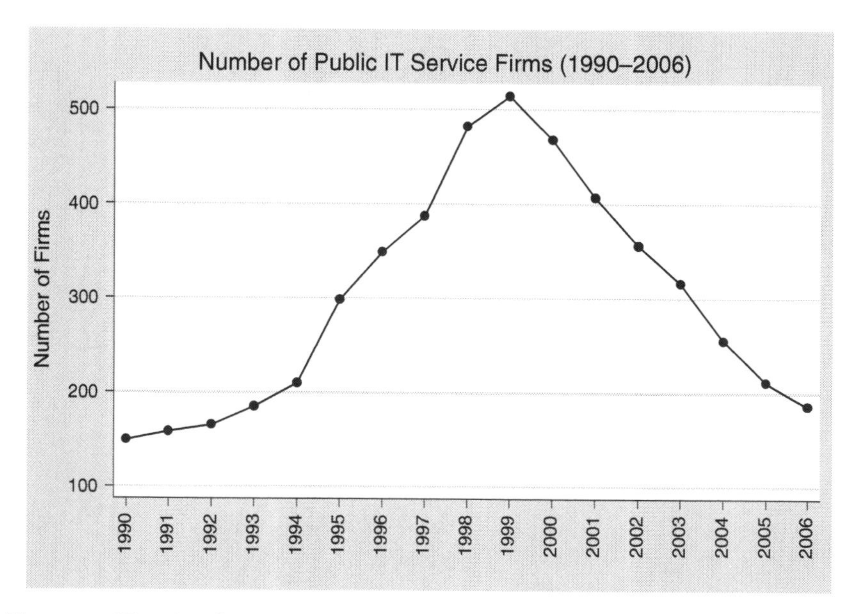

Figure 2.6. IT services firms, 1990–2006

Source: Suarez, Cusumano, and Kahl (2008).

enterprise software companies, especially in 2005, using a variety of new business or pricing models.[33] His survey of 108 companies competing in Web-based enterprise software (about 20 percent of the companies were publicly traded) indicated that monthly subscription fees are the most popular pricing model. A minority of companies also offered free software or advertising-based software (Google falls into this category). A few others charged the traditional license fee.

Figure 2.7 shows a model that Goeldi and other MIT students made to categorize the variations now occurring in revenue or business models, delivery models, and target customers. Until around 2000, nearly all software product companies sold software through the up-front license fee and did local installations on the customers' hardware. Now we have many different business models—subscription, advertising based, transaction based, and several kinds of "free, but not free." Software delivery models can be remote and Web based or bundled as hardware products. This trend toward potentially cheaper software, combined with less costly ways of delivering software over the Web, has made it possible for firms to target not only mainstream customers but small businesses and leading-edge early adopters. In addition, some

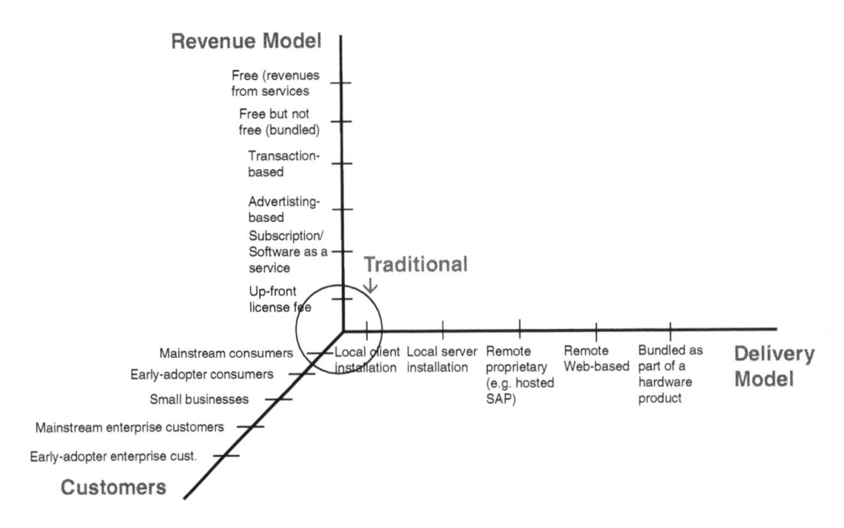

Figure 2.7. Dimensions of business models

Source: Cusumano (2008a: 81) (from MIT student team Krishna Boppana, Andreas Goeldi, Bettina Hein, Paul Hsu, and Tim Jones).

software-based companies are now turning into hardware vendors through the "appliance model" strategy.[34] The idea here is to make the technology more convenient for customers to install and use. But there is another effect: if you put software in a box, it is less likely that the price will fall to zero. People will usually pay for a box they can see and feel, even though they might not want to pay very much for software or digital media on their own.

There is yet another important element behind this entrepreneurial activity: now it seems to require less money to start a software company aimed at enterprise customers. Of course, it was always possible for "two guys in a garage" to launch a software firm or even a personal computer company, and many (such as Microsoft and Apple) started that way. But today, critical components that large organizations use—such as the operating system, database, Web applications server, customer management application—are available as free and open source software. Entrepreneurs can take these free components, write some applications code, and then hire other firms to host the software. With relatively little expense, they can launch an enterprise software company. Data from Goeldi's survey suggest that entrepreneurs funded about 37 percent of the new Web-based enterprise start-ups and only 36 percent relied on venture capital.

But, as we look at change in the software products business, a question occurs: is this commoditization trend and increase in services and new business models temporary or permanent? Permanent in my mind refers not necessarily to "forever" but to a trend lasting decades rather than years. One possibility is that we are now merely in between platform transitions and in a plateau in terms of product revenue growth. If some major innovation occurs, such as for a new computing platform, then individuals and enterprises will again start buying new products, both hardware and software, in large numbers. For example, Apple's iPhone clearly re-energized purchases in one hardware market, primarily because of a novel user interface. The iPad has started another wave of purchases it will happen again—but how long will the demand surge last?

By contrast, the permanent argument says that software might have experienced what computer hardware did in the past: investments from Intel and other firms along the lines of Moore's law helped reduce the price of computing power dramatically and bring powerful computers down to the level of commodities. In other words, permanent suggests that much software now is commoditized, just like hardware, and prices will fall to zero or near zero for any kind of standardized product. In this scenario, the future is really free software, inexpensive SaaS, or "free, but not free" software, with some kind of indirect revenue model, like advertising—a Google-type of business. And it is possible that other commoditizing industries, especially those with significant value coming from software or digital content, are likely to follow. The value may actually be shifting back to the box or to some ongoing, automated service. It is unlikely in this scenario, however, that a large number of product companies can survive or afford large R&D and marketing costs simply by the sale of advertising.

What the Data Say

The data on publicly listed software product companies allow us to look at some aspects of this transition from products to services in detail. Our database contains about ten years of financial information for each firm (totaling over 3,200 annual observations). Excluding game-software firms and some others (mostly, they did not break out products from services and we could not confidently classify their revenues), the total number of software product firms peaked at 300 in 1997 and stood at merely 111 in 2006. As Figure 2.8 shows, software product firms in the sample had an average of 70 percent of their revenues coming from product sales in 1990 and have had less than 50 percent since 2002, when the crisscross first happened for the industry as a whole. (We exclude video-game software producers, since they do not break out products from services and have nearly 100 percent product revenues.)

We did not separate maintenance from other services because less than 10 percent of our sample broke this out. Firms treat maintenance as a type of service because, unlike with product sales, companies can recognize these revenues only over time as they deliver patches, upgrades, or technical support. Some firms, such as SAP and Oracle, are now trying to relabel maintenance fees as "product license renewals" to make their total product business look healthier. This makes some sense because maintenance has profit margins closer to product sales (though a bit lower because of the routine technical support costs usually included in the agreements). But maintenance revenues still come from the installed

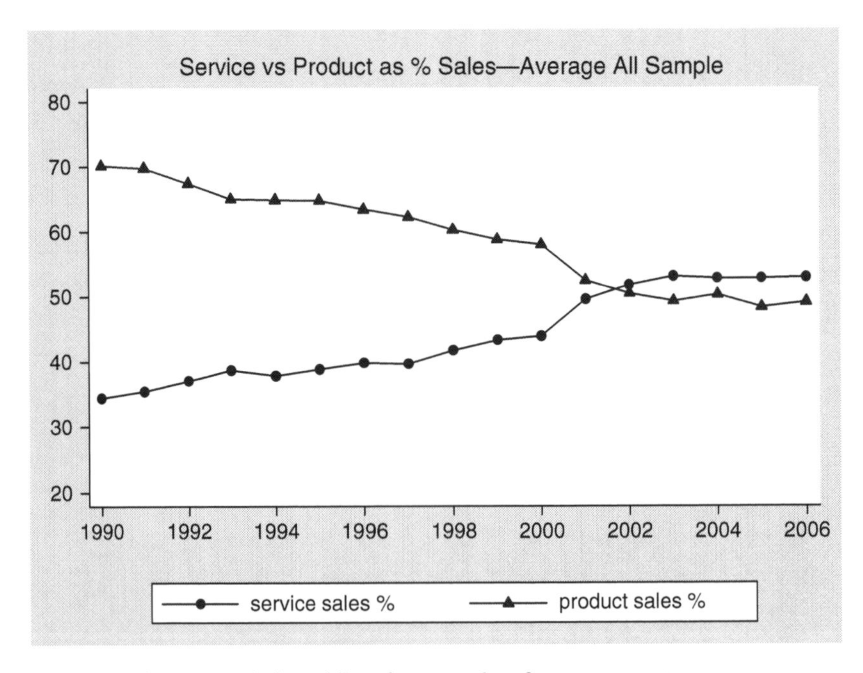

Figure 2.8. Crisscross graph for public software product firms, 1990–2006

Source: Cusumano, Kahl, and Suarez (2008).

base of customers, not from new product sales, and firms can recognize them only over time, like professional services.

The data indicate that product revenues have dropped but have not fallen to zero. Rather, they have stabilized at just under 50 percent of total revenues. So perhaps software product companies have reached a sort of equilibrium point as a business: There are more service (including maintenance) revenues from their existing customers than new-product revenues, but products are still holding significant value, at least for the surviving publicly traded companies.

A more refined analysis indicates that the degree of the shift toward services varies with the market segment. For example, in business applications, the percentage of services rose from just under 40 percent around 1990 to nearly 70 percent in 2006. In addition, specific segments where one or two platforms dominate sales tend to show more "resilient" product revenues. These slices of the market (operating systems and multimedia tools) do not show a crisscross point. Leading companies here (Microsoft and Adobe) have retained high levels of "standardized" or "packaged" product revenues (or prices) and thus have seen less pressure to resort to services such as consulting, integration, customization, or maintenance. They contrast to companies in segments such as business applications (led by firms such as SAP) and networking (e.g., Novell and Symantec). Business intelligence (e.g., Business Objects and Cognos, now owned by SAP and IBM, respectively) and databases (e.g., Oracle) follow a similar crisscross pattern.

We can also look at how common it is for software product companies to sell only products as well as have different hybrid mixtures of products and services. In *Microsoft Secrets* (1995), Richard Selby and I held Microsoft up as the ideal model for a software company: it had 100 percent product revenues and those wonderful gross margins. The trick for Microsoft and similar pure product companies is still to keep R&D, marketing, sales, and administrative expenses—all not included in the gross margin calculation—relatively low. But the data suggest that these kinds of companies are rare historically and are disappearing with time, even if we include the pure SaaS firms. In 1992, for example, there were twenty-eight publicly listed software product firms (apart from video-game firms) that had 90–100 percent of their revenues coming from software products. In 2006, only seven such firms remained (Microsoft, Adobe, Symantec, Salesforce.com, SCO Group, Echelon, and Smith-Micro).

The data also indicate that, whilst the average level of product revenues has dropped to less than 50 percent for the software product companies, the optimum mix for operating profitability (again, excluding games and some other firms) is still very product oriented—about 70 percent products and 30 percent services. Most software product companies make little or no money because of high sales and marketing costs as well as R&D expenses. This is another reason why so many firms are disappearing or agreeing to mergers with Oracle, Microsoft, Cisco, IBM, Hewlett-Packard, EMC, et al. There are also some companies in our database that have reported 100 percent service (including maintenance) revenues in a given year and no new product sales, even though they are nominally software product companies. Companies in this category are especially likely to be weak performers and candidates for takeover or bankruptcy.

The shift toward services for software product firms has been increasing at nearly 2 percent a year. The crisscross point by age is at twenty-two years for the whole sample—again, excluding game companies. In other words, if a software product company survives for twenty-two years (and does not sell software games), it is likely that service and maintenance revenues will at this point exceed product revenues. We also see new software product companies entering the industry with most of their revenues coming from professional services.

When we probe more deeply, the shift toward services for software product firms appears to have happened for two reasons. One is that product sales have continued to grow overall, but services have grown more quickly, perhaps because price levels or the number of new customers has fallen. This situation is still relatively healthy, and firms can easily survive as hybrid businesses. The other reason is that some firms have seen their products business collapse, and then they have crossed over from mainly product revenue to mainly service revenue. This second scenario is potentially disastrous, because it often means the firm must reorganize radically and quickly to accommodate declining product sales. This happened to Siebel. When this occurs, the product firm can no longer support large R&D groups or high marketing and sales expenditures. It must transition from designing products for a largely abstract set of users to building and servicing customized products for individual customers or sets of customers with very similar needs, such as in the same industry vertical (for example, computer-aided design software for automotive engineers). Many product-oriented firms do not make this transition or make it poorly and reluctantly, missing the opportunity to manage services as a new strategic area.

Impact of Services on Firm Performance and Value

In 2004, I assumed that rising services would have a negative impact on profitability and market value for a software product company. Services almost always have lower gross profit margins than software products, and their rise at the expense of product sales generally signals lower sales growth. For the first time, however, the data we collected on software product firms allowed us to examine that assumption, though the analysis suggests a somewhat more complex relationship than I had expected.

For most of the publicly listed software product companies in the United States, services contribute positively to profits but not in a linear manner.[35] There seems to be a "sweet spot" at the high end of the spectrum. Some firms do well with services at the low end of the spectrum. In general, though, services of all types create a drag on profitability until they reach about 60 percent of total revenues. At this point, the product firms seem to be gaining from economies of scope or learning-curve effects in their services business, or perhaps they are benefiting from service complementarities with their products business. Whatever the reason, when services reach a majority of sales, then the product firms seem able to manage the services side of the business more profitably than in the past. SAP and Oracle are great examples of this. Both are very profitable and have only about one-third of their revenues coming from new product license sales and another third from maintenance.

Market value, which generally tracks growth rates as well as profitability, seems to follow a similar nonlinear curve. Our analysis also indicates that, even in years when services positively contribute to profitability, market value can drop for a firm if services rise.

This suggests that investors still did not understand how important services had become to the revenues and profitability of software product companies such as SAP and Oracle.

The revenues of business application firms follow the crisscross pattern almost exactly as predicted, but firms in other sectors hit the crisscross later, and some segments (operating systems and multimedia tools) still maintain high levels of product sales. But the general argument seems valid that most software product firms should take more explicit advantage of professional services as well as maintenance opportunities, and not just let services "happen" to them as their product business declines. This means that software product firms—and many other firms in businesses ranging from computers and telecommunications to automobile manufacturing and dealership sales—should treat professional services and maintenance as strategic areas and targets of opportunity to increase revenues and profits. Services are especially valuable when the product business is suffering. By contrast, too many product firms seem to treat at least professional services as a necessary evil and manage them as a cost center, without much creativity or effort to grow that part of the business. Many managers have also tried to reverse the trend toward services, to no avail. In fact, most firms can look at their past trends and predict fairly accurately when they will hit the crisscross. They can take some actions to delay this, such as investing more or making acquisitions to rejuvenate the product line-up. Oracle has done this with dozens of acquisitions over the past several years. SAP followed suit with the acquisition of Business Objects in 2008. Or firms might accept this transition and launch a serious campaign to sell more professional services as well as maintenance contracts to each new product customer. SAP and Oracle also have done this very successfully over the past decade.

This trend toward services is not limited to the software business. However, it seems to be as much a strategic move as an inevitable industry life-cycle phenomenon in closely related sectors such as computers and telecommunications hardware and systems. For example, as Figure 2.1 shows, IBM services grew from 35 percent of revenues in 1996 to nearly 60 percent in 2008. Sun Microsystems, Hewlett-Packard, Cisco, and even Dell have shown major increases in services compared to a decade ago, a trend that seems to correspond to the commoditization of hardware. And their service revenues are likely only to increase because of their acquisitions of large service companies.

Software as a Service and Cloud Computing

To understand how services and service automation is affecting players in the computer software business and various other applications, we need to dive a bit more deeply into software as a service and cloud computing. Not only does SaaS represent a particular way to "servitize" a software product; it has also stimulated a new type of platform competition. In fact, all the concepts discussed in Chapter 1 that influence platform dynamics at the industry level—direct and indirect network effects, different uses of the four levers, differentiation and niche strategies, efforts to become a core technology or to tip a market—are shaping what has become a new industry. At the least, we see a new segment of the Internet services business.[36] SaaS platforms open to many different firms as well as neutral cloud-computing infrastructure providers now compete with owning your own personal computer and software product, or your own application server. Whether the cloud-computing market is "winner take nearly all" (like Microsoft in desktop operating systems) or "winner take most" (like Google in Internet search) remains to be seen. But now we know how to think about this question.

Delivering software applications over a network is actually an old idea. It goes back to time-sharing in the 1960s and 1970s, as well as application-hosting in the 1980s and 1990s. We also saw an increasing number of firms in the 1990s and 2000s deliver what used to be packaged software applications from a Web infrastructure, and usually for free—ranging from email to calendars, groupware, online publishing, simple word processing, and many other common consumer and even business applications. Advances in networks as well as virtualization technology have made Web delivery possible, regardless of the type of hardware. But only when vendors open their SaaS infrastructure to other companies do we have cloud computing and true industry platform competition.

A good starting point for the SaaS story is 1999, when Marc Benioff founded Salesforce.com to sell a new customer relationship management product. He offered this not as packaged software—as did Seibel Systems, his former employer and the company that largely invented this application. He offered the functionality delivered over servers and accessed through an Internet browser. When Salesforce.com did this, it created an in-house platform for delivering the software as a service to customers. Benioff priced the product like a service, with a fixed monthly fee that included some basic technical support and any functional improvements the company decided to make in the product. Professional services—such as customizing the application for a particular customer or providing strategic advice and training—were separate. Then, in 2005, Salesforce.com created AppExchange as an open integration platform for other software companies that wanted to build complementary applications utilizing some features in the Salesforce.com CRM product—or, more accurately, the Salesforce.com platform.

When the company did this, Salesforce.com created a new industry platform. Salesforce.com then extended the open platform concept with Force.com, a development and deployment environment using Salesforce's SaaS infrastructure. There are now competitors as well. Amazon (Elastic Compute Cloud, known as EC2) and Google (App Engine) have opened up their Internet infrastructures to host outside applications and to offer their own productized online services. By the end of 2008, Amazon was the most popular general-purpose cloud platform, with over 400,000 developers registered to use its Web services.[37] Amazon became so attractive because it has a rich infrastructure to support online retailing operations and made these services available to cloud users—data storage, computing resources, messaging, content management, and billing.

The cloud-computing platforms exhibit direct network effects to the extent that they have specific application programming interfaces (APIs) or Web services. These encourage developers to tailor their applications and can make it difficult for developers and users of the applications to switch platforms. The direct network effects do not seem as powerful as between Windows and applications written for PCs, or between particular smartphone operating systems such as Symbian, Blackberry, Apple, or Windows Mobile, and applications written for those environments. The cloud programming and Web service interfaces as well as technical standards for exchanging data and business rules are usually simple and standardized, relying on the Internet HTTP protocol. But some APIs and Web services are specific to individual platforms. For example, many real-estate companies or retail shops have built applications that incorporate Google Maps—tying the applications to Google's platform. Other companies have built e-commerce applications using facilities for handling payments provided by Amazon—tying the applications to Amazon's platform.

Cloud platforms also exhibit indirect network effects to the extent that the popularity of one over another with developers makes the platform more attractive to other developers or users. As more applications appear on a particular platform, they attract more application developers in a positive feedback loop. Cloud platform competitors also try to attract end-users by making use of their platforms free or very cheap, often subsidized with funding from advertisements. The application developers generally pay a fee depending on usage, data storage, or some other criteria. So what we see here is another instance of the "free, but not free" strategy that we saw earlier in PC and Internet software as well as Internet services.

Some firms, such as Salesforce.com (with Force.com) and Bungee Labs (with Bungee Connect), have taken the cloud idea further than just providing an environment to launch applications. They also provide services and program development tools that help companies build new applications within these platform environments. Developers can usually integrate with Web services from other vendors, even though some Web services or APIs may be specific to the platform. This is another kind of "open, but not open" or "closed, but not closed" strategy.

In short, SaaS and cloud platforms appear at multiple levels. First, the general technology of the Internet and virtualization make SaaS technically possible. Then some product companies utilize this technology in-house to offer SaaS versions of their products. Next, some firms not only offer SaaS versions of their products (or Web-based services) but open up their technology to allow other application developers to build and launch applications from these platforms. Finally, some companies offer general software-as-a-service platforms as a service—another way of describing cloud computing.

SaaS and cloud platforms appear to be efficient for both users and providers. Multiple customers can use the same facilities and thereby increase utilization rates for the hardware and the networks. For example, Amazon and Google have enormous data centers that they do not fully utilize. They can launch their own products (as standardized, automated services) whilst hosting applications from other companies, without sacrificing security of the different "tenants." Hosts such as Amazon, Google, and Salesforce.com generally guarantee security for their hosting customers through detailed service level agreements (SLAs).

On the other hand, there are some negative consequences of the SaaS and cloud platforms for traditional software product companies and users. As noted earlier, in the SaaS model, customers usually do not pay separate maintenance and product license fees, which account for about two-thirds of the revenues of major software product firms (Microsoft and Adobe are exceptions). Nor do the customers have to deal with upgrading versions of the software or installing patches—which have traditionally generated large service revenues for the product companies in the form of maintenance. That work is all done for them via the SaaS infrastructure and the one price.[38] There are sometimes issues of how well the applications perform over a shared, dispersed network. Some enterprise customers are also concerned about security of their data, proprietary knowledge in their applications somehow leaking to competitors, or the SaaS/cloud platform failing.

We also have potential conflicts of interest when application software companies move their products to a SaaS infrastructure for delivery and pricing but then open up their platforms to other application companies whose products are potentially complementary. There is less conflict when we have pure infrastructure provisioning from companies that have excess computing capacity on the Web but are not specifically application product vendors (for example, Amazon). But when a company tries to play both sides of this market, conflicts can occur. For example, Google's App Engine now includes Salesforce.com's APIs as part of its platform. A company can write an application, launch it on Google's App Engine, and use features from Google (such as Google Search or Google Maps) as well as features from Salesforce's CRM product. However, if Google decides to build its own CRM product, then we have a potential conflict of interest. Salesforce will have to rely on Google to maintain neutrality.

We will probably see more of these conflicts in the future. Microsoft has created a cloud platform service called Azure that competes with Amazon and Google as well as Salesforce.com. Azure seems relatively neutral to the extent that application developers can use various programming languages and not just Microsoft's proprietary .NET environment. They can incorporate features from other Web services platforms. But Microsoft is packaging its own online services and product features into Azure and clearly gives preference to its own products and services. Everything on Windows Live and Office Live will be available in Azure along with Microsoft SQL Server services, Microsoft CRM services, .NET services, and Sharepoint services. Microsoft is, therefore, making it possible for customers to use various Microsoft products as Web services rather than buying the packaged software. Customers should be able to integrate the Microsoft services with products of other vendors, but exactly how open the new platform will be in the future remains to be seen. The other issue for Microsoft is that usage of the Azure cloud will reduce demand for Windows desktop and servers. Microsoft expects that customers as well as some application companies will build applications using the Azure Web services and run the whole system on the Azure platform rather than purchasing more PCs or servers bundled with Windows.

Overall, we can say that SaaS and cloud computing are replacing some traditional software (and hardware) product sales, but will not eliminate them any time soon. It is relatively easy for a software product company to create a hosted version of its products. But delivering those products over an outside cloud platform such as Amazon, Google, Salesforce.com, or Windows Azure requires rewriting at least some and maybe most of the code to use the different interfaces and available Web-based services. Whilst the SaaS/cloud delivery and pricing model has many advantages, it also has disadvantages for product vendors and users. The transition, therefore, is likely to be gradual and partial, as companies create new versions of their products. Similarly, users have many customized applications and data stored in proprietary databases. They would all have trouble switching to a SaaS/cloud platform quickly but surely can do so gradually if the economics make sense.

Finally, as long as the SaaS/cloud vendors maintain some differentiation among their platform offerings, direct network efforts are not too powerful, and using multiple platforms is not too difficult or expensive for application developers or users, then we probably will see various cloud platforms coexist, just as we have seen in video-game consoles and the smartphone market, among others.

Lessons for Managers

First, managers need to realize—and find a way to leverage—the reality that most product companies today are hybrids or will soon become hybrids. By this term I mean that they combine products with complementary or essential services (including digital content provided as a service) delivered by themselves or ecosystem partners. Second, managers in hybrid firms need to understand how different the strategic and implementation challenges are for them compared to managers in dedicated product or service firms.

The first point—that most product firms are really hybrids—is true when we look at Microsoft, Apple, SAP, Oracle, Adobe, IBM, Sun Microsystems, EMC, Cisco, Hewlett-Packard, Sony, and similar high-tech firms. It is true when we look at General Electric, Toyota, Ford, General Motors, and similar firms with roots in manufacturing. We should also include banks, newspapers, or retail chains that sell standardized services and treat them as products but now want to offer personalized value-added services to make more money. In product industries subject to commoditization pressures, such as we have seen in enterprise software as well as automobiles and publishing, hybrid firms are likely to fight their ecosystem partners—the dedicated services companies—for the same pot of service revenues. Managers need to be aware of this risk and seek new kinds of partnerships and revenue-sharing models. Google, for example, is not only a platform leader. It is also a complementor and partner for all Internet service providers, all computer manufacturers, all smartphone producers, and all content providers. Whether we think of Google as a platform leader or a complementor, it must find a way to share revenues with other players in its vast ecosystem. Otherwise, the ecosystem will collapse.

With regard to the second point, not only must hybrid firms maintain both a product business and a service business. More importantly, these firms need to create deep synergies between the two businesses. Services are a great source of additional revenue as well as a great source of information on the product, its strengths and weaknesses, and what users may want in the future. But, for product companies, professional services and maintenance should never totally replace product revenues because products are the engines that drive these service revenues. The special challenge for managers in hybrid firms is to keep the product business healthy whilst making services into a secondary engine that drives additional revenue as well as new product ideas. General Motors made a mistake here. The company became far too dependent on financial services and forgot that people used its services only when they bought its products. Products and services are coupled for most firms, even though IBM and a few other companies, such as Hewlett-Packard and General Electric, have managed to become relatively neutral vendors of some services. But most product firms need to maintain strong product line-ups that keep customers paying for implementation, maintenance, and value-added or professional services.

It follows that the best mix of product sales, professional services, and personalized value-added services, as well as maintenance revenues, will vary for different firms and different business segments. Managers need to make these calculations carefully, with an eye to the present as well as to what is likely to happen in the future. Professional services, for example, are necessary in some businesses to sell products; in other businesses, they are mainly a drag on current profits but may be more important in the future if product prices fall. For platform companies, services may lose money but can be an important subsidy and help a market tip.

As one part of the implementation challenge, managers need to figure out how best to "servitize" their products. In some cases, this means changing the business model and devising how to deliver and price a product more like a service, such as Zipcar and SaaS vendors have done. This change brings the benefits of stable and recurring revenues, rather than relying solely on "hit" products and premium prices. But firms mostly need to create service offerings that add value and distinctiveness to their products. Services wrapped around products can make products less commodity-like as well as generate new revenues and profits, even as the product business declines. In some industries, there is evidence that services over the lifetime of the product can generate several times the initial profits on the product sale.[39] So some companies will choose to give away their products for free and just sell services. The cell-phone industry is well on the way toward this path—most contracts give the user the cell phone for free or at a heavily discounted price for a service contract of a particular length. Some service providers are even giving away small netbook computers in return for signing service agreements. The automobile industry might follow as well. It could be a good idea for automakers like General Motors, which made no profits from its product business for years, to "give away" automobiles in return for lifetime service contracts—not only loans or leases, but also insurance, maintenance and repair, and telematics offerings like OnStar (Table 2.2).

The second part of the implementation challenge is for managers to figure out how to "productize" their services so companies can design and deliver them more efficiently. Productization can come from component or design reuse, computer-aided tools, and standardized process frameworks and training, as seen in the software factory. Productization can also come through automating the service, following the example of Google, eBay, eTrade, Expedia.com, Lending Tree.com, Apple (iTunes), and other Internet-based companies. We also see firms automating services once done manually. For example, Eliza Corporation of Beverly, Massachusetts, discovered how to combine advanced speech recognition software with computers and telecommunications technology to deliver a variety of messages and hold interactive conversations. The company generates detailed analyses of how customers respond to

Table 2.2. *"Servitizing" the automobile*

- Financing (loans, leasing)
- Insurance (lifetime, term)
- Life-cycle management (warranty, maintenance)
- Repair (remote diagnostics)
- Customization (tailored or special features, from the factory or dealer)
- Telematics platform
  - Internet access (personal communications)
  - Navigation (GPS)
  - Communications (satellite radio)
  - Entertainment content (music, games, movies for back seat, etc.)

different communications, helping healthcare providers save money by communicating better with their customers. It is an important service used to tell customers when and how they can renew their benefits, get their health exams, receive follow-on care, or pick up their prescriptions.[40] There are many other applications for this type of automated service as well—any industry where companies want to communicate regularly with customers and not rely on one mode of communication. (Full disclosure: I have been a director of Eliza since 2008.)

In fact, fully automated services can potentially generate the same kind of gross margins as a traditional software product company. That is why Web-based delivery of software and other services is such an important innovation. SaaS and cloud-computing models have the power to change the way companies deliver both products and services, and continue to blur the distinction between the two.

Then there is the practical question: if focusing on services as much as or more than just products has such obvious benefits, why do not *all* managers in product firms embrace this principle? Again, the lower profit margins of services seem to dissuade many of these managers from taking a serious look at services and how to be more efficient and innovative in this part of the business. They also know that many services have low entry barriers, compared to products. So many competitors will continue to exist—another reason why margins can be so low in services compared to products. In some industries, though, managers now have little choice. In software application products, news publishing, commercial banking, and mass-market automobiles, products have already gone down the commoditization path. In these cases, personalized services that add value to the basic offerings are already more profitable than products. We can see this in strategic consulting, maintenance, and customization for technology companies; financial planning, individual asset management, and insurance for financial companies; and loans, leasing, maintenance, repair, and extended warrantys for manufacturing companies. This is true even though some pure services firms in the information technology business, such as Infosys and Accenture, make more money than the average software product firm. They have minimal R&D or marketing costs, charge a premium for their services, and leverage knowledge well—a form of scope economies.

In general, the reluctance of managers in product firms to build up their services business seems primarily due to a lack of appreciation for the fundamental changes going on in many industries today. Again, we are in an age of innovation and commoditization, as well as platforms and services. It is hard to make money on products, even high-tech products, that firms can produce cheaply anywhere in the world or replicate at zero marginal cost. Many products, like the automobile, are also platforms of a sort, which generate more ecosystem revenues over their lifetime from services than they do from the intial sale. Or the products have little value without services. In any case, managers in product companies need to invest in new types of capabilities—new people, processes, and knowledge—in order to innovate in both services and products, and to make money from them both. I will have more to say about the importance of capabilities in the next chapter.

Notes

1. See Cusumano (2004: ch. 1; 2008a).
2. Gadiesh and Gilbert (1998).
3. See General Electric Company (2009).
4. See the definition of services in Judd (1964), Rathmell (1966), Levitt (1972, 1976), and Bell (1973).
5. Mansharamani (2007).
6. See Mansharamani (2007: 85–92) for a review of the literature.
7. Barras (1986).
8. Thomke (2003a, b).
9. See also the discussion of this tension in Mansharamani (2007: 93–6).
10. See Mansharamani (2007), as well as public sources such as Change and Pfeffer (2004).
11. See Gerstner Jr. (2002), Lohr (2002), and Cusumano (2004: 102–8). For IBM's current services science initiative, see *IBM Systems Journal* (2008).
12. Gadiesh and Gilbert (1998).
13. This section is based on the unpublished working paper Cusumano, Kahl, and Suarez (2008), and Kahl (2008).
14. For a discussion of services, see Teece (1986), Potts (1988), Quinn (1992), Knecht, Leszinski, and Weber (1993), and Oliva and Kallenberg (2003).
15. Morgenson (2004).
16. Nambisan (2001).
17. Suarez and Cusumano (2009).

18. Carpenter and Nakamoto (1989).
19. See Gawer and Cusumano (2002).
20. See Potts (1988), Quinn (1992), and Wise and Baumgartner (1999).
21. See Bowen, Siehl, and Schneider (1991), Neely (2009), and Slack (2005).
22. The size of the global software industry is somewhat hard to track in that different groups include different segments of the industry. I usually rely on Standard & Poor's annual industry reports. India's NASSCOM (National Association of Software and Services Companies) also publishes excellent research reports, as does the International Data Corporation (IDC).
23. For an overview of how software became a business with an emphasis on the role of services and platform innovations, see Cusumano (2004: 86–127). See also Campbell-Kelly (2004).
24. This section is an updated version of Cusumano (2008a).
25. See Halligan and Shah (2010).
26. For these numbers and definitions of services, see Adobe Systems Incorporated, *Form 10-K*, annual.
27. Microsoft Corporation (2008: 5).
28. See Oracle Corporation (2009: 1, 8, 48–9) for how Oracle treats on-demand revenues. Oracle also lists Product License Updates and Support as product revenue, but lists the costs for these as services costs (p. 106). For a discussion of the complexities of how to treat SaaS and software products delivered over the Internet, see Campbell-Kelly and Garcia-Schwartz (2007).
29. Cusumano (2007a).
30. Cusumano (2004: 45).
31. Utterback and Abernathy (1975); Abernathy and Utterback (1978). See also Utterback (1994).
32. Utterback (1994); Christensen (1997).
33. Goeldi (2007).
34 Hein (2007).
35. Suarez, Cusumano, and Kahl (2008).
36. This section draws heavily on Bhattacharjee (2009). See also Hayes (2008).
37. See http://gigaom.com/2008/10/09/amazon-cuts-prices-on-s3/, cited in Bhattacharjee (2009).
38. Cusumano (2007a).
39. See Knecht, Leszinski, and Weber (1993).
40. See www.elizacorporation.com, as well as Scanlon (2009).

3

Capabilities, Not Just Strategy

The Principle

Managers should focus not simply on formulating strategy or a vision of the future (that is, deciding what to do) but equally on building distinctive organizational capabilities and operational skills (that is, how to do things) that rise above common practice (that is, what most firms do). Distinctive capabilities center on people, processes, and accumulated knowledge that reflect a deep understanding of the business and the technology, and how they are changing. Deep capabilities, combined with strategy, enable the firm to offer superior products and services as well as exploit foreseen and unforeseen opportunities for innovation and business development.

Introductory

The next four enduring principles have a longer history in management research and practice. Scholars in strategy and other fields as well as successful managers have long recognized that cultivating distinctive organizational capabilities, based on extensive domain knowledge, are usually as critical to success as making smart strategic decisions. In practice, it is difficult and usually unwise for managers to separate strategy (what they want to do) from implementation (how they will do it). One problem with "capabilities" as a concept is that this is another common yet vague term, like platforms, used in a myriad of ways. Nonetheless, most academics, consultants, and practitioners seem to agree that "distinctive" capabilities refer to specific skills necessary to design, build, and deliver products and services of significant value to customers and to do so better than the competition.[1] Capabilities also exist at the individual and organizational levels, but revolve around people, processes, suppliers, and accumulations of knowledge relevant to competing in a particular industry with specific technology or knowhow at a particular time and place.

We also need to define "strategy" more carefully. Here, we can follow Michael Porter of the Harvard Business School. He views "business strategy" as dealing specifically with how managers position the firm or a business unit in a particular market as well as how that organization chooses to compete. Distinctive organizational capabilities—such as how to manufacture an automobile with unsurpassed levels of productivity and quality, or how to design and mass produce a video recorder suitable for the home before anyone else—are the necessary complements to a successful business strategy. Then we have "corporate strategy." This deals with which businesses to be in and how the firm can generate synergies or scope economies (which also depend on capabilities) across those businesses.[2] A short review of the strategy field illustrates how we have come to this view of capabilities and their relationship to strategy and competition.[3]

Prior to the 1960s, rather than courses on strategy and competition as we know them today, the comparable subject taught in American business schools was "business policy." The key themes included how to organize the firm and what role the chief executive should play, such as how to lead the corporation and supervise operations.[4] There was a special need to manage growth in the 1950s because American companies expanded rapidly to service the domestic market and devastated regions around the globe. As US firms grew larger and more diversified, new managerial challenges emerged that went beyond the basic concepts of business policy, operations management, and executive leadership. For example, a widely cited study by the economist Edith Penrose in 1959 argued that the major constraint to further growth was what today we would call a resources and capabilities problem—the lack of enough talented managers.[5]

In the 1960s and 1970s, several consulting firms devised conceptual tools to help managers make strategic decisions and invest more effectively. The Boston Consulting Group (which popularized the term "cash cow" as well as the learning-curve concept) and McKinsey & Co.

(which popularized the notions of industry life cycles and different levels of industry attractiveness) counseled major corporations, such as General Electric, on how to introduce formal planning systems and apply financial concepts such as portfolio management to running a large, diversified company. The new goals included more "rational" resource allocation and tools to help managers select among mature versus new or growing businesses and take advantage of scale economies and learning-curve effects.[6]

This kind of portfolio thinking had its strengths—such as forcing more analysis of where to focus and invest. It also had drawbacks, particularly in that decisions not to put more money in a "mature" or "cash-cow" business often led to a self-fulfilling prophecy, even if the industry was about to enter a new period of growth. Managers interested in building long-term advantages quickly learned that firms are more than collections of financial assets to buy and sell similar to a stock portfolio. It also became apparent that scale has disadvantages as well as advantages. In any case, portfolio planning techniques and large formalized planning groups gradually fell into disfavor among practitioners as well as many academics. They tended to become too rigid and did not sufficiently emphasize how to build capabilities connected to long-term competitive advantage. Many academics also came to believe that managers in the trenches should take charge of making strategy, because they had direct knowledge of the customer, the business, the technology, and other skills the firm had to offer. In parallel, strategy courses generally followed the notions of James Brian Quinn in the late 1970s and then Henry Mintzberg in the late 1980s that managers should make strategy incrementally, through a trial-and-error process. The process more resembled "logical incrementalism" and "crafting" rather than "planning."[7] Mintzberg also dissected the pluses and minuses (mostly the minuses) of formal strategic planning, particularly in fast-paced or volatile environments, and found little empirical evidence that formal planning contributed to superior firm performance.[8]

The strategy field took another advance in 1980 when Michael Porter published *Competitive Strategy*. His "five forces" (power of buyers, suppliers, and substitutes; intensity of rivalries; degree of entry barriers) added more economic precision to strategic analysis and helped us understand why some firms in some industries made more money than others. The new concepts helped managers to distinguish how attractive one industry might be from another and to choose a particular "generic" strategy such as low-cost or differentiation, or focus on a market niche.[9] Porter added the idea in 1985 with his next book, *Competitive Advantage*, that some firms can manage their "value chains" more effectively than others and use operations and technology as sources of differentiation and sustainable advantage.[10]

Porter's work encouraged managers to view strategy in terms of the economics of different businesses and strategic positions, with some attention to organizational issues but perhaps not enough. Scholars also continued to disagree on how deliberately managers should try to make strategic decisions—how much they should rely on formal analysis of "the numbers" as well as detailed planning as opposed to intuition, experience, and qualitative judgment. Withal, by the mid-1980s, the strategy field had evolved considerably beyond ad hoc business policy and portfolio theories to a more economics-based foundation, with managers in the field taking the leading role in making strategic decisions, at least at the business-unit level.

Another milestone in the strategy field, at least for researchers, came in 1984. Following directly in Penrose's footsteps as well as building on Porter's work and that of others, Birger Wernerfeldt highlighted a concept that quickly became known as "the resource-based view of the firm." This is the notion that some firms have sustainable differences such as in production capacity or technology that can provide the basis of competitive advantage because of entry barriers or difficulty of imitation.[11] Other scholars such as Jay Barney developed this idea further.[12] Related ideas also became popular in the early 1990s, such as managing by "strategic intent" (rather than overly formal planning) and "core competences" (focusing on what the organization is really good at).[13] We also learned about the dangers of "core rigidities" that might take the place of core competencies if firms became too entrenched in their old ways of doing things.[14] Another generation of scholars wrote about this same problem from a different perspective—emphasizing that firms often become too attached to their existing capabilities as well as business models and customers, and leave themselves open to competition from new entrants and disruptive innovation.[15] Still other researchers added the concept of "absorptive capacity"—the argument that firms always need to be cultivating internal capabilities in order to understand new technologies, whether they develop technology internally or acquire it from outside, such as through acquisitions and strategic alliances.[16]

By the mid-1990s, it seems a consensus had emerged in the strategy field that building organizational capabilities that can evolve over time were at least as important as choosing a particular strategy and much more important than creating a formalized bureaucracy for strategic planning. The logic is that customer needs, industry structures, and competitive dynamics change, and often unpredictably. Different strategies may or may not work depending on implementation skills, responses of competitors, and whether the timing is right. It follows that, for firms to perform well over long periods of time, they need to re-create competitive advantages as conditions change, rather than locking themselves into particular ways of doing things. Richard D'Aveni described this idea in 1994 with the term "hypercompetition."[17] In 1997, David Teece, Gary Pisano, and Amy Shuen elaborated on the similar concept of "dynamic capabilities"—that firms can sustain competitive positions by reconfiguring critical knowledge assets and other factors of production that are difficult to imitate, thereby periodically recreating managerial and technical competences as markets change.[18] Other researchers went on to describe dynamic capabilities more precisely as specific processes for product development, alliance management, and strategic decision making.[19]

As noted in the Introduction, I illustrate this idea that firms need to focus on cultivating internal organizational and operational capabilities as much as on strategy itself by using three examples: a comparison of the different strategies that Nissan and Toyota followed to learn automobile design and manufacturing, and the long-term consequences of their specific approaches for internal capabilities; the two-decade-long competition to perfect a home VCR between JVC and Sony, with Ampex, Toshiba, RCA, and other companies joining the race; and the key strategic and organizational principles that guided Microsoft in its formative years and provided the basis for the company we see today.

The story of Nissan and Toyota is particularly important because it illustrates how firms can build and leverage internal capabilities.

My conclusion from this comparison is that there exists no good substitute for the deliberate cultivation of skills within the firm for product and process innovation. Companies can buy technology and assistance, and make all the acquisitions they want. But they must still possess and sustain those capabilities within the organization or very close at hand, and these normally come only through doing key design, or engineering, or manufacturing work within the firm. We will see a similar story in video recorders when it comes to explaining why JVC and Sony succeeded where other firms, including those with much deeper pockets, such as Ampex and RCA, failed. We also see this issue at play when explaining why IBM came to the tiny firm called Microsoft in 1980 to buy an operating system for its new personal computer.

The Japanese Automobile Industry

My story begins with Nissan, which pre-dated Toyota by several years in mass producing automobiles for the Japanese market. Most importantly, Nissan followed a dramatically different approach to entering the industry compared to Toyota: it imported technology and technical assistance directly from the United States in the 1930s and then from England in the 1950s. Nissan also ended up with markedly different—I argue inferior—innovation capabilities in both manufacturing and product development. Toyota has maintained this superiority through the present, even though Nissan, with the assistance of Renault and a president from outside Japan, Carlos Ghosn, has much improved its operations in recent years. The bottom line is that we can still see today the impact of the decisions Toyota and Nissan managers made in the 1930s and 1950s regarding how they would acquire the capabilities necessary to compete in the automobile industry.

Nissan and "Direct" Technology Transfer

Yoshisuke Aikawa (1880–1967), a mechanical engineer educated at what is now the University of Tokyo, founded the Nissan Motor Company in 1933.[20] After graduating from college in 1905, Aikawa went to the United States for two years and worked for an American iron-casting company. Soon after he imported American equipment and, using his connection to the wealthy Mitsui family, established his own firm, Tobata Casting, in 1910. Tobata would later merge with Hitachi, another firm connected to the Aikawa family.

Unlike his counterpart Kiichiro Toyoda, Aikawa was not so focused on building a great automobile company. He was an industrialist with ties to many different firms, many influential politicians, and the Japanese military. He was an engineer but did not do much engineering himself once he had started building companies—again, unlike his counterpart at Toyota. For example, Aikawa served as chairman of Hitachi from 1928 to 1941, in addition to being president of Nissan Motors in 1933–9 and then chairman in 1939–41 and again during 1944–5. Hitachi, which still provides many of Nissan's electrical and electronic components, was founded in 1917 as the equipment repair shop of Nippon Mining, another firm connected to the Aikawa family.[21] Aikawa also crafted the name "Nissan" as an abbreviation of Nippon Sangyo [Japan Industries], the holding company he used to control Tobata Casting, Hitachi, Nippon Mining, and several other companies. Aikawa invested heavily in developing Manchuria during the 1930s on the coat-tails of the Japanese invasion, served time in prison after the Second World War for war crimes, and later joined the upper house as a member of Japan's parliament in the 1950s. What is important to our story in this chapter is the strategy Aikawa used to enter the automobile industry and the impact on Nissan. Below is a basic chronology, which I will explain further in the text that follows.

- 1905 Yoshisuke Aikawa (1880–1967) graduates from Tokyo University, Mechanical Engineering Department, and goes to work in the United States.
- 1910 Aikawa founds Tobata Casting (later merged with Hitachi).
- 1920s Tobata starts making auto parts; hires William Gorham and other American engineers.
- 1928 Aikawa becomes chairman of Hitachi (established in 1917 from the machine repair section of Nippon Mining) and remains until 1941.
- 1931 Aikawa invests in DAT Motors and acquires rights to its small car.

1933 Aikawa founds Nissan Motors by spinning off the auto parts department of Tobata Casting and using funds from Nippon Mining stock sale. Remains as president until 1939; is then chairman during 1939–41 and 1944–5.

1934 Nissan begins production of small cars and trucks using Datsun technology.

1935 Nissan purchases Graham-Paige truck factory and truck technology, and imports to Japan.

Gorham sets up a modern mass-production plant for the Graham-Paige truck in Yokohama.

Nissan begins mass producing the Graham-Paige truck, with older side-valve engine; has to modify cab-over design for rough roads (not completed until 1939).

1937 Nissan passes 10,000 units per year production level.

1939 Nissan purchases Japan Ford's Yokohama plant.

1952 Nissan ties up with Austin to produce small cars in Japan, an arrangement that lasts to 1959.

Tobata Casting, Aikawa's original company, first moved from cast-iron parts into automobile parts production during the 1920s. Aikawa also hired three American engineers to work for Tobata in Japan, including William Gorham, who would later become a major figure in Nissan's history. In 1931, Aikawa then invested in a fledging small-car producer called DAT Motors and acquired rights to the "Datson" car (later renamed "Datsun"). DAT Motors made only ten vehicles during 1931–2, but this acquisition, as well as Tobata's experience in parts manufacturing, created the foundation for Nissan Motors.

Aikawa quickly realized that the real market was large trucks for the Japanese military, which needed vehicles to keep "advancing" into Manchuria. The Japanese army wanted local companies to provide the trucks and encouraged the Japanese civilian government to expel foreign automakers such as General Motors and Ford. Foreign auto companies were not allowed to return until the 1970s—giving invaluable protection to the local industry. Aikawa consequently decided to get big fast and buy rather than make his own truck technology. He sold

Nippon Mining stock for capital, spun off the auto parts department of Tobata Casting, and incorporated the new entity separately as the Nissan Motor Company in 1933. The company's main business for the next two years was making parts for Japan GM and Japan Ford and a few Datsun cars—good preparation for moving on to local mass production.

Nissan then followed a strategy of "direct" technology transfer. Aikawa knew there were distressed American automakers. To acquire new truck technology, he used Tobata contacts as a supplier to GM and Ford to gain introductions and visit different US companies. Then in 1935 he purchased (for $180,000) an entire factory from Graham-Paige, the fourteenth largest US auto company at the time, as well as designs for a 1.5-ton truck and a 6-cylinder side-valve engine. Twenty American engineers came with the Graham-Paige technology to help Nissan get its new factory up and running. Some of them stayed in Japan for as long as three years. In 1939, Aikawa also bought some of the factory assets from Ford's plant in Yokohama as well as hired many ex-Ford employees.

Gorham and other American engineers directed all of Nissan's key operations in the 1930s. They taught Nissan the latest American techniques for machine processing, stamping, welding, and forging as well as how to use steel in body and frame designs. Gorham also studied Ford's plant in Dearborn, Michigan, and used this as a reference when he set up Japan's first machine-paced moving assembly line at Nissan's Yokohama factory.

Demand for motor vehicles in Japan was still tiny in the 1930s, compared to the United States. This market reality made Ford-style mass production—expensive single-purpose machine tools, automated but fixed transfer equipment, specialized workers, assembly lines devoted to one model, and the like—difficult to justify economically in Japan. Nonetheless, Aikawa adopted the American production technology and used it to make the standard-size truck. Nissan became Japan's largest automobile producer practically overnight and held this position until 1951, when Toyota finally surpassed it. The Nissan truck and engine technology, based on the Graham-Paige technology with a modified cab design, served its purpose admirably for twenty-five years. Even after the Second World War, Nissan's production levels remained tiny. But, since it had focused on one model, Nissan's strategy maximized scale economies. Nissan and other Japanese automakers also got a boost from the Occupation troops and the Korean War, when the American army took over purchasing Japanese trucks for its own troops.

After 1950, as the demilitarized Japanese economy started its period of "miraculous" rapid growth, demand for passenger cars finally began to grow. Nissan had produced a small number of Datsun cars before 1945, and the company would resurrect this brand name. When it came to technological capabilities, however, post-war Nissan found that it again lacked engineers who understood how to design their own engines, bodies, components, and manufacturing systems. They had relied too heavily on Graham-Paige and American consultants. Looking only at the numbers, as of 1950 Nissan was by far Japan's most experienced producer of passenger cars, having made a total of 21,539 units since 1934, compared to only 2,685 at Toyota, the second most experienced producer. But most of Nissan's car production came before 1942 and relied on DAT Motors technology from the early 1930s. The Datsun engine technology was already twenty years old, as was the vehicle design, which was primarily a small truck fitted with a car body shell.

After the war had ended, Aikawa was no longer running Nissan. The next generation of Nissan managers, however, followed the same pre-war strategy of direct technology transfer. They wanted to make world-class small cars as quickly as possible, and the best place to go for such technology was Great Britain. The new Nissan president, Asahara Genshichi, who had negotiated the Graham-Paige deal in the 1930s, pushed through the decision to form an alliance with Austin in 1952. Nissan agreed to import 2,000 units annually of the Austin A40 car in the form of knock-down kits, assemble them in Japan, and market them under the Austin brand name. Over three years, Nissan would gradually switch to locally made components, with Austin's assistance and the right to use any Austin patents as long as it paid a royalty on each completed vehicle. Nissan then sold 20,855 Austin cars between 1953 and 1959, just over 23 percent of the Japanese company's total passenger car production during these years.[22]

The primary goal of the Austin alliance was to modernize Nissan's car design and manufacturing technology. Nissan based the 1959 Datsun 310 model and the 1960 Cedric, as well as the engine used in the Datsun line of small cars and trucks during this period, on technology from Austin. This infusion of new technology enabled Nissan to overtake Toyota briefly in passenger car sales (though not total vehicle sales) during 1960–2, after falling behind in the 1950s. Over the longer term, though, Austin technology did not provide a permanent advantage to Nissan. Toyota once again took over as Japan's largest car producer after 1962.

In retrospect, having the freedom to pick and choose the best of foreign technology, as well as cultivating the skills to design and make new automobiles on its own, enabled Toyota both to produce better products and to innovate more in production technology. For example, Nissan adopted Austin's old-style multi-section body for the Datsun cars. By contrast, Toyota introduced the latest technology—a more solid "monocoque" body—in its 1957 Corona. Two years later, Nissan followed. Nissan also copied Austin's transfer machinery and specialized machining and automation technology—all classic "push" systems. Meanwhile, Toyota continued to perfect its Just-in-Time "pull" methodology. Nissan also had to get outside help to move beyond the Austin product technology. It again hired the American engineer who had redesigned the Graham-Paige truck engine to modify the Austin engine for the new Datsun line.

During the 1950s and 1960s, as well as beyond, Nissan promoted itself as the Japanese leader in automotive technology. In reality, Nissan offered very good technology but with major gaps in internal capabilities. First, Nissan lacked in-house design skills, which is why it had to rely on foreign assistance. Second, and unlike Toyota, Nissan did not have any distinctive manufacturing philosophy that would set it apart from competitors. Nissan then overextended itself trying to match Toyota in model lines and Honda in engine technology. By 1999, Nissan faced bankruptcy before Renault came to the rescue. That alliance has worked very well, as CEO Carlos Ghosn forced Nissan to reduce costs throughout its supply chain, streamline product offerings, and emphasize more creativity in product design.

We should not dismiss direct technology transfer—in the form of alliances, acquisitions, or consulting—as a strategy to acquire new capabilities quickly. Rather, managers need to counteract the negative tradeoffs of the buy-versus-make shortcut. Interestingly enough, we see today major automobile producers in Korea, China, and India using similar strategies to Nissan to get big fast. Several Chinese and Indian automakers have already entered into technology tie-ups with GM, Ford, Suzuki, Chrysler, and Volkswagen. These arrangements usually begin by assembling mostly imported parts or even knock-down kits, and gradually move to local components production, with some eclectic reverse engineering and copying to design domestic models. We see different mixtures of production practices as well in these countries, with the Chinese and Indians emphasizing their relatively cheap labor compared to expensive automation, which is now more common in Japan, the USA, Europe, and Korea. In the long run, though, the winners in these markets will be the firms that find ways to produce the best products in terms of price, performance, and design. The Nissan and Toyota cases suggest a simple lesson: competitive differentiation requires managers to cultivate capabilities that are not easy to buy and are likely to take years of trial and error to evolve.

Toyota and "Indirect" Technology Transfer

The origins of the Toyota Motor Company, founded in 1936 (and named after a streamlined version of the family surname), go back to the family patriarch Sakichi Toyoda (1867–1930). Sakichi was an accomplished inventor and entrepreneur who founded Toyoda Spinning and Weaving in 1918 as well as Toyoda Automatic Loom in 1926. He licensed his automatic loom technology to the Platt Brothers of Great Britain for £100,000, then worth about $500,000 or one million yen. Sakichi had the original vision of creating an automobile company in Japan. He asked his eldest son Kiichiro (1894–1952) to make this happen and gave him the Platt money to get started. Perhaps most importantly, Sakichi served as a role model for Japanese ingenuity and self-reliance, rather than depending on the direct import or direct copying of foreign technology. There is no question that the mentality and technical capabilities Toyota began nurturing in the 1930s and 1940s, in both manufacturing and product development, provided the basis for its dominant position in the global automobile industry today. It may also be time for Toyota to move to another level of internal capabilities in order to design and test its advanced software and sensor-based systems for controlling braking, acceleration, vehicle stability, and power generation. But what follows is a basic chronology of how Sakichi and his son Kiichiro entered the automobile industry and developed the most important initial capabilities on their own. I will explain more of the details in the text afterwards.

1918 Sakichi Toyoda (1867–1930) founds Toyoda Spinning and Weaving.

1920 Kiichiro Toyoda (1894–1952) graduates from Tokyo University, Mechanical Engineering Department, and goes to work for Toyoda Spinning and Weaving.

1926 Sakichi founds Toyoda Automatic Loom (TAL); licenses technology to Platt Brothers for £100,000.

1929 Kiichiro heads the TAL delegation to England to monitor the Platt Brothers' contract; begins visiting modern auto factories.

1932 Kiichiro begins copying a Chevrolet engine.

1933 Kiichiro establishes an auto parts department in TAL; begins copying the American Smith motorcycle engine; buys a new Chevrolet car and dismantles it.

1934 TAL begins studying auto factory layouts in the United States and England.

Buys and dismantles Ford and Chrysler DeSoto cars.

Finishes copying the Chevrolet engine.

Begins importing modern machine tools from England then sets up machine-tool and steel-making departments in TAL headed by local professors.

Constructs a pilot factory, with the assistance of local Japanese experts.

1935 TAL buys and copies a Ford truck.

Opens first assembly plant.

Kiichiro hires Shotaro Kamiya (1898–1980) from Japan GM to head sales.

1936 Kiichiro founds Toyota Motors from the auto parts department of TAL.

1937 Toyota introduces its first truck, based on the best technologies selected from foreign models and utilizing a modern overhead valve engine design.

1939 Toyota passes 10,000 units per year production level.

1947 Toyota introduces its first small car, using the Volkswagen Beetle and other foreign cars as references.

Like Aikawa, Kiichiro studied mechanical engineering at the pre-war predecessor of the University of Tokyo, Department of Mechanical Engineering, and graduated in 1920. He worked for his father's company and became a specialist in casting technology, then took over manufacturing operations at Toyoda Automatic Loom in 1926. Kiichiro also headed the delegation to England for negotiations with the Platt Brothers during 1929–30. He took advantage of the opportunity to spend several months visiting automobile factories. Then he established an automobile department within Toyoda Automatic Loom in 1933 and spun this off in 1936 to form the Toyota Motor Company, a move done mainly to raise additional capital. The timing also corresponded to a new law the Japanese government passed to promote local automobile production and eliminate foreign automakers.

When founding Toyota, Kiichiro intended to create automobiles that were "cheaper and better than foreign imports."[23] But he concluded during the visit to England that he did not need to tie up formally with a Western automaker. Kiichiro's plan was to reverse engineer product designs with the help of Japanese university professors and other Japanese experts, and make most of the components using the skills of Toyoda Automatic Loom and machine tools that he would purchase. He also believed he could adapt Western manufacturing technology to the much smaller volumes of the Japanese market. At least part of the reason Kiichiro decided to proceed on his own was that, unlike Aikawa, he was short of funds. But Toyota passed the 10,000-unit production level in 1939, only two years behind Nissan, and with comparable or better levels of productivity (see Appendix II, Table II.5).

How Toyota learned automobile design is a classic story of reverse engineering. It also reflects great attention to detail and learning as well as hiring the right people—the essence of how firms build internal organizational capabilities that last. In 1933, Kiichiro and a former classmate and professor of automotive engineering at the University of Tokyo copied a 2-cylinder 60-cc motorcycle engine from the United States and made ten motorcycles. Meanwhile, a Toyoda Automatic Loom executive visiting England for the Platt contract began importing machine tools. Kiichiro then sent Toyoda Automatic Loom's in-house expert on engine casting to the United States to study factory layouts at Ford, General Motors, Chrysler, Packard, Nash, and Graham-Paige. This manager returned to Japan in summer 1934, two months after the modern machine tools arrived from England. Another friend of Kiichiro's who was a professor at the Tokyo Institute of Technology supervised construction of a pilot factory. Kiichiro also hired the chief engineer of a local steel company to set up Toyota's new machine-tool and steel manufacturing departments, both of which later became the important Toyota subsidiaries Toyoda Machine Works and Aichi Steel.

Necessity was also the mother of invention, or innovation. Kiichiro learned how to make his own machine tools because he did not have enough money to buy more than one of each kind. The quality of sheet steel available in Japan was inadequate for automobile use, so Toyota (as well as Nissan) had to make their own steel. As for electrical components, Toyota bought some locally and made many others in-house to save money and guarantee supplies. Kiichiro later separated the department for these components as Nippon Denso (NDC), today Toyota's largest subsidiary and one of the top automobile component makers in the world.

The list of everything Kiichiro did to build internal capabilities is very long. We can skip most of the specifics, but they illustrate Kiichiro's reasoning, level of commitment, and attention to detail. For example, when he realized that he needed his own design department, he hired four Japanese engineers from local companies that had made small vehicles. He asked a professor who was an expert in gear manufacturing to supervise machine-tool operations. Three other Japanese experts, including two with Ph.Ds, supervised steel production. Another friend joined the company and helped lay out the new assembly plant, opened in 1935. Equally foresighted, Kiichiro hired Shotaro Kamiya at this time from Japan GM; he would later gain recognition as a marketing genius whilst heading Toyota Motor Sales.

The engine work was critical because of the importance of this component. Kiichiro headed the design effort himself—something unthinkable for Nissan's Aikawa. He started by copying an American motorcycle engine (made by Smith) and then studied Ford and Chevrolet engines before deciding in 1932 to copy the Chevy engine, because this seemed easier to manufacture. Kiichiro and his team at Toyoda Automatic Loom also bought a new Chevrolet in 1933 and began systematically dismantling and studying its components. They ordered replacement parts from GM and local Japanese suppliers, including Aikawa's Tobata Casting. Toyoda Automatic Loom engineers then finished copying the 6-cylinder Chevy engine in 1934.

Toyoda Automatic Loom purchased a Chrysler DeSoto in 1934 and dismantled this as well as a Ford vehicle to study their frame and axle designs. Ultimately, Kiichiro's designers ended up in 1935 with a hybrid product. The body came mostly from the DeSoto, the frame and rear axle from Ford, and the front axle and engine from Chevrolet. We are told in the company history that engineers were careful to avoid violating patent laws, but there was also innovation in this first Toyota product. Kiichiro's team added a free-floating suspension system to improve stability on bumpy roads, which were common in Japan and Asia at the time and proved to be a problem for Nissan's Graham-Paige truck. They chose a wheelbase length that would enable Toyota to use the same frame and chassis to make either a car or a truck merely by sliding the engine forward. The company's engine experts also redesigned the Chevrolet engine's cylinder heads and widened the overhead valves to burn fuel more efficiently and produce more horsepower—62 compared to 50 before the redesign.

Like Aikawa, Kiichiro realized that the biggest market in Japan was for trucks. So he bought a Ford truck in spring 1935 and copied it by the summer. This became the Toyota G-1 standard-sized truck, equipped with the Model-A engine. This technology then became the basis for eight new truck models introduced between 1937 and 1956.

These various efforts proceeded through trial-and-error learning—another key feature of how to build internal capabilities. By Toyota's own account, the work did not go so smoothly. There were many problems with the designs and components, forcing company engineers to make constant changes and re-set manufacturing equipment. There was also no guarantee that Toyota would be able to manufacture a reliable product in volume. Because they were changing designs so frequently to improve the product, Kiichiro and his chief engineer decided to buy universal machine tools and easily adaptable specialized machines, rather than the single-function machine tools, large stamping presses, and expensive transfer machinery that Nissan had acquired and that were standard in Western mass-production facilities. Kiichiro even asked a local machine producer, Komatsu, to make him small, flexible stamping presses. In later years, the use of universal and adaptable machine tools, as well as small stamping presses that were easy to set up for different tasks, became hallmarks of the flexibility in Toyota's renowned "Just in Time" production system.

Kiichiro's decision to copy the Chevrolet engine proved fortuitous in another way. This engine used the latest design concept—overhead valves. Putting valves over the cylinders made it easier to modify the shape of the engine's combustion chambers and achieve higher compression ratios and thus more fuel and power efficiency. Toyota was able to continue tweaking the engine design for the next twenty years, without relying on foreign experts, and continually improved performance and in-house learning about the physics of internal combustion engines. By contrast, Nissan's Graham-Paige engine used an old-style side-valve design that was more difficult to modify and improve. Not until 1959 did Nissan introduce an overhead-valve design for its truck engine, and to do this it had to hire an American engineer to redesign the Graham-Paige engine (in addition to the Datsun engine).

Like Aikawa, Kiichiro did not have much role in the post-war company. He was ready to focus on passenger cars but had to resign in 1950. He did this to help settle a crippling strike and labor problems that plagued many Japanese companies at the time. Kiichiro died prematurely in 1952 at the age of 57. Nonetheless, Kiichiro's legacy grew after the war. In particular, the tradition of self-reliance and indirect technology transfer again provided an advantage as the company quickly mastered passenger-car production without a foreign tie-up. Unlike Nissan, which used the old Datsun technology from the mid-1930s, Toyota waited until 1947 to introduce a line of small cars. As they had done with trucks in the 1930s, the Toyota engineers used the latest foreign models such as the Volkswagen Beetle as their reference and consequently ended up with newer and better technology. Toyota also designed new vehicles so that small cars and trucks could utilize the same frames, chassis, engines, and other components, as well as the same manufacturing lines and equipment, until demand increased.

When explaining how well Toyota has done in Japan for so many years, it is difficult to overestimate the importance of the company's marketing and sales capabilities. Most of the credit goes to Shotaro Kamiya (1898–1980), hired in 1935 from Japan GM. Kamiya began his career in Toyota by recruiting Japan GM dealers, giving Toyota a national base of dealerships covering every Japanese prefecture. In 1949, Toyota separated the sales department as Toyota Motor Sales, which Kamiya then took over as president until the firms merged in 1982. By this time, global marketing had become well established at Toyota. After the Second World War, however, there was a danger that automobile companies would focus too much on technological catch-up; having a separate sales company with a strong executive meant that Toyota would never neglect the sales and marketing side of the business.

After the war, Kamiya right away set out to recruit Nissan dealers over to Toyota, and expanded the general dealership base through franchising. He set up specialized dealerships for each new product line and did not allow any one dealer to carry a full line of Toyota products. This approach forced salespeople to focus their efforts on particular models and, throughout the network, sell everything Toyota made with equal emphasis. Nissan, by contrast, set up large mega-dealerships and only later moved to specialized dealers. Perhaps showing his greatest foresight, Kamiya decided to invest heavily in driver education. Toyota Motor Sales bought its first driving school in 1957 and then continued to buy up schools around the country. As more Japanese became affluent and took up driving, their first exposure was often to a Toyota car. Many of these people became lifelong Toyota customers. In the 1950s, Kamiya also introduced services: installment plans for new vehicle purchases and training schools for repair technicians. He also started formal customer surveys and other measures that complemented Toyota's growing lead in product and production technology.

The outcome of this story is now common knowledge around the world. Both Toyota and Nissan began exporting their best small-car models in 1966—the Toyota Corolla and the Datsun Sunny/Nissan Sentra. It still took another decade to get their volumes high enough, costs low enough, and quality good enough to make real progress in the United States and then other overseas markets. Both companies did well, until Nissan's weaknesses in manufacturing, supply chain management, and sales resulted in losses by the 1990s. Honda also entered the Japanese automobile market in the 1960s and established itself as a true innovator in engine technology and other aspects of automobile design. But, at least through 2010, and despite some serious lapses in design quality, Toyota has remained the leader among the Japanese automakers in manufacturing quality and productivity, as well as long-term financial performance. The source of the company's remarkable strength, at least in the past, has been its unique production system and philosophy of learning, both of which we can trace back to the 1930s and 1940s. In addition to its manufacturing innovations, Toyota also introduced a new process for product development in the 1950s that became widely copied by other automakers, in Japan and around the world. I will say more about Toyota's manufacturing innovations in Chapter 4 and its product development system in Chapter 5, as well as how to interpret these systems in light of the quality problems that surfaced in 2009–10.

The Japanese VCR Industry

When Richard Rosenbloom and I first wrote about the VCR industry in 1987, we were trying to bring together two sets of ideas and research—on strategy and innovation.[24] We saw the key issue as how to help managers translate "technological capability into competitive advantage."[25] The strategy literature, exemplified in writings by Michael Porter and David Teece, generally talked about *timing* of market entry (first mover or follower), product *positioning* (segmentation, pricing), and organizational *implementation* (make or buy, supply chain, marketing channels, and the like). We argued that decisions made by both general managers and engineers with regard to how they would develop a new technology—even if they were not the inventors but were more aptly described as "pioneers"—shaped what firms were able to do later on. Many of the differences between Nissan and Toyota, for example, stem from decisions their founders made in the 1930s. Moreover, having the right strategic objectives and the ability to create a design in an R&D setting was not enough. Firms had to acquire the necessary engineering and mass-production capabilities. For the home VCR, these were not possible to buy or create in a short period of time.

The competition to produce a viable home VCR is a complex story—appropriate because this was probably the most advanced high-technology product ever offered to consumers until that time. Color televisions were highly complex as well but nowhere near as difficult to design and mass produce. Below is a basic chronology of the actions of key players, which I will explain in more detail in the text that follows.

1951 Chairman David Sarnoff of RCA challenges his research lab to create a video recorder.

1953 Sony and Toshiba begin studying video recording technology.

1954 Toshiba files patents for the helical scanner.

1955 JVC begins studying video recording technology.

1956 Ampex demonstrates a $50,000 four-head reel-to-reel video tape recorder (VTR) the size of a refrigerator.

1958 NHK (Japan National Broadcasting Co.) purchases an Ampex VTR and shows it to Sony, JVC, Toshiba, and Matsushita; the Japanese government offers R&D subsidies; Japanese companies, led by Sony and JVC, abandon the four-head design as too big for consumers and start designing two-head helical scanners.

RCA terminates video recorder development, before attempting to re-enter in 1969.

1960 Ampex and Sony exchange technology; agreement ends over patent dispute.

1962 Sony begins marketing a small transistorized reel-to-reel VTR, using a two-head helical scanner.

1964 Matsushita markets its first VTR, using JVC's two-head scanner.

1965 Sony introduces a $1,000 VTR with ½-inch-wide tape; fails with consumers but opens institutional market.

Ampex introduces $1,000 VTR with 1-inch-wide tape; fails with consumers and decides to focus on broadcasters.

1967 JVC introduces $630 reel-to-reel VTR that fails with consumers.

1969 Sony introduces 1-inch-wide tape cartridge VCR that fails with consumers; JVC counters with ½-inch-wide tape cartridge that also fails with consumers.

RCA announces Holotape and Videodisc design efforts.

1970 Ampex announces Instavideo home VCR; fails to outsource manufacturing to Toshiba joint venture.

1971 Sony, JVC, and Matsushita adopt common ¾-inch-wide tape cartridge—U-Matic.

RCA designs a VCR; fails to outsource manufacturing to Bell and Howell.

1974 Sony completes the Betamax design; begins mass production in 1975.

1975 JVC completes the VHS design; begins mass production in 1976.

The Video Recording Pioneers

With the spread of television and live coast-to-coast shows after the Second World War, broadcasting companies clearly saw the utility of recording (rather than filming) their programs. Innovation in audio recorders using magnetic tape provided the inspiration. Video contained far more information than audio, so any viable product would require considerable advances in various related technologies. Several companies undertook the challenge, beginning with RCA and Ampex in the United States and Toshiba in Japan. These three companies made considerable progress and eventually cross-licensed their inventions.

RCA in the United States began research in 1951 when its chairman, David Sarnoff, challenged his labs to develop a video recorder within five years. Company engineers then created a prototype that used fixed magnetic heads and moved a narrow tape past these heads at very high speeds. Toshiba in Japan, led by Dr Norikazu Sawazaki, tried a different approach whilst working on a contract for NHK, Japan's national broadcasting company. He allowed the heads to rotate at high speed and moved a much smaller amount of tape past the heads at a slower speed in a diagonal direction, using a device later called a "helical scanner." In 1954, Sawazaki filed for patents on this new device. Meanwhile, in 1956, Ampex in the United States demonstrated a $50,000 machine that used a "transverse scanner" to write and read on 2-inch-wide tape moving past four recording heads attached to a rapidly rotating drum. Ampex patented the technology and licensed it to RCA in exchange for color TV technology. Ampex also began to collaborate with Sony for access to transistor technology, which was necessary to shrink the size of the recording device (which initially was as big as a full-size American refrigerator).

There was speculation during 1956 in both the *Wall Street Journal* and the *New York Times* that this kind of machine would one day be available for the home consumer to record broadcast TV programs or play rented movies.[26] However, senior managers at RCA, Toshiba, and Ampex decided to focus on the broadcast market rather than the home consumer. RCA had many other businesses, including computers, and terminated video recorder development in 1958 before trying to return, unsuccessfully, to this market in the 1970s. Sawazaki had the best design, but Toshiba management decided to form a joint venture with Ampex in 1964 and adopt its technology. Ampex produced a small transistorized video tape recorder using a helical scanner in 1962, which it sold for use with closed-circuit TV monitors. But top management preferred to concentrate on the high-margin professional broadcast market as well as diversify into recording products for the nascent computer industry. In contrast, senior executives and managers at Sony, JVC, and Matsushita (Panasonic) decided to pursue the long-term goal of a home-use video recorder. Here, we can see that the vision and positioning adopted by these three firms in the 1950s would later shape the capabilities they developed by the 1970s—and influence who would succeed or fail. Only firms that targeted the consumer market and then incrementally refined their design and manufacturing skills for this market were able to mass produce the VHS and Betamax designs that eventually won over consumers.

As with automobiles in the 1930s and 1950s, the Japanese government played a role in this industry. In 1958, NHK imported an Ampex machine and invited Sony, Toshiba, Matsushita, JVC, and other Japanese firms to study it. Japan's Ministry of International Trade and Industry (MITI) organized a study group and provided some R&D subsidies. Through the study group and personal ties, engineers at JVC and Sony learned of Toshiba's helical scanner idea and imitated it.

Sony, established in 1946 by an electrical engineer, Masaru Ibuka, and a physicist, Akio Morita, was probably best positioned to create a home video recorder. The founders were committed to applying advanced electronics such as the transistor to consumer applications, even though they would produce many products for professionals. The founders also did not believe in market research for innovative new products and preferred to rely on their own judgment. Sony had the necessary technical expertise as well. It had produced Japan's first audio magnetic tape recorder in 1950, a TV camera in 1953, and a home-use stereo tape recorder in 1955. Its most talented engineer, Nobutoshi Kihara, began studying video recording at Ibuka's insistence in 1953. Ibuka then decided to focus more resources on video recording after seeing the Ampex machine at NHK in 1958. It took Sony only three months to duplicate the Ampex device, persuading Ibuka to give Kihara a specific challenge—create a $500 home machine. Sony and Ampex cooperated in 1960 but then fell into a dispute over patent rights, forcing Sony to proceed on its own. Sony then demonstrated a transistorized video tape recorder in 1961 and marketed its first product in 1962, using a two-head helical scanner that Ampex engineers had believed would never succeed commercially.

The second firm that had advanced capabilities in both television and magnetic recording technologies was JVC, founded in 1927 as RCA's Japanese subsidiary for making phonograph records and audio equipment. In 1955, JVC began studying video recording at the suggestion of Managing Director Kenjiro Takayanagi, who had pioneered television broadcasting in Japan during the 1920s whilst working at NHK laboratories. Takayanagi had joined JVC in 1946 with twenty other NHK television engineers. He heard of Sarnoff's challenge in 1951 and saw the Ampex recorder when he visited RCA in 1956. At this time, he asked one of his best young engineers, Yuma Shiraishi, to take on video recording. JVC had already produced an audio recorder in 1956. The JVC engineers studied the Ampex machine at NHK during 1958–9, decided the four-head design was unworkable for a home machine, and began devising a two-head helical scanner with circuitry that avoided the Ampex patents. JVC announced a prototype reel-to-reel design in 1960 and sold its first product in 1963 to the Tokyo police department.

The third Japanese firm to pioneer the home video market, Matsushita (Panasonic), dates back to 1918. Konosuke Matsushita founded the company to make electrical plugs and then a variety of other electrical devices and home appliances. It started making audio tape recorders in 1958 and TV sets in 1960. It is also relevant to this story that, in 1953, Matsushita acquired a half-ownership of JVC but chose to operate it as an independent company focused on consumer electronics innovation. In 1957, Konosuke encouraged his central labs to take up video recording. The task fell to Dr Hiroshi Sugaya, a physicist who had joined Matsushita three years earlier and was already studying audio heads. With help from NHK and the MITI subsidy, Sugaya built a prototype modeled after the Ampex design but then in 1960 decided to adopt JVC's two-head helical scanner. Matsushita introduced its first commercial video tape recorder in 1964.

Towards a Consumer Product

Ampex, RCA, and Toshiba all attempted to re-enter the consumer market in the 1970s but failed because they lacked the precision engineering and mass-production skills. By contrast, Sony introduced the successful Betamax in 1975, whilst JVC, supported by Matsushita, followed in 1976 with VHS (Video Home System). VHS eventually won this platform competition, as discussed in Chapter 1. Timing, positioning, and implementation skills relied on technical capabilities nurtured over decades, supported by clear goals and a specific vision of the future—a viable home market, where everyone with a TV set was a potential customer. Of course, the main market in the 1960s was for broadcasters, and selling to consumers remained a distant and risky goal.

The VCR story, like the Toyota story, illustrates how firms can develop distinctive capabilities—again, mainly through directed but incremental trial-and-error learning. In the VCR case, the early products aimed at consumers all failed to catch on in the marketplace. But the repeated efforts provided lots of ideas on what to do next. For example, in 1965 Sony came out with a $1,000 reel-to-reel machine that used ½-inch-wide tape. The product did not sell but served as an experiment to shrink the overall size of the machine, which would eventually be essential for consumer acceptance. Ampex countered with a machine at a similar price but used 1-inch-wide tape—resulting in a machine that was much too big for the home. Both the Sony and Ampex products were large, expensive, and hard to use, with the reel-to-reel format common on audio tape recorders of that era. Consumers did not buy them. Nonetheless, some schools, hospitals, companies, police departments, and other institutions did purchase the machines. These early video recorders thus opened up a new market and led to a temporary shift in strategy—targeting the audio-visual business.

For the home consumer, though, the best way to improve ease of use seemed to be cartridges or cassettes, which were being adopted in audio tape recorders and players. Sony in 1969 did this using 1-inch-wide tape; JVC countered with a ½-inch-wide tape cartridge. Sony, JVC, and Matsushita then decided to cooperate and adopt a ¾-inch-wide tape cassette as a new standard for the industry. Sony introduced the first machine in 1971 at a price of $1,000, dubbed the U-Matic. This design was relatively successful and dominated the audio-visual market for the next several years. Once more, consumers found the machines too large and expensive for home use. The competition continued.

Meanwhile, back in the United States, Ampex attempted to re-enter the consumer market in 1970 with a breakthrough product it called Instavideo. The company was only able to make a prototype, so it assigned the manufacturing task to its joint venture with Toshiba in Japan, called Toamco. But Toamco had never manufactured in volume and failed to mass produce the Instavideo design. This product, in fact, never shipped to customers. Ampex soon killed the Instavideo project entirely and withdrew from the race. Toshiba then tied up with Sanyo and came to market in 1974 and 1976 with its own machines. These were inferior to Beta and VHS and soon disappeared from the market. In 1969, RCA management decided to develop alternative technologies for home video, initially focusing on laser-based "holotape" and then a videodisc (forerunner to the modern CD and DVD). In 1971, RCA engineers also designed a video cassette recorder. Like Ampex, it did not have mass-production skills and tried to outsource the manufacturing to a leading US maker of movie projectors, Bell and Howell. But this outsourcing effort failed as well. RCA then adopted JVC's VHS technology in 1974.

By the early 1970s, the race to produce a successful home VCR had turned into a marathon, and only a few companies had the staying power to remain in the competition: Sony and JVC. Sony Chairman Ibuka kept pushing to develop smaller and cheaper machines than the U-Matic. His chief engineer identified ten different possible designs and set up parallel teams to pursue each one. Sony managers chose the best design and then spent eighteen months preparing for mass production. In December 1974, Sony showed the Betamax to Matsushita as well as executives at JVC and RCA, and hoped they would adopt this technology as a standard, as they had done with the U-Matic. But the initial Betamax had only a one-hour recording time and Sony was unwilling to change its design. The other companies declined to adopt Beta as well.

JVC, like Matsushita, had been continually evolving its own technology. In 1967, it introduced a video tape recorder for $630 that failed because of the awkward reel-to-reel format and lack of color recording. It introduced a cartridge model in 1969, which also flopped with consumers. But the U-Matic inspired Shiraishi, JVC's chief video engineer, to launch another effort in 1971, which five years later resulted in the VHS. His goal was to develop a cassette machine that would sell for $500 and use as little tape as possible. At this point, JVC management's commitment to the consumer industry was faltering. The video division was losing money, and some executives wanted to halt VCR development. For this reason, Shiraishi kept the new project secret from the board of directors and put only two of his ten engineers on the effort, before adding a third in 1973. Their first models used ½-inch-wide tape cassettes but recorded for only one hour and only in black and white. The team then came up with a matrix plan outlining desired features for the consumer market and technological difficulties and options. They followed this roadmap to produce the VHS, which by 1975 recorded for two hours and in color. Shiraishi revealed the new design to Matsushita in April 1975, a few days after the Betamax went on sale.

Winners and Losers

To recap: Ampex, RCA, and Toshiba led the way in developing video recording technology but failed to translate this advantage into a viable home product. Sony, JVC, and Matsushita, as well as their licensees (mostly of the VHS platform), eventually took nearly 100 percent of this blockbuster market. (Philips in Europe temporarily gained a small percentage with a competing format before deciding to license VHS.) Ampex had about $68 million in revenue in 1960, compared to about $55 million for JVC and $52 million for Sony. It had grown only to about half a billion dollars in revenue by 1985, compared to nearly $3 billion for JVC, where two-thirds of its revenues came from video sales, and over $5.3 billion for Sony, where more than one-third of revenues came from video (Table 3.1).[27]

To explain who won and who lost this competition, it is important to realize that the VCR is a sophisticated "system" technology. It is also a platform to the extent that the basic technology (the player-recorder), whilst useful to record broadcast TV programs, requires complementary innovations (pre-recorded tapes as well as movie cameras) to be particularly valuable for consumers. Moving down to the consumer from a broadcasting machine the size of a refrigerator and priced at the cost of a house in 1960 necessitated many advances—in magnetic recording materials (the tape), precision machining (heads and scanners), solid-state circuitry, television and FM encoding (for the video signals), and the automatic tape-handling mechanisms (a mechanical assembly that reached into a cassette, pulled out the tape, and threaded it diagonally around the scanner). Several firms, including Ampex and RCA, were able to combine these diverse technologies in a laboratory setting; but only Sony and JVC succeeded in putting them together in a mass-production setting, with Matsushita close behind.

Table 3.1. *Revenue comparison of VCR competitors, 1960–1985 ($ m.)*

	1960	1970	1980	1985
Ampex	68	296	469	480*
JVC				
Total	55*	293	1,810	2,941
Video	nil	2	1,068	1,912
Matsushita				
Total	256*	2,588	13,690	17,120
Video	nil	6	800	3,000
Sony				
Total	52	414	4,321	5,357
Video	nil	17	973	1,982
Toshiba				
Total	390*	1,667	7,738	12,598
Video	nil	n.a.	n.a.	400*

Notes: Yen converted as 360 = $1.00 in 1960 and 1970; 200 = $1.00 in 1980 and 1985. Revenues do not include unconsolidated subsidiaries. Numbers with * indicate authors' estimates.

Source: Rosenbloom and Cusumano (1987: 66).

The reason has everything to do with capabilities, and not just strategy. We saw how Sony, JVC, and Matsushita started their home video design efforts in the mid-1950s and stayed committed to the business, learning along the way how to design, engineer, and mass produce the critical components, especially the complex heads and tape-handling assemblies. Each company brought various models to market during the 1960s and early 1970s that failed commercially but provided a technical agenda for improvement. The failed products, in a real sense, became prototypes of the successful designs. But putting together the necessary technologies required an incremental, iterative, and experimental approach to R&D and manufacturing lasting over two decades. Top management at these three firms supported the long-term efforts, though the VHS project at JVC did proceed secretly for a few years. By contrast, Ampex, RCA, and Toshiba management saw the market potential and eventually wanted to go after the home market. But they did not put in the time and investment necessary to implement that strategy.

Sawazaki of Toshiba, who invented the helical scanner that became the key component in modern VCRs, moved into research management in the late 1950s before Toshiba decided to tie up with Ampex. Alex Maxey, Ampex's chief designer, left the company largely unappreciated in 1964. RCA also moved into and out of a variety of businesses and largely divested its consumer electronics manufacturing business during the 1960s and 1970s.

One can also detect a breakthrough mentality in both Ampex and RCA that differs from the kind of thinking behind deep capabilities development. Managers and even some engineers at these companies believed they could duplicate the great success of the first Ampex machine with a bold R&D effort in the laboratory. In a sense, Ampex did this with the Instavideo prototype, and RCA did this with its holotape and videodisc designs. But neither company, nor the Ampex–Toshiba joint venture in Japan, paid much attention to developing the requisite knowhow to transfer such complex designs to manufacturing. It was not possible simply to change strategies and shift from the professional or audio-visual market to the consumer so quickly. Superior design and manufacturing capabilities, nurtured over two decades and sustained by the vision of a future mass market, ultimately separated JVC and Sony from the losers in this competition.

Microsoft "Secrets"

My last example of capabilities takes us back to the founding of Microsoft—a story that is often recounted.[28] Many observers also describe Bill Gates and Microsoft as "lucky" or at least in the right place at the right time. This is true. But we should acknowledge that Microsoft was truly a pioneer in a new industry—with a new technology that would soon require novel strategies and marketing skills based around a mass-market platform. The company already had distinguished itself as having the best PC programming skills when IBM came by in 1980 looking for an operating system. So, whilst Microsoft's history surely includes some fortuitous events, it is also a history of remarkable vision and distinctive capabilities.

The highlights of the story go as follows. In 1975, whilst Bill Gates was a junior at Harvard University, his friend and fellow entrepreneur from high school, Paul Allen, saw an ad in *Popular Electronics* magazine for the Altair personal computer kit. Gates and Allen had been fortunate to attend a private high school that gave the students access to a computer. The programming skills they acquired enabled them to found a small company—whilst in high school—to write applications to monitor traffic data and do other tasks. By the mid-1970s, Allen was working as an engineer for Honeywell in Massachusetts but was still on the lookout for another entrepreneurial opportunity. He showed the article to Gates and suggested they make hardware insert boards for the kit and bundle this with a rudimentary programming language.

Gates immediately saw a different future. The ad convinced him that someday all desktops would have personal computers on them, and he wanted to make the software for them. So he used the computer center at Harvard to write a version of the Basic programming language for the Altair. Allen brought the new product out to New Mexico, location of Altair's manufacturer, and ran it on the test machine. The software worked perfectly and the rest is history. Microsoft the company began as a 60–40 partnership in favor of Gates, reflecting his greater role in creating the first product. Allen would leave the company a few years later because of illness, though he recovered and remains on Microsoft's board of directors. He is still a major shareholder as well as involved in numerous other ventures.

Paul Allen was the first to see the magazine ad, but Bill Gates's insight was particularly remarkable for 1975. Most companies in the computer industry at the time, led by IBM and Digital Equipment Corporation, and even a new firm founded the next year, Apple Computer, were focusing on hardware. Gates recalled his thinking in a 1994 interview:

> I thought we should do only software. When you have the microprocessor doubling in power every two years, in a sense you can think of computer power as almost free. So you ask, why be in the business of making something that's almost free? What is the scarce resource? What is it that limits being able to get value out of that infinite computing power? Software.[29]

Microsoft and its two employees (Gates and Allen) generated only $16,000 in revenue for 1975 and $22,000 for 1976, when they expanded to seven employees. They had no strategy other than to build programming languages for the Altair and other computers that might come along. Gates, however, knew he needed more talented people like himself. He spent most of his time recruiting friends and other people who he knew could master software development for personal computers. This was a rare technical skill at the time, just as knowing how to sell software products to the masses would become a rare and valuable marketing skill a few years later. Even in 1980, just before IBM decided to launch a PC, Microsoft had only thirty-eight employees and about $8 million in revenues (Table 3.2). But, because of the new company's reputation, IBM executives asked Gates to produce an operating system, even though Microsoft was a languages company at the time and had not built any other products. Microsoft was again very fortunate to know about and then acquire a rudimentary operating system (for $75,000) from a local Seattle company, which it modified and then licensed to IBM (royalty-free) as DOS. Microsoft received only some development funds from IBM but retained the rights to sell DOS to other computer manufacturers, which then Microsoft helped to build compatible machines.

Let us pause here for a moment. Microsoft sort of did what Nissan did—buying technology as a short cut into the business, in this case, the nascent PC operating systems market. It became a financial bonanza for Microsoft as the IBM PC "clone" market took off in the early 1980s. Over the next decade, Microsoft also built Windows as a graphical layer on top of DOS. But, as Nissan experienced, there was a tradeoff. Microsoft engineers never got the chance to build an operating system from scratch, at least, not until Gates hired new people to create Windows NT in the mid-1990s. Moreover, though DOS was fast, light,

Table 3.2. *Microsoft company data, selected years*

Fiscal	Revenues ($000)	Employees (approximate)	Operating profit (%)
1975	16	3	n.a.
1980	8,000	38	n.a.
1985	140,417	1,001	n.a.
1990	1,183,000	5,635	33
1995	5,937,000	17,801	35
2000	22,956,000	36,582	48
2005	39,788,000	61,000	37
2008	60,420,000	91,000	37
2009	58,437,000	93,000	35
2010	62,484,000	90,000	38
2011	69,943,000	90,000	39

Note: Microsoft began publishing data on operating profits after going public in 1986.
Source: 1975–90: Cusumano and Selby (1995: 26); 1990– : Microsoft's annual *Form 10-K* reports.

powerful, and cheap, it had many technological limitations, not the least of which was the lack of a graphical user interface and networking functions, or the ability to do multiple tasks simultaneously. Microsoft would fix these and other problems, as well as diversify its products and skill set, but only after acquiring a new set of capabilities to add to its knowledge of programming languages.

For example, at the time of the IBM deal, Gates decided to recruit several programmers and managers from Apple as well as other companies to build desktop applications, beginning with a word processor and a spreadsheet. He believed these were new emerging mass markets and he wanted Microsoft to be in them with products compatible with the IBM PC platform. Then, to learn graphical programming, Gates agreed to provide the basic applications to Apple that later helped make the Macintosh, introduced in 1984, a successful product. These new skills learned for the Apple contract then enabled Microsoft to produce Windows and stave off future competition from Apple. The result: by 1985, Microsoft had grown to 1,000 employees and $140 million in revenues, and topped a billion dollars in sales in 1990.[30]

For many years now, Microsoft has been the largest software product company in the world. In fiscal 2009, it had 93,000 employees and nearly $60 billion in revenues, and made over $20 billion in operating profits. It has expanded from that single Basic programming language product to dozens today, including Windows and Office (see Table 1.2 in Chapter 1 and Appendix II, Table II.3, for more financial details). In the 1990s, Microsoft managed to fend off the challenge from Netscape, largely by "embracing and extending" the Internet and working this new technology into Windows and Office, though it also violated antitrust rules. Microsoft has done less well competing with a newer dominant firm—Google—in Internet search. The Internet, however, has not eliminated Microsoft's enormous revenue and profit streams, at least not yet. Microsoft also trails Apple by a wide margin in the consumer electronics, cell phone, and digital media businesses, though it remains much more profitable, for the time being.

I have already commented on the competitive dynamics and prospects for Microsoft, Apple, and Google in Chapter 1. In Chapters 4 and 6, I also discuss how Microsoft refined the development process that is now common at PC and Internet software companies, and that it still uses to churn out a continuing stream of "good-enough" products for the PC mass market, the enterprise software market, and Internet customers. To conclude this chapter, though, I want to summarize the key strategies that Richard Selby and I identified in our book *Microsoft Secrets* (1995). These categorize what Bill Gates and his colleagues emphasized as they built the company during its critical formative period, roughly 1976 to the mid-1990s.[31]

The first strategy dealt with how Gates organized and managed the company: *find "smart" people who know the technology and the business.* Microsoft's success story began with the talents and insights of its founders, but the company's staying power depended on refining a rigorous process for selecting and screening new programmers, managers, and other employees. Gates in particular looked for people who deeply understood the technology and the business. In the late 1970s, this meant knowing how to program in Assembler whilst using minimal memory and processor power, and knowing how to translate this knowledge into products and platforms that would one day generate revenue for the company. Many programmers at the time, especially for the new personal computers, wrote and shared software as a hobby and did not see this as a profession for which they should get paid.

The second strategy dealt with how to nurture creative people and technical skills: *organize small teams of overlapping functional specialists.* Gates and his managers early on decided to empower people and organize work in small multi-functional teams in order to avoid narrowly skilled employees and overly bureaucratic compartmentalization. They cultivated the basic functional skills needed to design, build, test, and support Microsoft products, which grew enormously in size and complexity during the 1990s. Following IBM, Microsoft also created formal career paths and "ladder levels" to recognize and compensate people for achievements within their technical specialties.

The third strategy dealt with how to compete by creating product portfolios and setting industry standards: *pioneer and "orchestrate" evolving mass markets.* In today's language, we would call this a strategy for platform leadership. This powerful strategy, backed by programming talent and evolving market power, emerged after the introduction of DOS in 1981. Microsoft has continued to enter every major PC software mass market since 1975 and focuses on incrementally improving its products rather than seeking to invent new technologies or pioneer new products.

The fourth and fifth strategies dealt with how Microsoft managed product development for the mass market: *focus creativity by evolving features and "fixing" resources*, and *do everything in parallel with frequent synchronizations.* Microsoft went through a remarkable transformation from 1989 through the mid-1990s that completely revamped how teams managed development as well as treated feedback from customers. The transition from the relatively simple DOS operating system to the more complex Windows nearly brought all new product development to a standstill and could easily have bankrupted the company. To handle the more complex graphical programs and rapidly changing PC technology, Microsoft teams adopted "vision statements" to guide the design process but without trying to determine everything in advance. They learned to structure projects in "milestones" or subprojects of prioritized features to make big tasks easier to manage and have multiple checkpoints to rethink what they were doing. They adopted the practice of creating daily "builds" (a kind of working prototype) to keep everyone synchronized as they made their changes. The programmers also began listening carefully to data gathered from customer support as well as usability labs. Overall, these and other practices made it possible for many small feature teams to experiment, evolve their designs, and work in parallel, but still function as one large team, able to build larger and more complex products. This change in process was essential not only to build Windows but also to create new graphical applications. Microsoft would again run into trouble in the early 2000s with what became Vista, which was several times the size and complexity of earlier versions of Windows. But the company would again find a way to revise its development process and shipped a highly successful Windows 7 in 2009.

The sixth strategy dealt with building a learning organization: *improve through continuous self-critiquing, feedback, and sharing.* Selby and I noted: "Companies filled with smart people can easily degenerate into a motley collection of arrogant and fiercely independent individuals, teams, and projects that do not share knowledge, learn from past mistakes, or listen to customers."[32] Again, this has been an ongoing process at Microsoft, with each project writing postmortem reports, usually after each milestone, and focusing on what went well, what went poorly, and what they should do in the next milestone or the next project. The product groups learned how to share components (economies of scope!) as well as listen much more carefully to customers and improve the usability of products. In the later 1990s and through the present, the learning challenge also included how to adapt to the Internet and software as a service.

We ended our book with another Microsoft philosophy that we saw embedded in the company at the time: *Attack the future!* True, we wrote that Microsoft exhibited several weaknesses in basic software engineering, such as less attention to quality and architecture than we thought appropriate. These issues became much more serious as Windows got larger and more complex, and as security and stability became more important to users. The Internet gave Microsoft its most serious challenge, and the company is still adapting. Both Bill Gates and Steve Ballmer, who took over as CEO in 2000, worried constantly about these challenges and often behaved as if all their markets were about to collapse. Indeed, Microsoft now has strong competition, principally from IBM, Apple, Netscape, Intuit, Adobe, Google, Nokia, RIM, Oracle, SAP, Sun Microsystems, and the free and open source communities (mostly Linux). Government officials in the United States, Europe, and Asia have also forced Microsoft to modify several practices to allow more open competition in browsers, media players, and enterprise operating systems and servers. Overall, though, Microsoft is an exemplar of staying power: it has preserved the Windows and Office platform franchises whilst exploring, albeit reluctantly and incrementally, how to evolve its products, platforms, services, and business models.

The great strength of Bill Gates as an entrepreneur was the ability to identify market movements early enough to mobilize resources and exploit those trends. He also learned how to leverage the platform positions of DOS and then Windows.[33] But Microsoft was not much of a technological pioneer whilst Gates was in charge, except for adapting programming languages from bigger computers to the tiny PC. Gates also made sure that Microsoft cultivated mass-market engineering and marketing skills. But he let other firms be the product innovators, such as in desktop applications (led by WordPerfect and Lotus 1-2-3), graphical computing (led by Apple), Internet computing (led by Netscape), online services (led by AOL), Internet search and online content (led by Yahoo, Google, and others), multi-platform computing (led by Sun Microsystems), open source computing (led by the Linux community and best exploited by Red Hat and IBM), or the transition from products to services in the computer industry (foreseen best by IBM) and software as a service (pioneered by Salesforce.com) and cloud computing (led by Amazon, Google, and IBM).

I have written elsewhere that Gates took too long to adjust to the position of power that Microsoft attained and never quite understood that the rules of competition differ for a firm with a monopoly market share (in this case, two monopolies—Windows and Office). He did not realize early enough that it was unnecessary as well as illegal for Microsoft to cross the antitrust line to battle Netscape.[34] Nonetheless, no individual did more than Bill Gates to grow the PC software business by bringing cheap, powerful software products to the masses. This is his business legacy, and it is the result of both remarkable personal vision as well as an unmatched understanding of what it took to succeed in PC software. And the world is now benefiting greatly from Gates's enormous wealth and second career as a philanthropist, to which he has devoted the same kind of intelligence, commitment, and foresight that he used to build Microsoft.

Lessons for Managers

First, it should be obvious at this point that most firms are better off evolving strategy and capabilities together, and incrementally. Not only is strategy worth little without the appropriate implementation skills. But finding the right strategy is also usually a process of trial and error, and risk-taking. Companies should run experiments, such as in production management or product design, and see how customers and competitors respond. Second, there are unforeseen benefits to having talented managers and engineers very close at hand—within the firm—who deeply understand both the technology and the business and can innovate in products, processes, services, and strategy as opportunities appear.

On the first point, investing in capabilities such as precision engineering and manufacturing, Just-in-Time production and supply-chain management, software engineering, or platform marketing, seem at least as important as having the "right" strategy on paper. Of course, it helps for managers to have a general vision of the future that is on target and can guide them when building capabilities for the future. We saw this in the Japanese automobile industry, the global video recorder industry, and the American PC industry. Chance and good fortune played a role in determining the winning firms, as they usually do. But there also were significant differences in how managers approached the coupling of strategy and capabilities. Toyota, Sony and JVC, and Microsoft took one course: they cultivated distinctive in-house skills that eventually put them at the top of their industries for a generation or more. Nissan, Ampex, Toshiba, RCA, GE, and IBM took another course: they went outside for critical skills and ended up at a long-term competitive disadvantage.

On the second point, focusing on capabilities development, though not itself a business strategy, can have enormous strategic implications and benefits. It is impossible to imagine Toyota's long-term success without the self-reliance philosophy of Sakichi Toyoda or without his son Kiichiro laying the groundwork in the 1930s by hiring the best managers and engineers he could find in Japan. JVC and Sony, as well as Toshiba and Matsushita, also developed impressive skills in audio and video technologies during the 1950s, with brilliant engineers and highly trained Ph.D.s running their R&D efforts. These kinds of people help generate the product and process innovations—planned and unplanned—that truly differentiate firms over the long run. Their expertise is usually critical for absorbing new technology and scientific knowledge from outside the firm, such as from acquisitions, universities, government laboratories, suppliers, partners, and even competitors. They also can help CEOs craft strategy—such as to know when to emphasize manufacturing, design, or marketing, if that is where the firm's advantage turns out to be. With the right people close at hand, even random events or chance information can inspire unplanned innovation. This happened when Japanese firms learned of RCA's challenge to create a video recorder. It also occurred when Taiichi Ohno by chance read about American supermarkets and aircraft companies, and decided to experiment with a similar kind of supermarket "pull" system in automobile production—as we will see in the next chapter.

Of course, many firms successfully buy technology or technical assistance. Nissan did this in the 1930s and 1950s. Microsoft did with DOS and continues to make acquisitions as it diversifies beyond PC software. Apple recently bought a firm that provides it with online music storage and streaming technology. The "buy" strategy can save time and money. Some firms—Cisco is a great example—become masters of acquisitions and successfully acquire smaller firms to renew and expand their product lines, rather than investing heavily in internal R&D. Bringing in new people and new technology can also help people in an established firm think differently and respond better to change or opportunities to innovate.

But we also know that as many as two out of three acquisitions fail to create value for the acquirer.[35] By analogy, firms without adequate internal capabilities to absorb outside or new technology and then innovate on their own are likely to fail or end up disadvantaged. Moreover, relying too heavily on outside assistance often reveals a fundamental internal weakness. We saw this with Nissan's lack of skills in automobile design and manufacturing. Internal weakness also lay behind the failed attempts at Ampex and RCA to outsource manufacturing of their video device prototypes. Internal weaknesses (as well as politics—the mainframe people were still in control)—prompted IBM's decision to go outside for a PC operating system (from Microsoft) and a PC microprocessor (from Intel).

At the same time, capabilities, as well as strategy, must be dynamic and evolve as conditions change. Not all managers understand this, and many firms struggle to make the transitions. In recent years, Sony and JVC, as well as other Japanese consumer electronics firms, have been slow to develop or capitalize on new products. They must now also compete with lower-cost and dynamic competitors such as Samsung in Korea and other firms in China. Sony, for example, has been a major disappointment for most observers of Japan. After introducing the Betamax in 1975, the same chief engineer, Nobutoshi Kihara, invented the Walkman in 1979. This product is similar to Apple's iPod in many ways. Though a generation earlier, it makes you wonder why Sony did not later on find a way to connect the Walkman to a library of digital media and then the Internet. Unfortunately, Sony management seems to have remained too tied to hardware products. The company did buy movie and music producers following the Betamax debacle to get more control over content. It also went on to design new digital video recorders and many other successful audio and video products. But Sony's senior executives and engineers never seem to have fully understood the value of creating a new industry-wide platform or of investing more heavily in software programming skills and networking technology. This is a lapse in both strategy and capabilities. The Sony PlayStation has done relatively well in recent years, and Sony makes excellent PCs, high-definition TVs, and other audio and video equipment. But it would have done much better with a deeper appreciation for platform dynamics, software engineering, networks, and the changes occurring in consumer technology.

JVC is also a disappointment. It has always been an underdog of sorts, competing against much larger firms, including its parent, Panasonic/Matsushita. But it did not have another hit product or industry platform after VHS. Again, we can point to relatively weak capabilities in software, computers, and networks (there is a pattern here in Japan!). The VHS machine is a miracle of precision electronics and assembly. But it is a machine and not a computer or networked device. Like Sony, JVC went on to make excellent digital products for audio and video applications (displays, camcorders, home entertainment systems, car systems, projectors). Still, the firm struggled to make money from consumer electronics—tough to do without a platform strategy and without ownership of digital content or services such as iTunes. In 2008, Matsushita finally sold its holdings in JVC. This historic company then merged with another Japanese consumer electronics company and now operates under the name JVC-Kenwood.[36]

No company is immune to ups and downs or the impact of technological change. This is another reason why strategy and capabilities must evolve together. If they do not, good and great firms will surely decline unless they can reinvent themselves or recover. Even Microsoft saw its revenues fall for the first time in 2009, though it remains highly profitable and probably has the resources and flexibility to continue adapting to the Internet and mobile computing. Then we have Toyota, passing General Motors to become the world's largest automaker in 2008 but suffering serious safety-related quality problems. No one should doubt that Toyota is here to stay and will rebound quickly. But to see even the mighty Toyota struggle should give pause to every manager. No advantage is permanent. No market is guaranteed. Companies must continually renew their capabilities and their reputations.

Finally, there is the practical question: if focusing more on capabilities rather than on the particulars of strategy and planning has such obvious benefits, why do not *all* managers and firms embrace this principle? This subject needs at least an entire book to explore. We quickly get into complex issues of how organizations learn, how leaders lead, and how change occurs or does not occur—including the problem of conflicting incentives for managers, employees, and all stakeholders who are not in it for the long haul. Then there is the problem of the future: it is rarely clear in advance what specific capabilities a firm will need five or ten years from now. But, if senior managers and technical executives are consistently wrong about the general directions of markets and technology, then there is little hope that the firm will survive or thrive over long periods of time.

We do know that building deep technical skills and business knowledge takes years of patient experimentation and investment. Firms then must learn from these efforts, figure out how to solve recurring problems in design or operations systematically, and accumulate knowledge at the organizational level and not just the individual level.[37] Many CEOs are in their jobs only for a few years or do not have the technical understanding to gaze into the future with any level of confidence. Some managers grow complacent or arrogant, especially if their firms have a long uninterrupted record of success. Again, this is a problem of both strategy and capabilities as well as organizational dynamics and leadership.

Nor can we say that executives who explicitly say they are in it for the long term are likely to generate more staying power. Firms need to manage both the short term and the long term, and do well in both. For example, Japanese companies tend to have a very long-term view because shareholding is mainly institutional and senior executives are generally promoted from within the firm. The bigger firms also used to practice lifetime employment. But Japanese firms often make bad investments because of the lack of pressure to earn quarterly profits or show more financial discipline. Japan is also at a long-term and short-term disadvantage because of the country's relatively weak investment in university research and the basic sciences, compared to the United States and Europe. As more industries become dependent on leading-edge scientific research and technology that firms can quickly apply—as we have seen in computer software, semiconductors, biotechnology, nano-materials, and optical technologies—then Japan will fall further behind.[38] This is not really a problem of strategy for Japanese firms; managers and even government policy-makers generally know what they should do. It seems more an issue of what they can do, given the limitations of financial resources and other societal constraints. These are problems of capabilities, not just strategy.

In the next chapter, I return to the story of Toyota and the concept of "pull" versus "push" in production management, including a discussion of the strengths and weaknesses of this approach. Then I return to Microsoft and describe how it evolved an iterative or agile process for software development, which I see as another application of pull rather than just push. It is surprising (or perhaps not surprising) to see such an important similarity in the processes followed at both Toyota and Microsoft. There are other similarities, as we shall see in later chapters.

Notes

1. This emphasis on building distinctive capabilities rather than focusing on strategy in isolation is the central theme in Kay (1993). My thanks to David Musson for pointing this book out to me.
2. See Porter (1987, 1996).
3. Kay (1993: 337–63) contains a detailed review of the strategy field through the early 1990s.
4. For the thinking behind the business policy course at Harvard Business School, which created a foundation for courses on strategy, see Andrews (1980). See also the classic text, Barnard (1938).
5. See Penrose (1959), republished by Oxford University Press in 1995.
6. See Hax and Majluf (1984) for a discussion of the consulting methods popular in the 1970s and 1980s.
7. Quinn (1978) and Mintzberg (1987).
8. Mintzberg (1994).
9. Porter (1980).
10. Porter (1985).

11. Wernerfeldt (1984).
12. Barney (1991) and Peteraf (1993). See also Barney and Clark (2007).
13. Hamel and Pralahad (1989) and Pralahad and Hamel (1990).
14. Leonard-Barton (1992).
15. Henderson and Clark (1990); Christensen (1997).
16. See Cohen and Levinthal (1990) and Zahra and George (2002). Detelin Elenkov and I noted as well in a 1994 article that other scholars had studied this same problem from the perspective of international technology transfer. See Elenkov and Cusumano (1994). See also Rosenberg and Frischtak (1985) and Lall (1987).
17. D'Aveni (1994).
18. See Teece, Pisano, and Shuen (1997) and Teece (2007, 2009).
19. Eisenhardt and Martin (2000).
20. This section is based on Cusumano (1985: 27–57).
21. For the corporate history of Nippon Mining, see www.shinnikko-hd.co.jp/english/corporate/history.php (accessed Apr. 14, 2009).
22. This section is based on Cusumano (1985: 73–112).
23. This section is based on Cusumano (1985: 58–72, 112–36).
24. This section is based on Rosenbloom and Cusumano (1987).
25. Rosenbloom and Cusumano (1987: 52).
26. Rosenbloom and Cusumano (1987: 55).
27. Rosenbloom and Cusumano (1987: 66).
28. The best account of Gates and Microsoft through the early 1990s is Manes and Andrews (1993).
29. "Playboy Interview: Bill Gates," *Playboy* (July 1994), 63, cited in Cusumano and Selby (1995: 5–6).
30. Data are from Cusumano and Selby (1995: 3).
31. This section is based on Cusumano and Selby (1995: 8–13).
32. Cusumano and Selby (1995: 12).
33. This short discussion is based on Cusumano (2009b).
34. See Cusumano and Yoffie (1998). Also Cusumano (2000).
35. See Porter (1987: 3), as well as Oster (1999: 232–6).
36. For more information on JVC-Kenwood Holdings in English, see www.jk-holdings.com/en.
37. This is a broad literature, but some classic works that influenced me include Senge (1990), Garvin (1993), and Kotter (1996). For some later research on organizational change and leader, see publications from the MIT Leadership Center http://mitleadership.mit.edu.
38. For more on the tightening linkages between scientific research and application, see Stokes (1997).

4

Pull, Don't Just Push

The Principle

Managers should embrace, wherever possible, a "pull-style" of operations that reverses the sequential processes and information flow common in manufacturing as well as product development, service design and delivery, and other activities. The goal should be to link each step in a company's key operations backward from the market in order to respond in real time to changes in demand, customer preferences, competitive conditions, or internal difficulties. The continuous feedback and opportunities for adjustment also facilitate rapid learning, elimination of waste or errors, and at least incremental innovation.

Introductory

The push–pull distinction is not only relevant to manufacturing and product development. At a higher level of abstraction, we can see this principle at work in how managers think about the future, and in how they operate in the present. If we look carefully, we can find push or pull concepts in nearly every function within the firm: from decision making to research and development as well as manufacturing, supply-chain management, marketing, sales, and services.

At one end of the spectrum, there are the "rational planners" who believe (or hope) the world is a highly predictable place. They prefer to create detailed plans and to try to implement them—often valuing persistence even when things are not going as anticipated. At the other end of the spectrum, there are the "incremental innovators" and "experimenters." They continually try to get feedback from customers or the sales force on prototypes and existing products. They listen closely to the market and find ways to adjust what they are doing and planning. Organizations and managers that behave this way tend to avoid detailed, rigid plans, and prefer processes that allow them to learn, adapt, and innovate, at least incrementally.

In mass production, firms generally have followed a push-style of management when market demand is relatively predictable and product variety limited—such as in the extreme case of Ford with the Model T in the early twentieth century. We know this approach broke down for Ford in the 1920s when General Motors introduced more product variety. If we jump to the 1970s, we can see firms learning how to automate the push-style of planning with the introduction of the first MRP (materials requirement planning) systems. Early versions of these software programs in the 1960s and 1970s required detailed advance production schedules and bills of materials. They had little or no flexibility to accommodate change once the plans had gone into action. By contrast, we have Toyota's manual pull system, which originated in the 1940s—before firms used electronic computers. In this approach, managers can adjust production volumes and even the product mix in very short intervals—daily, if necessary. Toyota's main process innovation was to treat each completed product as a signal back into the production system to draw in more materials, components, and labor in order to complete another product. This differs fundamentally from a push system, where the schedule "forces" the arrival of more materials and components. They in turn push the production system to make more products—whether the market needs them or not. The pull system can immediately expose manufacturing flaws or overproduction of unnecessary parts and finished goods. It is important to realize, however, that it cannot expose faulty designs or architectural flaws in a product unless they are visually obvious to workers or create some difficulty in the assembly process. Only much more extensive usage testing prior to mass production could have exposed the kinds of problems Toyota vehicles experienced with frame corrosion, sticking gas pedals, unintended rapid acceleration, and braking software controls.

In product development, the analogy to push-style mass production is the traditional "waterfall" style of development, initially associated with how NASA and IBM in the 1960s managed large-scale projects. The waterfall process starts with designers or analysts gathering information on customer requirements. Then the analysts create a detailed plan. The work proceeds to component construction, testing of the components, and later integration of the components and system testing, based on what functions the requirements documents said the product or system should perform. The waterfall implies a sequential process flow, with no specific provisions for the project members to go back to a prior stage and to redo their work or experiment with designs. In reality, nearly all software projects end up going back to fix designs and to redo testing.[1] Most other products, such as automobiles, are also nearly impossible to get "right" the first time, just from writing down requirements and talking with potential customers. This is why automobile designers first make drawings, then build clay models, and create computer simulations of their designs, before they start bending metal and molding plastic. There are also limitations to how well a computer simulation can truly represent the behavior of a complex product or component in the physical world. Over-reliance on computer-aided designs and simulations may have contributed to the difficulty Toyota engineers had in foreseeing problems with the materials used in its gas pedals as well as with how antilock braking software behaved under adverse road conditions.

Sticking to the plan in a waterfall project or limiting the amount of experimentation and testing in any project tends to push the development work forward, sometimes before it is ready. A strict adherence to the specifications may also result in higher quality in terms of fewer defects, as defined by any departure from the original specification or unintentional errors. In particular, introducing late product design changes within a process that does not easily accommodate them—such as a project lacking tools for continuous testing and retesting—does tend to insert new defects. The firm may not discover them before shipping the product to the customer. The waterfall may also result in products that no one wants to buy or features that do not meet customer needs very well. This is because the divergence among actual customer desires, early requirements documents, and the final product is usually unclear until the end of the development cycle.

In response to these weaknesses of push-style product development, many companies have introduced "iterative" or "incremental" processes, now commonly known as "agile" approaches. As in crafting strategy, agile projects should have fairly clear goals for what to build, much like Sony and JVC did for a home VCR, or like Kiichiro Toyoda did for his first automobiles. But the work should then proceed incrementally, relying on learning from trial and error experimentation as well as extensive testing. We usually see several working prototypes that generate opportunities for feedback and adjustments. The project team may reject or change early design decisions as they get more input from the creation process, such as through prototypes or beta versions of the product introduced into the market. In recent years, IDEO, the design company based in California, has made famous this type of prototype-driven or experiment-oriented design process, but it is now relatively common.[2]

In fact, many companies in a variety of industries now use prototyping as well as multiple short cycles of concurrent design, build, and extensive testing activities to improve responsiveness to user feedback as well as market uncertainty.[3] In the software community, since the mid-1970s, researchers and managers have talked about "iterative enhancement," a "spiral model" for iterating among the different project phases, and "concurrent development" of multiple phases and activities.[4] The "extreme" part of the spectrum is XP, short for "extreme programming" (which has about a 70 percent overlap with the Microsoft-style techniques described later in this chapter).[5] XP does away with detailed specifications and focuses entirely on incrementally writing, evolving, and testing code. Programmers create many "builds" or prototype-like working versions of the evolving product each day and then test new code with "quick tests" for each new feature or change. This development style is especially suited to software because the product consists of digital instructions that programmers can modify relatively easily compared to traditional "hard" products. Nonetheless, firms in the aircraft, automobile, machine tool, and many other businesses now design and test non-software products by creating virtual prototypes and synchronizing the work of many engineers and geographically dispersed teams by using computer-aided styling, design, engineering, and manufacturing technology.

There are times, though, when individuals and organizations find a push-style of management useful. This occurs in basic research when there is no market pull, and in product development and manufacturing when companies know exactly what they want to build and do not want to change their designs or production plans. NASA, for example, once it had finished the research stage for the Apollo projects or the space shuttle, wanted to have subsequent versions of their working systems with minimal changes to reduce the risk of errors. There was also only one customer—a US government agency.[6] Even when IBM and Hitachi built their tenth or so version of a similar real-time banking system, they could pretty much proceed in a sequential manner. Figure 4.1 maps out potential situations when a project might use a waterfall or agile process or something in between.

There also are cases when inventors come out with a product or service "new to the world" and need to push this out to market. Some large science projects, such as for nuclear technology and space exploration, clearly have done this successfully.[7] The computer and plain-paper copier innovations were also both cases of push. Outside consultants hired in the late 1940s and 1950s to evaluate their potential first thought there were no mass markets for these products, but the

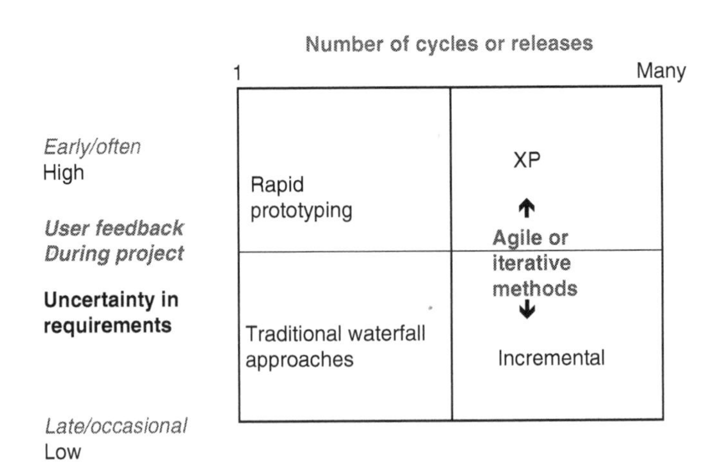

Figure 4.1. Spectrum of process approaches in software development

Source: Adapted from Bill Crandall (formerly of Hewlett-Packard).

companies (IBM and what is now Xerox) persisted. Yet, even in these famous instances, the researchers, engineers, and managers involved were responding to existing market needs—such as for calculating large sums of numbers, or easily creating many copies of documents. There is no doubt that some technologies, and the visions of their creators, will clearly be ahead of the market and fail for that reason alone. At the same time, every technology that does succeed commercially must eventually find a market and, in a sense, become "pulled" by consumers demanding new and better features, higher reliability, or lower prices. Otherwise, we have the proverbial "technology in search of a market" or a "solution in search of a problem." The challenge for managers is to figure out when to push and when to respond to existing pull forces, as well as how to create pull-style linkages with the market.

Nor should pull-style linkages end with the delivery of a product. It is equally important for companies to continue responding in as close to real time as possible to feedback from customers, the sales force, and the service network. For example, had Toyota reacted more quickly to early negative reports on its gas pedals, throttle controls, or the braking software in the Prius, it could have avoided much of the negative publicity it received in 2010. More importantly, the company might have prevented some fatalities. Instead, Toyota chose to continue studying the problems and recalled vehicles only when pushed by government safety officials in the United States and then Japan.

The concept of pull versus push is so important that it has its own principle in this book, even though it requires a high degree of flexibility to implement—and flexibility, broadly considered, is my sixth principle. We also need more research on push versus pull. When Taiichi Ohno began to reorganize production at Toyota in the 1940s, there was no academic work on why a pull approach might provide an advantage. It was not possible for academics to study Toyota, however, because the company did not reveal much about what it was doing. Not until 1977 did we get a clear glimpse inside the Toyota production system. In that year, several Toyota managers trained under Ohno (including Fujio Cho, a future CEO of Toyota) published an article (in English) for an international conference finally explaining their system in some detail. Ohno followed with a book in 1978; this was translated into English in 1988.[8] Meanwhile, in the early 1980s, a

Japanese accounting professor, Yasuhiro Monden, who consulted for Toyota, began publishing articles in English describing the inner workings and the economics of the Toyota production system. He also published a book in 1982, *The Toyota Production System*, which came out in English and Japanese. Two American operations management professors followed with books for managers: Richard Schonberger's *Japanese Manufacturing Techniques* in 1982 and Robert Hall's *Zero Inventories* in 1983.[9] I benefited from all these publications whilst writing *The Japanese Automobile Industry* (1985).

As we look back, not until the 1980s did academics begin to research the pull system at Toyota in any depth. Since then, the pull-systems literature has remained dominated by operations management specialists working on lean production and a few operations researchers, who have modeled stochastic (random) versus non-stochastic systems.[10] The product development literature is generally less mathematical, with more cases and solid empirical work on relatively large samples.[11] In both literatures, the research finds that pull systems and iterative development processes are more flexible than push or waterfall approaches because *they are designed to accommodate change*—often unforeseen change. Using these approaches enables a firm to react more quickly to changes in the market or competition.

As noted in the Introduction, I illustrate the distinction between pull and push concepts by relying on two examples: the evolution of the just-in-time system for managing automobile production at Toyota and the iterative or agile style of product development for software at Microsoft. The Toyota example is much better known to the average reader and is probably easier to understand conceptually. Externally driven changes in market demand (such as orders coming in from dealers) as well as sales forecasts trigger the pull system in manufacturing plants and suppliers; managers adjust production schedules and component deliveries on a daily or weekly basis. The system works extraordinarily well in minimizing inventory and adjusting to demand changes—but only as long as suppliers and parts factories deliver components with zero defects, and assembly workers make no errors. If something goes wrong in assembly, then the entire production system comes to a stop. There are generally no stockpiles of components to substitute for the defective parts. The trigger in product development is more internal—an evolving product specification that designers or engineers compile. But a project team can also build working prototypes and make frequent changes in their designs based on feedback from internal usage (sometimes called "alpha testing"), usability labs, focus groups, in-process testing, and outside beta users. Product development does not have to proceed sequentially from concept to requirements to building the product as originally conceived, just as production does not have to follow a manufacturing plan based on an early market forecast if later information suggests that demand has changed or some components are defective.

Toyota: Just-in-Time ("Lean") Production

The story of the pull system in Japanese automobile production takes us back to 1948. This is when Toyota began deviating from probably the most fundamental manufacturing principle in the automobile industry at that time.[12] First in one department and then gradually through all of its production operations as well as in many suppliers, over a period of fifteen years or more, the company decided not to "push" materials and components but to have final assembly lines "pull" them through the manufacturing system and supply chain. If you looked at a production flow diagram of Toyota's assembly operations, you did not see this change because all the materials and components physically moved forward. But Toyota reversed the conceptualization and management of the production flow by allowing final assembly to control the ordering of more parts and materials as workers completed each product. Nothing was pushed forward because workers moved backwards to previous stations to take only the parts or materials that they needed "just in time" for their operation. Toyota also decided to attach small paper signs, called *kanban* in Japanese, to pallets or containers of components. The kanban accompanied all parts in transit and signaled machine operators as well as suppliers when and how many more components to provide (Figure 4.2). Workers learned to check for mistakes as they took the parts they needed as well as to stop production lines and correct problems as they found them or ask for help from supervisors. This new process soon eliminated the need

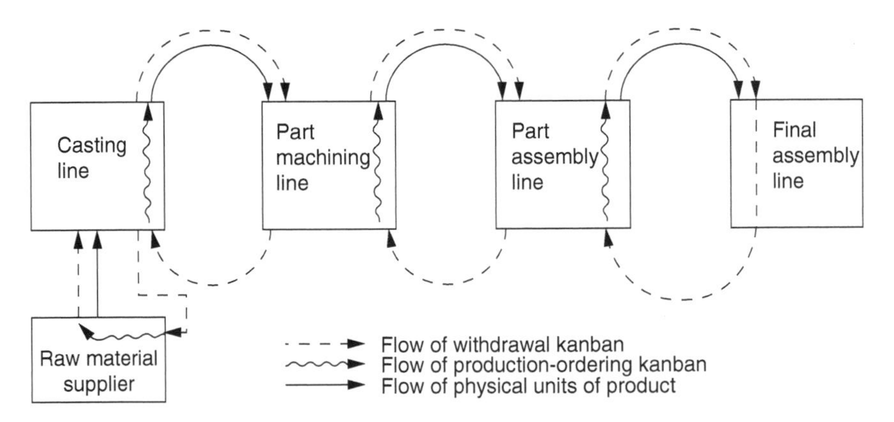

Figure 4.2. Kanban flow chart

Source: Monden (1981: 34). Reprinted with permission from *Industrial Engineering* (May 1981). Copyright © Institute of Industrial Engineers, 25 Technology Park/Atlanta, Norcross, GA 30092.

for large numbers of inspectors at the end of final assembly as well as large rework or reject piles.

For the system to work properly, it was necessary to make or receive components in batches (production lots) as small as possible. Toyota's objective was to have one worker or station provide components to the next quickly, with minimal stockpiling of in-process components. Small lots were uneconomical if machinery had long set-up times for different components. As a result, utilizing the general-purpose and small machine tools that Kiichiro Toyoda had insisted on buying in the 1930s, Toyota was able to reset equipment quickly for different jobs and make "small-lot production" economical. As sales increased, Toyota would eventually combine the benefits of flexible small-lot production (economies of scope!) with massive volumes (economies of scale).

Reversing the process flow to create a pull system, setting up flexible equipment rapidly, finding and fixing mistakes in production immediately, producing in small batches with minimal in-process and finished inventory, and utilizing fewer but more broadly trained workers, were *revolutionary* techniques in the automobile industry of the 1940s and 1950s. By the 1980s, these techniques had come to represent best practice worldwide, in nearly all manufacturing industries.[13] Toyota's process innovations altered the Western assumption that mass production of components in batches as large as possible with single-purpose, automated machinery and specialized workers—maximizing economies of scale—was the most efficient way to make an automobile or any mass-produced product. Yet, to a large degree the changes Toyota introduced were *evolutionary* adaptations to the company's circumstances and the unique needs of the Japanese market at the time. All Japanese automakers had to modify conventional mass-production techniques to accommodate much lower volumes, and relatively high model variety, covering both cars and trucks, than were common in the United States or Europe during the 1940s and 1950s. For example, the entire Japanese automobile industry in 1950 produced only about 30,000 vehicles—1.5 days' production in the United States!

Toyota pioneered the manufacturing techniques that would later be called "lean" production. Another MIT Sloan graduate student, John Krafcik (in 2010 CEO of Hyundai Motor America), coined this term in 1988.[14] The reality is that Toyota changed the concept of manufacturing productivity. Prior to 1948, Toyota workers tried to make as many parts and finished vehicles as possible each day. Once management had introduced the pull system and the notion of producing only what the next station needed, then too much production on any given day became "waste" in the form of unneeded inventory. Nissan and other Japanese automakers learned of Toyota's new approach in the 1950s through information shared among consultants and machine-tool suppliers. They gradually made similar modifications in their own production systems, for similar reasons. But none matched Toyota, as it relentlessly refined its processes and continued to train and retrain its managers, workers, and a tightly knit group of suppliers located around Toyoda City. As a result, in the Japanese automobile industry, Toyota achieved the smallest economical production batches, the shortest set-up times, and the lowest inventories. It also achieved the best manufacturing quality, as the pull system and related techniques led to fast identification of processing errors and encouraged rapid learning cycles. Toyota has also demonstrated the highest worker productivity and, over long periods of time, the highest profits of any automobile mass producer. This long record of excellence makes Toyota's recent transgressions in design quality and safety so difficult for company executives, employees, customers, government regulators, and industry observers to understand. But the long record of excellence has also made it difficult for company executives to acknowledge imperfections in their products—a weakness that has finally come to light. Correcting this problem will surely benefit Toyota and its customers in the long run.

The Role of Taiichi Ohno

One observation that jumps out at anyone who studies the history of Toyota is the importance of a few key individuals in establishing the early traditions and values of the company. Sakichi Toyoda and his son Kiichiro were obviously critical figures. But equally important was the engineer most directly responsible for Toyota's unique system of production management.

Taiichi Ohno (1912–90) was born in Dairen, Manchuria, the son of a Japanese ceramics technician who worked for Japan's South Manchurian Railway. His father moved to Aichi Prefecture in Japan (the home province of Toyota) at the end of the First World War. Ohno attended a technical higher school and graduated in 1932 with the equivalent of a college associate's degree in mechanical engineering. His father was an acquaintance of Kiichiro Toyoda, prompting Ohno to apply for a position in Toyoda Spinning and Weaving. He then joined the original Toyoda family company as a production engineer in cotton thread manufacturing. Ohno moved to Toyota Motors in 1943 when the automaker absorbed the parent company to concentrate more group resources on truck production. He managed machining and assembly shops from 1945 to 1953, was promoted to a company director in 1954, and then headed Toyota's major factories in the 1950s and 1960s, before retiring as an executive vice president in 1978. He then moved back to Toyoda Spinning and Weaving as chairman and continued to work with Toyota subsidiaries. (I had the great pleasure of interviewing Ohno in 1983 as part of the research for my doctoral thesis.)

Ohno first visited an automobile factory—a General Motors plant in Osaka—around 1930. He had no particular interest in automobile manufacturing at the time. But later on, at Toyoda Spinning and Weaving, he learned several principles for thread manufacturing that seemed useful for automobiles: how to control production costs, reduce defects, and utilize automation or mechanical devices to prevent mistakes but without hindering the ability of workers to adapt to different tasks as needed. Overall, Toyoda Spinning and Weaving had focused on reducing "waste" in its various forms, and this concept became Ohno's overriding objective when he rethought Toyota's automobile production system.

In retrospect, it must be significant that Ohno joined Toyota in 1943 with no experience in the automobile industry. He carried no prejudices in favor of American-style mass-production methods, such as Nissan had already adopted. After working in Toyota for five years, he became convinced that the American production methods, based largely on Ford's system of producing a limited product line in massive quantities to gain maximum economies of scale, contained two fundamental flaws.

First, as in Ford's case, only the final assembly line achieved anything like a continuous process flow, which Ohno thought should be the ideal for all mass-production factories. Ford made months of parts inventory in single production runs and stored the components for the assembly lines. But the massive inventories of unfinished goods wasted huge amounts of space and operating capital—both in short supply in post-war Japan. Moreover, if workers or machines were producing defective parts, often these went undetected for months, until they appeared in the finished products—another major source of waste. Second, the Ford style of mass production worked fine only as long as product variety remained limited (any color was fine as long as it was black) with predictable demand. General Motors had exploited this weakness in the Ford system during the 1920s by introducing more variety in final assembly but with standardized components underneath the different car body shells. Ohno took this idea of flexibility a giant leap further, introducing it throughout Toyota's production system, and not only in final assembly.

The timing was also right for Ohno to introduce change. In 1948, Toyota management adopted a five-year plan to reverse the large financial losses and piles of unfinished and unsold product that had accumulated after the end of the Second World War. Ohno, then head of a machine shop that made engines, convinced his superiors to experiment with his shop in the effort to raise productivity whilst reducing costs and personnel. He used time-and-motion studies to take a fresh look at cycle times and job routines, and eliminate waste in the form of unneeded motions, defects, parts production, and in-process inventory. At this time he also conceived of using a pull system. He had first read about this in a Japanese newspaper that described the practice being used in the US aircraft industry during the Second World War and in US supermarkets, where customers bought goods only when they needed them and stores replenished goods on shelves only when they ran out of inventory—just in time.

This is not to say that Ohno planned everything in advance and Toyota encountered no missteps along the way. They relied as much on trial and error or experimentation as on enlightened direction from Ohno and managers he trained.[15] Clearly, when Ohno began the pull measures in 1948, he had only a vague idea of how successful they would be—otherwise he probably would have implemented these changes much more quickly. Below is a rough chronology of how the Toyota production system evolved as Ohno moved up in management and incrementally introduced this approach into more departments and suppliers.[16]

1948 Ohno began the pull system in Toyota's engine machining shop by asking each worker to move back to the previous station to retrieve work-in-process, just when it was needed for the next processing step.

1949 The pull system in the engine shop enabled Toyota to end the stockpiling of finished engines. Ohno also asked workers in his shop to operate several machines, rather than specialize on one, because demand was not large enough to keep all machines operating continuously. He then asked workers to conduct their own inspections and reduced the inspection staff.

1950 Toyota extended the pull concept to marketing, limiting production to orders received from dealers. Toyota synchronized engine and transmission machining with final assembly to reduce in-process inventories further. It also introduced indicator lights on the engine lines to alert supervisors to problems.

1953 Ohno introduced a kanban system into the machine shop, using the backward exchange of these paper tags to signal when to begin the next processing operations. To simplify procurement, Toyota also began a standardization program for car and truck components.

1955 Toyota synchronized its body and final assembly shops to eliminate more in-process inventories. It introduced controls on parts deliveries from suppliers to cut inventories further.

It then started to mix the loading of components in small lots for machine tools and to mix model runs on final assembly lines to raise equipment utilization as well as to lower inventories. Toyota also introduced line-stop buttons on assembly lines and gave workers the authority to halt production if they noticed a defect or other problems.

1957 Toyota installed indicator lights on all production lines to alert supervisors outside the machine shop to problems.

1959 Toyota cut in-process inventories further and reduced waiting times by introducing a kanban control system for internal and in-house-to-outside parts conveyance.

1961 Toyota introduced the kanban system to some outside parts suppliers.

1962 Toyota extended the kanban system to all in-house shops, placing the entire company on a small-lot, pull system. Toyota also introduced "foolproof" devices to machine tools to help prevent defects and overproduction. By this time, Toyota had reduced stamping press set-up times for dies to approximately 15 minutes, compared to 2–3 hours in previous years. Toyota achieved this by automating as much of the changeover process as possible, doing set-up preparations whilst machines were running, and training teams to specialize in set-up.

1963 Workers were now operating an average of five machines each, compared to only two in 1947 and one in prior years.

1965 Toyota extended the kanban system to all outside parts deliveries, further reducing in-process inventories.

1971 Toyota adopted the practice of moving workers on the assembly lines to different positions as needed. Die set-up times for stamping presses dropped to approximately 3 minutes.

1973 Toyota began allowing suppliers to deliver directly to assembly lines, fully linking them with the in-house parts conveyance system and cutting more inventory throughout the production network.

In his 1978 book, Ohno gave some credit to Kiichiro for first promoting the idea of what later became just-in-time production and parts delivery. In 1937, Kiichiro set up a preparations office to coordinate production planning so that the machine shop received materials for only one day's needs at a time, and the assembly line followed. But Toyota stopped restricting production to one day's needs when military orders quickly increased. In 1940, Kiichiro added a lot number system to keep track of components and their assembly dates. But workers continued to produce as many components and finished vehicles as possible each day, raising in-process inventories to as much as two months' worth of production when Ohno joined Toyota in 1943. Ohno recalled that he observed these stockpiles and bottlenecks, and concluded they were caused by the lack of coordination among parts and subassembly production as well as suppliers and final assembly. Over the next several years, he contemplated a solution, which emerged in 1948: reverse the process flow with a pull system from final assembly.[17]

It is ironic that the most important inspiration Ohno received came from American aircraft manufacturers and supermarkets. The critical concept of fast set-up times for stamping presses and other machinery also came from the United States. Ohno recalled that he first saw American Danley stamping presses using rapid set-up techniques on a trip to the United States during the mid-1950s. He bought several of the machines for Toyota. He noted that American auto companies were still taking several hours to change dies because they produced in large lots and had become accustomed to resetting equipment in between shifts. Americans took hours to do a job that needed only a few minutes. The Japanese automakers in the 1950s and 1960s, with much smaller volumes and growing demand for different car and truck models, had no such luxury. Toyota also benefited by hiring a talented consultant, Shigeo Shingo, to study die set-up, to cut set-up times further, and to teach the techniques to Toyota suppliers.[18]

Another change at Toyota and other Japanese automakers was the increasing tendency to "de-verticalize" or outsource more components production and even final assembly. As discussed in Chapter 3, in the 1930s Nissan and Toyota had to make the most of their own steel materials and components because good-quality inputs were unavailable in Japan. After 1945, however, many Japanese manufacturers of components for aircraft, bicycles, and other machinery switched to auto parts to accommodate rising demand. When sales began surging in the 1960s as Japan encountered its "miracle" rapid growth period, both Nissan and Toyota recruited many of these small and medium-sized firms. Toyota in particular quickly built a tightly knit supplier network. By 1965, the percentage of components and assembly work done in-house (by cost) stood at 41 percent for Toyota, compared to 32 percent at Nissan and 50 percent at General Motors. However, the Toyota group (Toyota plus subsidiaries in which it held at least 20 percent equity) accounted for 74 percent of production at this time, compared to only 54 percent for the Nissan group. This level continued for the next two decades (and beyond). By the mid-1980s, both Toyota and Nissan accounted for merely 26 percent of components and manufacturing costs in-house whilst their lower-wage but high-quality first-tier suppliers accounted for another 50 percent. The outsourcing pyramid gave the leading Japanese companies a far higher level of group integration and a lower cost structure than American automakers, which relied on independent suppliers for 50–75 percent or so of their components and manufacturing costs. By contrast, the numbers indicate that Toyota and Nissan relied on independent suppliers for no more than 25 percent of their components and related costs.[19]

Because of its reliance on an internal and external supply chain of components and materials, the kanban system had some limits to its flexibility. It worked best when variations in monthly production demand fell between 10 and 30 percent. Toyota could handle this level of variation merely by changing the frequency of kanban exchanges and using worker overtime. More than 30 percent variance required more kanban tickets and more workers, as well as more capacity at suppliers. The system also required internal departments and suppliers to operate at close to zero defects, otherwise Toyota workers would be stopping the production lines constantly—the equivalent of continually "breaking the build" in agile software development.[20]

It is also significant that Ohno never did like computerized production control systems or too many robots; he considered both too inflexible, and flexibility has remained a key benefit of the Toyota production system. (Krafcik also found little or no correspondence between the use of robots and productivity when he compared Japanese, American, and European auto assembly plants.[21]) The MRP systems of the 1960s and

1970s in particular, according to Ohno, were push-style. Managers tried to follow the computer-generated schedules rather than to respond to demand. Moreover, MRP required precise real-time information to work accurately, and this was nearly impossible to achieve when he was running Toyota's operations. The manual kanban system, by contrast, always provided the equivalent of real-time information on needs and capacity in the production system. Nonetheless, Toyota did not shy away from using computers to calculate its materials and parts requirements or to make short-term forecasts. From 1953, Toyota began installing the latest IBM machines for inventory control, cost computations, parts ordering, equipment scheduling, and other functions, and continued to be a leader among Japanese companies in introducing advanced computer technology. Around 1970, Toyota also began using computers to coordinate (but not to control or push) the production schedules as well as to calculate the optimal number of kanban tickets. In the 1980s and 1990s, Toyota became a leader in the use of flexible robots and computer systems, adeptly mixing these technologies with its continued reliance on pull concepts and highly trained workers.

Performance Improvements and Comparisons

My analysis in *The Japanese Automobile Industry* (1985) relied on public data and measured aspects of the pull system only indirectly, such as inventory turnover at the firm level. At the parent Toyota company, for example, this increased from 11 times per year in the late 1950s to 36 times per year in 1983. Meanwhile, inventory turnover went from 7 times per year at Nissan to 19 times, and from about 8 times per year at GM, Ford, and Chrysler to 11 times per year. In other words, by the mid-1980s, annual inventory turnover at Toyota had become twice that of Nissan and about 3 times that of the Big Three American automakers, at least at the parent-firm level.[22] During this same period, both Toyota and Nissan experienced substantial improvements in the number, type, and quality of products introduced. Physical unit productivity and value-added productivity both continued to increase, at the firm level and the group level (parent firm plus subsidiaries).

Toyota's continual refinements of its production system corresponded to impressive improvements in productivity. The data presented in Appendix II, Table II.5, show that Nissan and Toyota each produced between two and three vehicles per employee during 1939–42. Vehicles manufactured per worker in gross numbers tripled at Toyota between 1955 and 1957 and rose another 60 percent by 1964 and then tripled again in the mid-1980s.[23] Even adjusting for vertical integration (the number of parts made in-house versus purchased) as well as capacity utilization and labor hour differences between Japan and the United States, Toyota was still 1.5 times more productive than GM, Ford, and Chrysler in 1965 and 2.7 times more productive in 1979. The ability of the Toyota-style production system to reduce the number of required workers and other inputs (working capital, inventory, space, tooling) roughly by half inspired Krafcik to dub it the "lean" production system in 1988. Nissan, which copied aspects of Toyota's pull system and some other process innovations, also reached double the US productivity levels in 1970 as its production volumes soared (Figure 4.3).

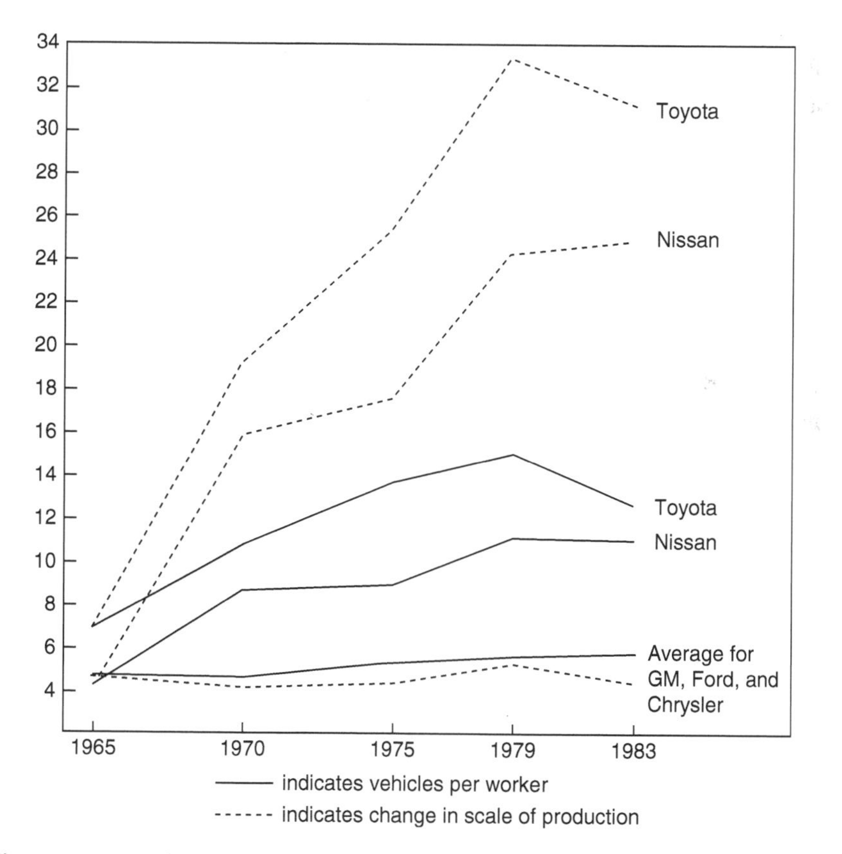

Figure 4.3. Automakers' adjusted vehicle productivity and change in production scales, 1965–1983
Source: Cusumano (1985: 201, figure 4).

In terms of value-added productivity, in 1983 the levels were similar at Nissan and the Big Three in the United States, whereas this productivity measure was twice as high at Toyota. Even if we factor in the Toyota and Nissan subsidiaries, value-added productivity in the Toyota group at this time was still 50 percent higher than Nissan or the American automakers. (Productivity was similar for Nissan and the Nissan group, whereas Toyota the parent company was about 50 percent more productive in value-added than its group overall.)[24]

Toyota did not achieve these productivity levels through the magic of the pull system alone: its workers had much more equipment to help them than US workers, at least from around 1970. By this time, Toyota had 30 percent more fixed assets per worker, adjusted by labor hours, compared to the US companies, and 150 percent more by 1983. Nissan also had double the assets per worker in 1983 compared to US competitors. But these figures are misleading. If we look at capital investment per vehicle produced, we see the real story. In 1983, Toyota had only 13 percent more investment per vehicle than the US automakers and 10 percent more than Nissan.

In other words, by the time the Japanese automobile industry passed the US industry in production around 1980, it took roughly the same amount of capital investment to make an automobile—in the United States or Japan. Toyota workers, however, using the pull system and other techniques, made more than twice as many vehicles per person each year as US autoworkers (approximately 13 to 6 in 1983 on an adjusted basis). So the reality is that Toyota used half the number of workers that American companies did to produce the same number of products with comparable levels of investment for their production volumes. This performance translated into double the value-added (economic) productivity for Toyota the parent company and still 50 percent higher economic productivity at the Toyota group level (Toyota plus subsidiaries). By the early 1980s, Nissan was not far behind the Toyota parent company in physical productivity (vehicles produced per worker), though its group value-added productivity was about the same as the US automakers. A major part of value-added is operating profits, so these numbers suggest that Toyota in the early 1980s was already highly efficient or charging a premium for its products. Another major part of value-added is employee compensation, though this was about 80 percent the US level at Nissan and Toyota in 1983. Toyota also paid its workers slightly less than Nissan did.[25]

For a more precise analysis of productivity and the relationship to manufacturing flexibility, or to quality, we need factory-level data. John Krafcik collected this for his master's thesis and published the results in an *MIT Sloan Management Review* article in 1988. Krafcik and then John Paul MacDuffie, a professor at the Wharton School of the University of Pennsylvania, continued this work for the International Motor Vehicle Program. In 1990, our program directors published the results in the international best-seller, *The Machine that Changed the World*.[26] These data and other analyses confirmed that the average Japanese auto plant had roughly a 70 percent productivity advantage over American automakers. The average Japanese assembly plant around 1990 took less than 17 hours to assemble a finished vehicle in Japan and about 21 hours in North America. The average American-owned factory took 25 hours and the average European plant nearly 36 hours. Moreover, there was no tradeoff between productivity and quality. The average Japanese plant produced cars with only 52 defects per 100 vehicles in Japan and 55 in the Japanese North American plants. By contrast, American cars had an average of 78 defects and the Europeans 75.[27]

The Japanese automakers proved they could transfer their production techniques and performance levels outside Japan. There was nothing "cultural" about their achievements in manufacturing. Toyota, in particular, achieved similar productivity and quality levels at New United Motor Manufacturing, Inc. (NUMMI), the joint venture with General Motors in California established in the mid-1980s (where Krafcik worked as a quality engineer before coming to MIT) and then at other American Japanese factories (the "transplants"). The Japanese also transferred their practices for managing suppliers. In 1991, I published data on this with another former MIT Sloan master's and doctoral student, Akira Takeishi, in 2010 a professor at Kyoto University. We compared GM, Ford, and Chrysler with Toyota, Nissan, Honda, Mazda, and Mitsubishi as well as six transplant factories owned by the Japanese firms in the United States. The analysis showed lower costs and higher quality in the Japanese firms and their transplant factories, even when the Japanese received parts from US suppliers.[28]

American and European companies reduced the gap in costs and quality during the 1990s as they learned the Toyota-style techniques and improved supplier management. But the damage to their reputations and cost structures, particularly for the American-owned companies, has been difficult to overcome. The result was the bankruptcy of GM and Chrysler in 2009 and their financial reorganizations, including Fiat's takeover of Chrysler. Again, it is ironic that, in 2010, Toyota is now fighting to preserve its reputation for quality. The gas pedal, brake software, and other problems in Toyota vehicles have created an opportunity for GM, Chrysler, and every other automaker to win back or win over customers from Toyota. At the same time, however, no one should underestimate Toyota's ability to learn from its mistakes.

The Tradition of Learning and Process Innovation

Toyota's past accomplishments in production management, including quality control, have inspired considerable research from academics and consultants, in Japanese, English, and other languages.[29] Several authors have even analyzed how to apply Toyota principles, such as pull concepts and elimination of waste, to other industries and activities, such as software engineering and other forms of product development. Indeed, many of the changes that personal computer and Internet software developers championed in the 1980s and 1990s very much resembled the just-in-time process improvements that Toyota championed.

But the pull system is just one example (albeit a powerful one) of what can happen when a company establishes a tradition of learning and process innovation. Taiichi Ohno made various efforts to teach pull concepts and other ways to reduce or eliminate waste in operations to workers in different Toyota departments and then to suppliers. As he did this, Toyota as an organization gradually evolved an in-house corps of consultants and teachers as well as specific training methods and a philosophy of learning. The company embraces what Stephen Spear and Kent Bowen have called "rigid" process rules, whilst simultaneously teaching workers how to test the assumptions inherent in those rules. Trainers show employees how to use the scientific method and run their own experiments, such as checking whether a recommended standard time for performing a specific task makes sense or not. Done consistently, the result at Toyota is a carefully standardized production process divided into many small, rationalized steps, which are then subject to scrutiny and continuous improvement by employees working with trained supervisors. When changing processes or learning new tasks, such as in assembly, which may require half a dozen discrete steps for each job, Toyota also follows a specific method: it focuses on getting workers to master one highly refined step at a time before moving on to the second step, and so on, rather than asking workers to learn an entire multi-step task at once.[30]

The introduction of new products into mass production is another area where engineers and workers are continually challenged to learn new tasks as well as to solve problems quickly without compromising ramp-up time, production efficiency, or quality. In 1999, Paul Adler and colleagues published a detailed study of how Toyota managed this part of its operations.[31] Their study builds on earlier work by Kim Clark and Takahiro Fujimoto.[32] According to Adler et al., during the late 1990s, American automakers took between 60 and 87 working days to introduce new models. Toyota was able to do this in 5 days even at NUMMI in California, which employed unionized former GM workers (but Japanese managers). The American companies also experienced major increases in defects that persisted for months after introducing new models, whereas NUMMI and other Toyota plants did not.[33]

The authors' explanation of how Toyota accomplished this feat is somewhat complicated because of their mixture of empirical observations with theoretical explanations, but several factors appear to have been critical. Most importantly, they cite Toyota's use of highly stable "meta-routines" (defined as "standardized procedures for changing existing routines and creating new ones"). These principles facilitated problem solving and made non-routine tasks more routine. Toyota also switched employees between production and improvement tasks on a regular basis, and carefully partitioned tasks, such as creating a separate pilot team to explore rapid new-product introduction techniques before introducing these into factory practice. As usual, Toyota relied heavily on extensive training of employees and suppliers, and cultivated high levels of trust with suppliers that enabled it to offload numerous tasks to them. Some of these measures explicitly enhanced flexibility, whilst others contributed more directly to efficiency. All measures increased operational effectiveness, though we must acknowledge that efficiency and effectiveness do not matter to the customer if there is a design flaw in a component or subsystem.

We must also be careful not to underestimate how difficult it was for Toyota to create the capabilities that have made it one of the most admired manufacturing companies in the world, despite the problems that surfaced in 2009–10. In particular, in his book *The Evolution of a Manufacturing System at Toyota*, Professor Takahiro Fujimoto of the University of Tokyo (and a former doctoral student of Kim Clark at Harvard Business School) cautions against oversimplification. He describes Toyota's approach to both production management and product development as somewhat less deliberate or "rational" than it appears in retrospect. Fujimoto insists it was "evolutionary" in nature, resulting from years of directed and ad hoc trial and error, with some random good luck.[34] But Toyota combined this learning-by-trying approach with impressive problem-solving skills and systems thinking, and with outcomes shaped by unique constraints imposed by the Japanese market—tiny production volumes in the 1950s but relatively high demand for product variety.

Microsoft: Agile ("Synch and Stabilize") Product Development

We have just seen how Toyota, followed by Nissan and other Japanese automakers, devised a novel manufacturing process in the 1940s and 1950s to suit the Japanese market. Post-war Japan required much more flexibility than traditional mass-production systems offered. Similarly, we can see process innovation occurring in software product development, again to accommodate the unique requirements of a new market. This time, it was the personal-computer business in the United States of the 1980s and 1990s. Compared to mainframe-oriented software development, Microsoft and other PC software companies turned to a less formal "hacker-style" of development. At least in part, this was because they competed in a fast-paced, more rapidly changing and less predictable business than the mainframe computer market. PC and then Internet software teams usually wanted to make design improvements or fix problems in their products *until the last possible moment.* This style of engineering required an agile process with few time-consuming rules and bureaucratic procedures.

I became convinced in the early 1990s that what we now call agile development represents best practice not only for *most* software projects but also for *many different types* of product development.[35] The principle that underlies agile is, again, pull rather than push, though not as directly as Toyota's just-in-time system. In agile or iterative software development, the pull comes from continuous feedback during the project—from testing the daily builds as well as results from usability labs, in-house "alpha" testers (eating their own "dog food"), and outside beta testers. Project teams begin with a rough plan but then modify the product design as they go along. The challenge, especially for teams of more than a few people, is to keep everyone's changes synchronized. It is the same issue when you have more than one person editing the same document. There needs to be a check-in process for editing changes, and someone needs to keep control of the "master" document. The alternative is to plan all the writing work in advance, to assign specific parts of the document to different people, not to allow anyone to edit or change the writing done by other members of the team, and then to put the pieces together at the end.

In general, when a large team works on the same product, and especially when team members need to coordinate because their components are interdependent, then an incremental process has important advantages. It seems inherently useful for project members frequently to *synchronize* what they are doing whilst working in parallel so that components remain compatible (or terminology remains consistent). And it seems inherently easier to coordinate and periodically to *stabilize* the components under development when managers break up a large complex project into smaller subprojects and thus smaller chunks of the product. Most professional engineers and academic researchers in the product development field, especially software, now agree it is an outdated practice to design monolithic systems without clear demarcations of subsystems and components, or to follow initial plans too rigidly and then to wait until the end of the project before trying to integrate components and to see if all the different parts work together properly.

Steps toward the New Process

Microsoft did not introduce a structured agile process easily or quickly, but the change was deliberate and precipitated by a series of crises. For the first decade after being founded in 1975, Microsoft had no defined process for product development and relied on the skills of a few talented programmers to design, build, and test new products. These products began with programming languages such as BASIC and then the DOS operating system, introduced in 1981. DOS applications such as Word and Multiplan, the predecessor to the Excel spreadsheet, followed in the early 1980s. For larger, more complex new products that contained a graphical user interface, such as Word for Windows, Microsoft loosely tried to follow a waterfall style. But a series of events set in motion some major process changes by the late 1980s. Similar to Toyota, Microsoft ended up relying on a critical practice that stopped work if the engineers detected a problem—the daily build. In *Microsoft Secrets* (1995), Richard Selby and I dubbed this type of structured iterative process "synch and stabilize." As noted below, several turning points and key actors helped shape its evolution at Microsoft.[36]

1984 Microsoft establishes separate testing groups in the Systems and Applications Divisions.

Recall of Multiplan 1.06 spreadsheet for the Macintosh.
Microsoft establishes the main technical specialties (functions) in the company—Program Management, Product Management, and Testing.

1986 Project teams begin to write postmortem reports to identify quality and project management problems as well as potential solutions.

1987 Microsoft recalls the Macintosh Word 3.0 word processor.

1988 Mike Maples arrives from IBM and establishes separate business units for each product group within the Systems and Applications Divisions.

The Publisher 1.0 project is divided into several "milestone" subprojects.

1989 The May retreat and November "Zero-defects code" memo highlight the importance of daily builds and milestones.

1990 Excel 3.0 project finishes only 11 days late on a 14-month schedule, led by Chris Peters, using key elements of the synch-and-stabilize process (subproject milestones and daily builds).

1992 Microsoft centralizes the Systems and Applications Divisions under the Worldwide Products Group, headed by Executive Vice President Mike Maples.

1993 Microsoft creates the Office Product Unit, unifying the separate Word, Excel, and PowerPoint product teams, headed by Vice President Chris Peters.

Not until there was a lot of pressure from customers and partners such as IBM did Microsoft managers and senior engineers decide to change the way they built software. Again, this market feedback is a kind of pull phenomenon, responding to Microsoft's habit at the time of pushing products out of the door with minimal testing. In fact, the company shipped several late and buggy products during the early and mid-1980s that prompted loud complaints from both PC manufacturers and individual retail customers. Retail customers were especially unhappy because they bought Microsoft products directly in stores, which meant that the software had not undergone the independent testing that IBM and other PC manufacturers usually performed.[37]

In 1984, Microsoft's senior executives decided to set up an independent testing group, following the lead of IBM and other more established software producers. They did not ask managers to review all software project artifacts, such as product specifications documents or code and test plans. Nor did they require executives to "sign off" on documents at the various development stages. These were "bureaucratic" practices common in software production for mainframe computers and defense applications but still rare in the PC world. Instead, Microsoft selected what seemed to be a few good techniques, such as a separate testing group and automated tests, and code reviews for new people or critical components. Then they promoted these as "good practices" for projects to adopt voluntarily. Microsoft teams also now began documenting their experiences through written postmortems and emphasized the need to "learn from mistakes."

Quality improved, but not enough. The company continued to ship buggy products as developers came to rely too much on the separate testing group. Most notable was Word 3.0 for the Macintosh, released in February 1987 (and originally scheduled for July 1986). Mac Word 3.0 had approximately 700 bugs, several of which destroyed data or "crashed" the program. Microsoft sent a free upgrade to customers within two months, costing more than $1 million.[38] This incident forced managers to realize that developers needed to look for bugs along with testers. They also concluded that only developers—the workers creating the product—can prevent errors from happening in the first place.

By now, it had become apparent even to skeptics within the company that Microsoft would have to become more systematic. Gates himself took over the Applications Division, but several key projects remained in chaos. None of the new applications for Windows, except for Excel, was progressing. A database program (dubbed the Omega project, which evolved into Access) and a project-management application for Windows were in serious trouble. The Opus project, later renamed Word for Windows, coined the infamous phrase "infinite defects." This describes a situation where testers are finding bugs faster than developers can fix them, and each fix (which requires changing lines of code) leads to yet another bug. Under these conditions, predicting the schedule and eventual ship date becomes impossible. After testing the product once during the "late and large" integration periods that Microsoft used to attempt, developers had to return to the old code, whose details they had largely forgotten or whose authors had disappeared. They had to rewrite much of the code to fix the innumerable bugs, and tended to add as many new errors as they repaired.[39]

About this same time, program management (which took charge of writing specifications and managing the schedule) began to emerge within Microsoft as a function distinct from product management (which took charge of feature prioritization and communicating with marketing) and software development (which wrote the product code and performed the first level of technical testing). Also important was the arrival of Mike Maples in 1988 from IBM and his decision to create smaller business units that were more focused and easier to manage. Maples also encouraged each group to define a repeatable development process suitable for its individual products.

Over in the Systems Division, Microsoft was having severe troubles with Windows. It shipped two versions in 1985 and 1987, but not until version 3.0 in 1990 was the product reasonably stable and well received by users.[40] Meanwhile, delayed products and recalls continued to frustrate customers. Microsoft settled one shareholder suit in 1990 for $1.5 million because of its failure to ship Word, which then accounted for 20 percent of sales. The database project was also years late and finally cancelled.[41]

Then Microsoft held a retreat for senior managers and developers in May 1989 to discuss how to reduce defects and make processes more systematic, but without overly structuring what they were doing. The basic concepts that came out of this meeting became the essence of the development process all Microsoft teams would adopt going forward. The first idea was to break up projects into subprojects or milestones, which Publisher 1.0 did successfully in 1988. The second idea was to create daily builds of the evolving products, which several groups had done successfully. A widely circulated memo summarizing the discussions galvanized the product groups. Chris Mason, a development manager in the Word group, wrote the memo, titled "Zero-Defects Code":

Microsoft Memo

To:	Application developers and testers
From:	Chris Mason
Date:	6/20/89
Subject:	Zero-defects code
Cc:	Mike Maples, Steve Ballmer, Applications Business Unit managers and department heads

On May 12th and 13th, the applications development managers held a retreat with some of their project leads, Mike Maples, and other representatives of Applications and Languages. My discussion group investigated techniques for writing code with no defects. This memo describes the conclusions which we reached... *There are a lot of reasons why our products seem to get buggier and buggier. It's a fact that they're getting more complex, but we haven't changed our methods to respond to that complexity*... The point of enumerating our problems is to realize that our current methods, not our people, cause their own failure... Our scheduling methods and Microsoft's culture encourage doing the minimum work necessary on a feature. When it works well enough to demonstrate, we consider it done, everyone else considers it done, and the feature is checked off the schedule. The inevitable bugs months later are seen as unrelated... When the schedule is jeopardized, we start cutting corners... *The reason that complexity breeds bugs is that we don't understand how the pieces will work together.* This is true for new products as well as for changes to existing products... I mean this literally: your goal should be to have a working, nearly-shippable product every day... Since human beings themselves are not fully debugged yet, there will be bugs in your code no matter what you do. When this happens, you must evaluate the problem and resolve it immediately... Coding is the major way we spend our time. Writing bugs means we're failing in our major activity. *Hundreds of thousands of individuals and companies rely on our products; bugs can cause a lot of lost time and money. We could conceivably put a company out of business with a bug in a spreadsheet, database, or word processor. We have to start taking this more seriously.* (Emphasis added).[42]

Several projects adopted these two ideas and other practices, most of which Microsoft people had debated in the past but not applied consistently or together. The project that received the most attention after the retreat was Excel 3.0. Not only was this a major revenue generator for Microsoft, but it was the first big project to use milestones and daily builds together. It had thirty-four developers at the peak, not including testers, working for 14 months, with 2 months of planning and the rest of the time spent programming and testing. Most importantly, the product shipped only eleven days late—astounding in that most Microsoft projects during those years, including Word for Windows, were *years* late.

Applying the Key Ideas More Broadly

As implied in the May 1989 memo, Microsoft managers and senior engineers wanted an approach that was fast and minimalist but still resulted in better quality and more predictable schedules.[43] Following an arbitrary schedule that some executive created, or trying to eliminate bugs completely, were not viable business goals for a PC software company in the 1980s and 1990s. Speed to market with "good-enough" products (to get ahead of Lotus, WordPerfect, and Apple) and backward compatibility with DOS (to keep the installed base of customers) were more important to Microsoft. Microsoft's solution worked well when first adopted, and it has continued to work well in most projects. Of course, the company has had some failures and did make modifications for very large, highly complex systems such as Windows Vista. But the two principles that emerged from the May 1989 retreat have remained fundamental to Microsoft's development process.

Figure 4.4 shows an example of a build chart for one of the milestones in an Excel project during the mid-1990s. Each day, as programmers checked in their work, they generated new bugs but fixed nearly all of them very quickly. The goal was to bring the number of active bugs to a low level (but not to zero, which was impractical from a business perspective) and eliminate the most serious defects. This team, led by Chris Peters, set new standards within Microsoft for project management as well as architectural elegance. Projects that followed these principles (and not all have—such as the Windows Longhorn/Vista project in the early 2000s) would avoid doing years of coding and feature development before trying to integrate modules and see if the whole product actually worked. They were also able to make late changes more easily.

There are two important examples of how this more agile development process enabled Microsoft to adapt quickly to the unpredictable

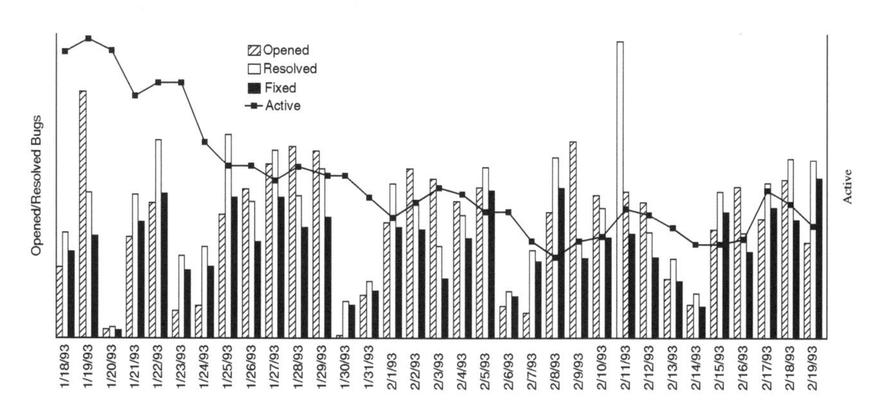

Figure 4.4. Daily build bug chart for Excel 5.0 Milestone 2 integration

Source: Cusumano and Selby (1995: 318).

world of the 1990s, with enormous business consequences. The original Windows 95 product specification, completed in early 1993, did not include an Internet browser—which was becoming an important new technology by 1994. And the original specification for Windows NT, done in the late 1980s, did not include a graphical user interface—which users demanded in the 1990s.[44] Microsoft made major changes to both designs, midstream, and ended up with very successful products. Microsoft won the browser wars with Internet Explorer, bundled into Windows 95 (albeit with what I believe were unnecessary unfair play and illegal actions). And Windows NT opened up an entire new opportunity for Microsoft to sell enterprise operating systems to corporations and other large organizations, as well as to replace the aging Windows code base built as a layer on top of DOS.

The initial task in the synch-and-stabilize process is for projects to begin their work by creating a short "vision statement" of a few paragraphs or pages (Figure 4.5).[45] This document defines the goals for a new product and describes as well as prioritizes the user activities that individual product features need to support. Product managers

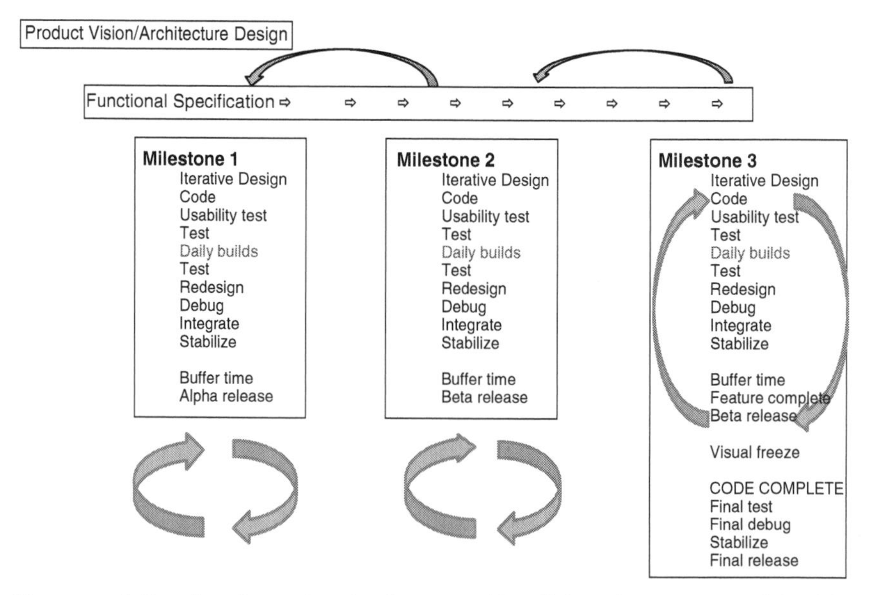

Figure 4.5. Agile or iterative product development, with pull from the customer usability labs, prototypes and builds, and an evolving spec

Source: Adapted from Cusumano and Yoffie (1998: 241).

(often MBAs who are marketing specialists) usually take charge of writing up the vision document whilst consulting program managers as well as other members of the product team, key executives, customer representatives, and other important stakeholders. The program managers are more technically oriented and usually write up functional specifications of the product and coordinate project work. They are not actually the "managers" of the programmers, but work with one or two feature teams of about five or eight developers. There is also usually a parallel team of "buddy testers" assigned to work with each developer on a one-to-one ratio (Figure 4.6). The developers report to their feature team leaders and a development manager. In addition, the program managers consult with developers when they write up the functional specification. Developers generally have veto power over feature suggestions because, after all, they have to estimate the effort required and write the code.

The functional spec should outline the product features in sufficient depth for project leaders to organize feature teams and estimate schedules with the programmers. In contrast to the typical waterfall process, Microsoft's specification documents generally do not try to decide all

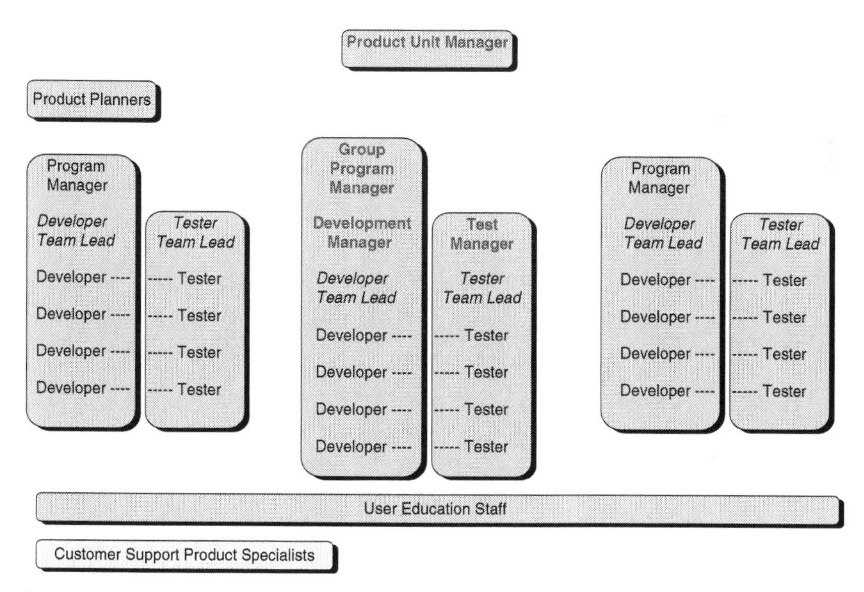

Figure 4.6. Microsoft project and feature team structure

Source: Cusumano and Selby (1995: 75).

the details of each feature, or lock the project into the original set of features. During product development, the program managers and developers will revise the feature set and functional details as they learn more about what should be in the product. Our interviews at Microsoft in the mid-1990s indicated that the feature set in a typical specification document often changed 30 percent or more by the end of a project. Again, this kind of change can be *good*, not bad, in that it gets the team closer to a better product—as long as the company can avoid tradeoffs in cost and quality (see Chapter 6).

Microsoft projects are usually led by a committee consisting of the product unit manager, the development manager, the test manager, and the head or "group" program manager. Together, with input from other team members, they decide the initial feature set. Then they divide the product and the project into parts (features and small feature teams), and break up the project into three or four milestone subprojects. The milestones represent completion points for major portions of the product, usually represented by clusters of prioritized features. All the feature teams go through a complete cycle of design, development, feature integration, usability testing, integration testing, debugging, and stabilization in each milestone subproject. Moreover, throughout the whole project, the feature teams synchronize their work by building the product, and by finding and fixing errors, on a daily and weekly basis. At the end of a milestone subproject, the developers are supposed to fix most of the bugs and all of the truly serious defects (those that "crash" the system) before they move on to the next milestone or ship the product. This debugging or error correction work stabilizes the product and enables the team to understand which features are done and which need more work.

At a higher level of abstraction, Selby and I grouped the Microsoft practices into two strategies: (1) *focus creativity*, and (2) *do everything in parallel with frequent synchronizations*. The first was important to define the product's features and place some limits on the project. The second was important to manage the actual process of designing, building, testing, and shipping products with an acceptable level of quality and functionality.

We identified five specific guidelines that product groups followed to focus creativity. Microsoft had no specific list or handbook of these practices, but this is what we observed:

1. Divide large projects into multiple milestone cycles with buffer time (about 20–50 percent of total project time) and no separate product maintenance group.
2. Use a vision statement and outline specification of features to guide projects.
3. Base feature selection and prioritization on user activities and data (to minimize arguments over what to do).
4. Evolve a modular and "horizontal" (that is, prioritized) design architecture, with the product structure mirrored in the project structure.
5. Control by individual commitments to small tasks and "fixed" project resources.

These practices can bring discipline and structure to a project in several ways. The basic idea is that, whilst having creative people is important, it is often more important to *direct* their creativity. Managers can do this by getting team members to think about features customers will pay money for, and by putting pressure on project personnel by limiting their resources, such as number of people and amount of time. If managers fail to create limits, then projects run the risk of never shipping anything to market. This risk becomes especially high in fast-moving market segments, when individuals or teams have unfocused or highly volatile user requirements, frequently change interdependent components during a project, or do not synchronize and stabilize their work adequately or frequently enough.

Microsoft got around these problems at least in applications groups like Word and Excel by breaking up large projects into several smaller subprojects, and by prioritizing features. Managers also introduced buffer time between each subproject milestone and the promised ship date to allow time to respond to unexpected delays or to add unplanned features. Projects used vision statements and outlines rather than complete specifications and detailed designs before coding because managers and programmers realized they could not determine *in advance* everything that the team would need to do to build the right product for the market. This incremental or iterative approach left the team room to innovate and adapt to unforeseen competitive opportunities and threats.

By the mid-1990s, most of Microsoft's major applications products as well as Windows NT also had modular and "horizontal" architectures. (This was not true of Windows 95. Like earlier versions of Windows, it superimposed a graphical user interface layer almost magically on top of DOS, with a lot of clever but non-modular code making the product appear to work as a graphical operating system.) Modularity allowed teams to add, cut, or combine features incrementally and independently in a prioritized fashion, like working off a horizontal list of most to least important components. Particularly for applications products, development teams also tried to design features that mapped directly to activities typical customers performed. This mapping required continual observation and testing with users during development. In addition, many Microsoft managers allowed team members to set their own schedules, but only after the developers had analyzed tasks in detail (half-day to three-day chunks, for example) and committed personally to the schedules they set. Managers then tried to "fix" resources by limiting the number of people allocated to any one project or by limiting the time teams could spend. Managers enforced these limits more in applications projects, where teams could more easily delete less-important features if they fell too far behind the schedule or if competitive pressures changed. Cutting features to save schedule time is less practical with operating systems because reliability is usually much more important than new features. In addition, many operating system features are closely coupled in subsystem layers, and developers cannot easily delete them individually.

Most important, prioritizing features, fixing resources, and modularizing components simplified project management. Instead of creating a long wish list of features and then guessing at how many people and how much time would be necessary to do the job, Microsoft's application projects were generally scheduled in a simpler way. For example, the Office product unit might want to deliver a new release in 12 months, and might have 100 developers and 100 testers. The problem then becomes how many new features can a team of 200 engineers build, debug, and stabilize in 10 months or so, with some time set aside as buffer. If they have done their prioritization and design work correctly, they can cut features if the project runs short of time.

The second strategy of *do everything in parallel with frequent synchronizations* required another set of guidelines to implement. We also broke these down into five main practices:

1. Work in parallel teams, but "synch-up" and debug daily.
2. "Always" have a product you can ship, with versions for every major platform and market.
3. Speak a "common language" on a single development site.
4. Continuously test the product as you build it.
5. Use metric data to determine milestone completion and product release.

This approach imposed discipline on projects but in a subtle, flexible way—without trying to control every moment of every developer's day. Many small teams and individuals have enough freedom to work in parallel yet still function as one large team so they can build large-scale products relatively quickly and cheaply. But the feature teams and individual engineers must adhere to a few rigid rules that enforce a high degree of coordination and communication.

For example, one of the few rules developers had to follow is that, on whatever day they decide to check in their pieces of code, they must do so by a particular time, such as by 2.00 p.m. or 5.00 p.m. This allows the team to put available components together, completely recompile the product source code, create a new build of the evolving product by the end of the day or by the next morning, and then start testing and debugging immediately. Another rule is that, if developers check in code that "breaks" the build by preventing it from completing the recompilation or quick test, they must fix the defect immediately. Again, this aspect of Microsoft's build process resembles Toyota's just-in-time production system, where factory workers stop the manufacturing lines whenever they notice a defect in the product or component they are assembling.

In agile development, the frequency of the builds is like a "heartbeat" that sets the pace of the project, much like the pace of kanban exchanges and cycle times set the pace of production at Toyota. The builds pull together all the changes occurring in the evolving product. I have seen Microsoft operating system projects do builds multiple times a day as well as once a week. I have seen Japanese computer manufacturers and European telecommunications equipment producers move from building their software products only in the last phase of a project—which can be after one or two years of development—to once every two weeks. The general principle is that more frequent synchronizations are better than less frequent synchronizations. Otherwise, developers can accumulate errors or incompatibilities as well as pile up changes in components that later on become nearly impossible to integrate. But the actual frequency is less important than the concept of doing continuous integration and testing.

The Daily Build

The typical build process introduced at Microsoft remained similar in the 2000s, though the company added much more sophisticated automatic testing and verification tools. The process described here is for software, but can be useful for many types of product development, especially projects with distributed teams or work done through computer-aided design and engineering tools. In fact, the translation team in Japan for *Microsoft Secrets*, published in 1996 by Nihon Keizai Shinbunsha, was so influenced by this process that they adopted it for the translation work. Every night they synchronized—using faxes!—by sharing a list of uncommon English terms encountered in their work that day and how they had translated them. The senior editor then figured out which terms were best. The team used each chapter as a milestone point to stabilize their changes and make sure everyone was translating terms (including "daily build") in the same way.

Back to Microsoft and software: as a first step, in order to create new features such as cutting and pasting text, developers checked out private copies of source code files from a centralized master version. They evolved their features by making changes to their private copies of the source code files. The developers then created "private builds" or copies of the evolving product that contained the new or more completed features, and tested the build on their own, with feedback from their testing buddies. Once they got the build to work, developers checked the changes from their private copies into the master version of the source code. The check-in process included an automated regression test to help assure that their changes to the source code files did not cause errors elsewhere in the product's basic functions. Developers usually checked their code back into the master copy at least twice a week, but they sometimes checked in daily or more frequently, especially toward the end of a project and during the final destabilization phase.

Regardless of how often individual developers checked in their changes, a designated developer or a dedicated person on the team, called the project build master, generated a complete build of the product on a daily basis using the master version of the source code. Generating a build for a product consisted of executing an automated sequence of commands called a build script. This created a new internal release of the product and included many steps that compile source code. The build process automatically translated the source code for a product into one or more executable files. The new internal release of the product built each day is the daily build. Projects generated daily builds for each platform, such as Windows and Macintosh, and for each market, such as US and the major international versions.

Product teams also tested features as they built them by bringing in potential customers from "off the street" to try early versions or prototypes in a usability lab. In addition, Microsoft tried to concentrate work on particular products on a single physical site, using a common set of programming languages, coding styles, and development support tools. This goal changed as Microsoft distributed some development work, such as to India and China, but it remains an important strategy for critical components. Concentrating development on a single site with common tools helps teams communicate, debate design ideas, and resolve problems face-to-face relatively easily and quickly. Project teams also used a small set of quantitative metrics to guide decisions, such as when to move forward in a project or when to ship a product to market. For example, as seen in Figure 4.4, projects rigorously tracked progress of the daily builds by monitoring how many bugs were newly opened, resolved (such as by eliminating duplicates or deferring fixes), fixed, and active. Developers and testers tested the builds until they reduced the number of bugs to a relatively low and stable level before moving on to the next development milestone or shipping the product.

Of course, Microsoft had variations and lapses in its product teams, especially within the Internet groups. For example, if managers of new projects wanted to move *really fast*—such as on the early versions of Internet Explorer and virtual meeting software—developers usually took the lead in proposing features and writing up outlines of specifications. In these cases, Microsoft program managers came on board later and focused on managing project schedules, writing up test cases with testers in parallel with development, working with interface or Web page designers, and building relationships with outside partners and customers. As in Extreme Programming, the best-written representation of the specification in fast-moving projects was the set of test cases, rather than a more formal and separately written functional specification.[46]

Microsoft deviated somewhat from these basic principles when building what eventually shipped as Windows Vista in 2005, after several years of delays and wasted efforts. As I have written elsewhere, I believe that Microsoft adopted a political strategy in the late 1990s and early 2000s to tie as many functions together into Windows so as not again to appear to violate antitrust law.[47] Netscape and the US government sued Microsoft in 1998 for bundling Internet Explorer with Windows; Microsoft lost the case in rulings that came out in 2000–1. Though the company avoided a break-up, antitrust authorities and competitors have continued to monitor Microsoft and occasionally challenge it in court, such as for bundling a media player with Windows or linking certain functions in its Windows servers to Office.

The Windows Longhorn/Vista project began with the Windows NT code base, designed by David Cutler (hired from DEC in 1988) and other software veterans who joined Microsoft in the late 1980s and early 1990s from several established firms. However, the desktop version of the product was supposed to contain many new features, and quickly grew to more than 50 million lines of "spaghetti" (that is, nonmodular) code. It became impossible to test thoroughly and stabilize. The poor state of the code forced Microsoft to abandon a couple of years of work on new features, go back to the 2003 Windows server code base, and make some refinements to its engineering and design approach. In particular, Microsoft took on the risks of product bundling and violating antitrust laws by breaking up Windows into different branches. This approach resembles how Microsoft engineers have treated Word, Excel, and PowerPoint as "branches" or subsystems of Office and built them separately and then integrated the branches periodically, such as weekly. The branching strategy makes the coding and daily builds for Windows and Office much more manageable than continuing to build these as single or "monolithic" products. They really now are complex systems. For Vista, Microsoft Research also introduced a new generation of testing tools to check automatically for a wider variety of errors (code coverage and correctness, application programming interfaces and component architecture breakage, security, problematic component interdependencies, and memory use) and automatically reject code at desktop builds and branch check-in points.[48]

Despite the problems Microsoft encountered with operating systems, at least prior to the streamlined Windows 7, released in fall 2009, Office and other applications groups best demonstrated the effectiveness of modular architectures and agile or iterative development. During the 1990s and 2000s, these kinds of techniques became much more common around the world, with some variations. In a global survey of 104 projects, conducted with Alan MacCormack, Chris Kemerer, and Bill Crandall, and published in 2003, we found that 64 percent of the projects surveyed used subcycles and thus were iterative and not waterfall.[49] Use of subcycles was most common in India (79 percent of projects) and Europe (86 percent), relatively common in the United States (55 percent), and least popular in Japan (44 percent). Most projects (73 percent) used beta releases, which became a useful testing and feedback tool with the arrival of the Web. Over 40 percent of the projects paired testers with developers—a Microsoft-style practice especially popular in India (54 percent of projects). Thirty-five percent of the projects used the XP practice of pairing programmers. Again, this was especially popular in our Indian sample (58 percent). More than 80 percent of the projects used daily builds at some time during their work and 46 percent used daily builds at the beginning or middle of the project, which is close to the Microsoft style of development. More than 83 percent of projects ran regression tests on each build. This important practice was most common in the Japanese (96 percent) and Indian (92 percent) projects.

Lessons for Managers

First, pull versus push is really a fundamental difference in management philosophy. The former emphasizes continuous adjustments to real-time information and the latter emphasizes detailed planning and control. As a manager, do you promote decentralization and empowerment to allow people and processes to respond directly to new information? Or do you try to anticipate everything in advance? This is the difference. Second, managers can use the pull philosophy to set their own company "clock speed"—that is, the pace they want to see for responses to feedback from customers, manufacturing facilities, the supply chain, product testing, or marketing and sales channels. Both the kanban exchange and the daily build have served as the "heartbeats" of change and innovation processes at Toyota and Microsoft. Both depend on fast cycle times and real-time information. Both are "lean" relative to conventional mass production or waterfall-style product development: they minimize use of people, work-in-process, inventory, bureaucratic procedures, or complex control systems. If pull systems work properly, they immediately and ruthlessly expose waste and inefficiency. As we have seen, though, a pull system is a processing system; companies must still make sure they get the product design right and respond to market feedback.

I recognized whilst studying Microsoft that aspects of the Toyota production system resembled the software company's development approach, but I never did a detailed comparison.[50] Here, I offer some observations on how certain Toyota and Microsoft processes are similar, point by point (Table 4.1). The analysis should help managers think about how to reduce or eliminate bottlenecks in their production systems or development processes, at least after the stage of initial concept design.

At Toyota, the kanban cards sent back into the factory and supply chain with each completed product enable the company to manufacture components and finished goods in small lots, just as the market requires them. This approach reduces in-process (and finished) inventories to a minimum. It works particularly well when suppliers and factories are geographically nearby, so that they can deliver components just when needed. The system also works best when the cycle times for

Table 4.1. *Process comparison of Toyota and Microsoft*

Toyota-style "lean" production	Microsoft-style "agile" development
Manual demand-pull with kanban cards	*Daily builds with evolving features*
JIT "small lot" production	Development by small-scale features
Minimal in-process inventories	Short cycles and milestone intervals
Geographic concentration—production	Geographic concentration—development
Production leveling	Scheduling by features and milestones
Rapid setup	Automated build tools and quick tests
Machine/line rationalization	Focus on small, multifunctional teams
Work standardization	Design, coding, and testing standards
Foolproof automation devices	Builds and continuous integration testing
Multi-skilled workers	Overlapping responsibilities
Selective use of automation	CA tools but no code generators
Continuous improvement	Postmortems, process evolution

Source: Toyota list adapted from Cusumano (1994: 28).

each worker are similar and machines produce comparable numbers of components—which Toyota called production leveling. It was possible to produce small lots of components just when needed, because Toyota redesigned its machinery for rapid set-up and changeover for different components. The reverse flow and minimal inventory also expose excess production and wasteful activities, and encourage work standardization. Foolproof automation prevents production of unnecessary and defective components. Multi-skilled workers make it possible to reduce the total workforce and eliminate idle time. Selective use of automation—in particular, avoidance of inflexible robots and too much fixed machinery—facilitate just-in-time adaptations to product mix and scheduling changes as well as immediate identification and correction of processing errors. Toyota managers, especially in the fast-growth decades of the 1970s and 1980s, also encouraged workers to make suggestions for continuous improvement—the exact opposite of what Frederick Taylor had taught as part of Scientific Management. Taylor had pressed factory managers to "freeze" procedures once the industrial engineers had analyzed the work flow and optimized the production process at a given point in time.

Again, at the center of what Microsoft does is another just-in-time process with a pull effect—the daily build, created as programmers test their evolving features each day and then try to synchronize them with changes made by other members of the team. The daily builds are akin to assembly lines bringing together components. The build does not work without modular architectures and development of new functionality (software instructions in code) in small increments, otherwise project engineers must integrate too many interdependent components each day, with too great a risk of errors causing "the line to stop" (that is, the build to break). Microsoft's focus on short development cycles (daily, weekly, monthly, and at milestone intervals) is similar to minimal in-process inventories of components that the team has to test and retest. Geographic concentration of teams facilitates fast communication among developers, testers, and program managers. Microsoft and other firms following agile processes also relied heavily on automated build tests to examine and integrate new code—rather like Toyota's rapid set-up times as well as foolproof automation techniques.

The focus on small, multi-functional teams with overlapping responsibilities, coordinated through the daily builds as well as techniques such as design reviews, reminds me of how Toyota used multi-skilled workers and reduced unnecessary staffing and bureaucracy. The processes at both companies require individuals and teams to follow a small number of iron-clad work rules. Microsoft also used automation selectively; for example, unlike the Japanese, the American company did not generally deploy automatic code generators. Finally, Microsoft introduced the idea of holding postmortems and writing reports after each milestone as well as at the end of each project in an effort to make improvements during projects as well as over time. Major changes came with major disasters, such as Windows/Longhorn, but there has clearly been a mechanism in Microsoft for continuous process improvement. Windows 7 is evidence of Microsoft's ability to redesign the product as well as its development process.[51]

Pull techniques can also help generate or refine product concepts. Microsoft, for example, is not very good at inventing things; it is much better at quickly exploiting ideas that already have some traction in the marketplace. In this sense, product managers working with program managers to suggest product concepts and feature ideas generate another kind of pull, rather than relying on, say, R&D to push out new technologies as products. When Microsoft has resorted to pushing out new technologies, rather than relying on its usual development system, it has often introduced products and features that were clumsy to use or not what the market wanted at the time. We can cite examples such as Microsoft Bob (an awkward "wizard" deploying some simple artificial intelligence that helped users with Office, introduced in the mid-1990s and then withdrawn), WebTV (introduced in 1997 and later withdrawn), Tablet PCs (introduced in 2002 to weak market acceptance), and the wireless-enabled SPOT (Smart Personal Objects Technology) watch (introduced in 2004 and discontinued in 2008).[52] The highly acclaimed Windows 7 is a positive example of pull versus push. Customer feedback drove nearly the entire development agenda for this latest version of Windows. In particular, users found Vista too slow, unstable, and riddled with awkward security features. Vista even drove many Windows users to switch to the Apple Macintosh. Many enterprises refused to buy Vista and continued to buy the older Windows XP until Microsoft released Windows 7.

Finally, the practical question: if pull concepts have such obvious benefits, why do not *all* managers and firms embrace this principle? The pull approach, in either manufacturing or product development, requires several elements to work properly. Perhaps most important is the concept of performing work in small chunks—short cycle times and small batches of components made just when needed or small increments of functionality added daily to the evolving product. The production "systems" should also generate a pull effect driven by new information—on changes in market demand or in customer preferences, or from feedback on quality of the components or the evolving product. The systems need to avoid too much rigidity—in automation and processes. Therefore, managers should structure around small semi-autonomous teams in engineering and production but insist on a few rigid rules to enforce communication and coordination. Both engineers and production workers need the freedom to respond to new information, adapt desirable changes, make other improvements, and innovate as opportunities occur.

Some managers have debated the complexity of managing such processes and stayed with push-style systems. Even back in the mid-1990s, it was clear to Toyota that so many frequent deliveries of components exacerbated urban congestion, creating traffic gridlock and the risk of not delivering on time.[53] Toyota's approach also placed a special strain on overseas operations, where it was more difficult to replicate the just-in-time system and control quality with widely dispersed suppliers. Suppliers and factories overseas generally had to hold more in-process inventory from distribution centers and then deliver components on a just-in-time basis, to maintain the continuous flow of the assembly lines. As came to light in 2009–10, Toyota was unable to increase overseas production rapidly while maintaining its traditional quality standards, though the most serious problems seemed the result of poor designs and inadequate testing, which Toyota, not suppliers, ultimately controls. Japan since the 1990s has also been suffering from an increasing shortage of blue-collar workers as well as growing product variety and unnecessary product variations. These conditions have further strained suppliers and factories. Toyota and other Japanese firms have adapted by slightly increasing in-process inventories, using electronic data transfer rather than physically exchanging kanban cards, standardizing more components, and reducing product variations.[54] Toyota has continued to upgrade the automation in its factories as well as to use more technology to assist workers, while still adhering to Ohno's basic principles of production management.[55]

Microsoft also encountered problems with its development approach, particularly when building Windows. The general observation is this: if the underlying product becomes too large and complex, then it becomes extremely difficult to add new features or make changes without introducing new errors. And the new errors are often equally difficult to find and fix. The problem is that it is usually impossible to predict how one change will affect all other parts of a large system with many interdependent components. Moreover, the range of potential scenarios to test (combinations of features or modules, data, user applications, hardware, and so on) quickly becomes an extraordinarily large number. Therefore, developers must take the time to break up large products into smaller, more manageable subsystems, and these should contain as many "modular" rather than "integral" or tightly coupled components as possible. Projects also need sophisticated and automated testing tools when the product and the project team get very big—such as hundreds of thousands or millions of lines of code, and hundreds or even thousands of engineers. Otherwise, the daily build process most likely will come to a grinding halt most of the time—and delays in terms of years can result, as Microsoft experienced with Windows Longhorn/Vista.

In short, a pull system—in manufacturing or product development—is not a panacea for all the kinds of problems that can occur with complex technologies. It is also tough to implement and manage. That is why pull systems, properly executed, can be such an advantage to firms that master this type of process innovation. The advantages are many: both Toyota and Microsoft, and many other firms using similar approaches, have demonstrated that simply making a pull system work in manufacturing or product development forces immediate improvements. The pull approach exposes manufacturing defects (though not design defects unless the faults are visually obvious) and reduces rework or wasted work, bringing about higher productivity and manufacturing quality. The reduced rework comes not only from detecting defects early and learning how to prevent them, but also from building the right assortment of products because of better information on what customers are buying. A pull system can also improve scope economies by facilitating production in small lots or engineering in small, customer-driven features that different products can reuse, though this strategy can backfire if the reused technology is faulty. There are other relationships between pull concepts, scope economies, and flexibility, as we will see in the next two chapters.

Notes

1. Another doctoral student of mine, the late Nancy Staudenmayer, compared how Bell Labs and Microsoft managed design changes. The former used a much slower, bureaucratic process, making changes only reluctantly, even in applications code. See Staudenmayer (1997).
2. On IDEO, see Kelley (2004) and Nussbaum (2004). Also Thomke (2004).
3. For examples of this style of product development, see Wheelright and Clark (1992), Eisenhardt and Tabrizi (1995), and MacCormack, Verganti, and Iansiti (2001).
4. See Basili and Turner 1975; Boehm (1988), and Aoyama (1993).
5. See Cusumano (2007b).
6. See Smith (1993) and Cusumano and Smith (1997).
7. See the discussion in Dodgson, Gann, and Salter (2005), esp. pp. 27–38.
8. See Sugimori, Kusunoki, Cho, and Uchikawa (1977) and Ohno (1978, 1988).
9. See Monden (1981a, b, 1982). Also Schonberger (1982) and Hall (1983).

10. There are a series of articles in the *International Journal of Operations and Production Management* that have analyzed push versus pull systems in some detail. See, e.g., De Toni, Caputo, and Vinelli (1988), Lee (1989), and Slack and Correa (1992). For the notion of tying the pull system to ordering products from customers and implications for the supply chain as well as factory management, see Holweg and Pil (2004). See also a new book that includes details from interviews with Ohno recounting his experiences and thinking: Shimokawa and Fujimoto (2009).
11. For reviews of the academic work on product development, see Brown and Eisenhardt (1995), Krishnan and Ulrich (2001), and Ulrich and Eppinger (2004).
12. This section is based on Cusumano (1985: 262–307).
13. See Schonberger (1982), as well as Womack, Jones, and Roos (1990).
14. See Krafcik (1988a, b). For a detailed account of the history of research on lean production, see Holweg (2007).
15. See the discussion of how the Toyota system evolved in Fujimoto (1999).
16. This chronology is derived from various sources. The text here follows Cusumano (1988).
17. Cusumano (1985: 276–7).
18. Cusumano (1985: 285–6).
19. Cusumano (1985: 190, table 46).
20. Cusumano (1985: 292–3).
21. Krafcik (1988b: 50).
22. Cusumano (1985: 302).
23. Cusumano (1988: 36, table 3).
24. Cusumano (1985: 199–203).
25. Cusumano (1985: 207–8).
26. Womack, Jones, and Roos (1990).
27. Womack, Jones, and Roos (1990: 85–6, figs. 4.3 and 4.4).
28. Cusumano and Takeishi (1991).
29. Again, for English publications, see Holweg (2007). Japanese business and academic writers have probably written enough books on Toyota to fill a small university library. These are of varying quality, though some are excellent. The most prolific author in the West not trained in Japanese studies but writing extensively on Toyota has become Professor Jeffrey Liker of the University of Michigan. See, e.g., Liker (2003) and Liker and Hoseus (2007). A new and somewhat more critical but still laudatory treatment of Toyota that focuses on how the firm manages seemingly opposite qualities is by three Japanese: Osono, Shimizu, and Takeuchi (2008).
30. See Spear and Bowen (1999) and Spear (2008).
31. Adler, Goldoftas, and Levine (1999).
32. Clark and Fujimoto (1991).
33. Adler, Godoftas, and Levine (1999).
34. Fujimoto (1999).

35. See Cusumano and Selby (1995: 14–18) and Cusumano (1997: 19). Also Cusumano (2004: 130).
36. This section is adapted from Cusumano and Selby (1995: 35–45). The chronology is adapted from p. 37.
37. See Manes and Andrews (1993: 205, 231, 317, 366).
38. Manes and Andrews (1993: 329).
39. Mason (1989), discussed in Cusumano and Selby (1995: 42–3); Gill (1990).
40. Manes and Andrews (1993: 398).
41. Manes and Andrews (1993: 373, 398–9).
42. Mason (1989: 1–4), reproduced in Cusumano and Selby (1995: 43).
43. This section is adapted from Cusumano (2004: 144–60).
44. Zachary (1994). Also, Cusumano and Selby (1995: 223).
45. This section relies on material, initially from Cusumano and Selby (1995), in Cusumano and Selby (1997).
46. Interview with Max Morris, Program Manager, Internet Applications and Client Division, Microsoft, Oct. 31, 1997, cited in Cusumano and Yoffie (1998: 250).
47. Cusumano (2006b).
48. See Larus et al. (2004).
49. Cusumano et al. (2003).
50. See the reference to Toyota in Cusumano and Selby (1995: 12). I did give some thought to the comparison in an interview with a Japanese magazine in 2006, which helped get me started on this chapter. See Cusumano (2006a).
51. See Mossberg (2009) and Pogue (2009).
52. My thanks to Andreas Goeldi for these examples.
53. See Nishiguchi and Beaudet (1997).
54. Cusumano (1994: 29).
55. See Fujimoto (1999), Liker (2003), and Spear (2008).

5

Scope, Not Just Scale

The Principle

Managers should seek efficiencies even across activities not suited to conventional economies of scale, such as research, engineering, and product development as well as service design and delivery. Firms usually pursue synergies across different lines of business at the corporate level. But scope economies within the same line of business can be an important source of differentiation in markets requiring efficiency and flexibility, and responsiveness to individual customer requirements. These deeper economies of scope require systematic ways to share product inputs, intermediate components, and other knowledge across separate teams and projects. Firms can also eliminate redundant activities and other forms of waste, and utilize resources more effectively.

Introductory

Every manager knows the power of scale economies. This concept, facilitated by the division of labor, standardization of product designs, components, and production processes, as well as factory mechanization and automation, has been behind many of the productivity and quality improvements in manufacturing since the beginning of the Industrial Revolution more than two centuries ago. The logic is that costs should drop as the size of operations increases, at least until a firm has to duplicate equipment or personnel, or when "diseconomies" of scale set in because of organizational bureaucracy and other difficulties.

We have already seen several examples in this book where massive economies of scale have been important in manufacturing (or product replication) as well as in marketing, sales, and distribution: Toyota with automobiles, JVC with VHS video recorders, and Microsoft with software products, among others. But none of these companies really relied on scale to succeed. Rather, they put themselves in positions to grow and eventually benefit from scale because of what they achieved when relatively small. The Toyota story is primarily about in-house capabilities that led to the just-in-time process innovations as well as independent product engineering skills. The JVC story is also about distinctive engineering and manufacturing capabilities supported by clever strategic maneuvering in what became an enormous platform market. The Microsoft story is mainly about having the right technical skills at the right point in time and then leveraging these capabilities to exploit a powerful platform position in a rapidly growing industry.

Scope economies are less commonly part of the management toolkit, perhaps because—like platforms and capabilities—they are relatively vague to define and difficult to measure. Researchers have trouble with this concept as well. Most MBA students and managers do learn about scope economies, but from the perspective of corporate strategy: diversification into related but separate lines of business should provide more opportunities to share resources or generate "synergies" compared to investing in unrelated lines of business. Most diversified firms, though, find that corporate-level synergies are hard to achieve in practice, and grow more elusive as new businesses become increasingly distant from the core expertise of the firm.[1]

The type of scope economies highlighted in this chapter are much more practical and operational, though still difficult to achieve in practice: sharing *within* the same line of business or technology domain. My broader argument is that scope economies are potentially even more important to differentiation and competitive advantage than scale economies precisely because they are difficult to achieve, within or across lines of business. Research, design, and engineering, for example, are critical activities for most manufacturing or technology-based firms; managers and other employees who can identify ways to share these activities have a potential edge. Even more important are service firms or service departments of product firms: they do not readily benefit from scale in many of their activities. Often they see performance or customer satisfaction suffer as their size increases unless they can find ways to share expertise effectively across customer engagements.

Within the business, ignoring scope in favor of scale has led some great companies astray. One of the more famous cases is that of Ford with the Model T. It is well known that Henry Ford tried to maximize production of one model. Indeed, the company experienced continual productivity increases from 1913 through the mid-1920s. In the later years, sales declined and then collapsed when customers began to prefer more product variety, such as offered by the smaller and seemingly less efficient General Motors.[2] GM soon became the largest automobile producer in the world until passed by Toyota in 2008. If scale economies alone were so important to performance of the business, then GM would have remained the most profitable automobile company. Yet for decades it has badly trailed Toyota in productivity and profitability as well as much smaller niche producers such as BMW. GM also wasted years of management time and billions of dollars when, during the mid-1980s, it diversified into non-core areas by acquiring EDS, the information technology services firm, as well as Hughes Aircraft—both of which it later sold. GM finally went bankrupt in 2008–9, and then re-emerged as a smaller version of itself. GM remains one of the larger automobile producers in the world. But it should be obvious to GM's new management team—as well as to executives at Toyota and other companies—that scale is generally far less important than variety, quality, reputation, marketing skills, or timing—having the right product or service at the right time. In fact, as we look back in 2010, it is ironic that Toyota seems to have followed GM down the path of placing too much emphasis on scale—becoming number one in the industry—while neglecting the more important issues of product design and customer safety. Perhaps we have been too hard on GM. The lure of becoming number one may be so powerful that it can persuade even the very best companies to compromise their most cherished values.

That scale is less important than other goals, including scope economies, is hardly a new idea. Alfred Chandler titled one of his major books *Scale and Scope*. He treated scope mainly from the point of view of production and related operations—the ability of a firm to make more than one type of product within the same facility at a lower cost than producing the two products in separate facilities.[3] But scope can extend far beyond manufacturing, as I have indicated, and not only to operations such as distribution. (Pepsi, for example, once touted the scope economies of selling snack foods along with soft drinks, and justified acquisitions along these lines, because they could utilize the same distribution channels.) Ford with the Model T failed to exploit scope economies in product development as well as in production. Engineering and design work can benefit from scale of operations, which are helpful to fund investments such as in product and process R&D or tools development. But, in the 1920s, Henry Ford's major mistake was that he did not design multiple products that looked different to the customers but still shared major components. Nor did he design a factory that could handle more variety in final assembly.

Several scholars have demonstrated empirically that large, diversified firms can achieve economies of scope in areas beyond manufacturing and distribution such as research and product development, mainly through knowledge "spillovers" or intentional sharing across projects.[4] Particularly in activities such as R&D, it seems that economies of scale are very limited or non-existent—simply increasing the size of research programs does not make them more productive, such as in generating patents. But, for example, the economists and strategy researchers Rebecca Henderson and Ian Cockburn found that increasing knowledge sharing across pharmaceutical research projects does result in higher performance.[5] Other scholars have shown how services firms can achieve economies of scope through consolidations and sharing across related product areas or business units.[6]

In the software industry, many firms have found that, with carefully designed processes, tools, and training, as well as reusable or easily modifiable components and design frameworks, they can achieve scope economies across multiple projects for different products or customers. The benefits are obvious when the technology is innovative and the quality impeccable; the risks to a wide set of products are also obvious if designs and quality are poor (as Toyota managers learned the hard way). The entire free and open software movement, as well as efforts to promote software reuse in commercial settings, also center on the principle of economies of scope: sharing software components and whole products in different contexts where project teams or individual programmers might otherwise re-create the technology from scratch. This kind of sharing is especially important in custom software development, such as practiced by major IT services firms. They have many customers with similar but not identical needs. Other service businesses, like management consulting, also can benefit from systematic efforts to share knowledge across separate customer engagements.[7]

As noted in the Introduction, I illustrate the idea of scope economies within the same line of business using two examples of product development. One deals with the software factory, as implemented most famously in Japan but also in the United States and India. The other deals with how automobile companies such as Toyota built products around common in-house platforms and quickly transferred technology to multiple projects. There are many similarities between multiproject management in the automobile industry and software factories or even open-source software development efforts that exploit reuse.

Again, I must emphasize that pursuing scope economies, if not implemented skillfully, can have tradeoffs in cost and quality, innovation, or general customer satisfaction. Software factories and automobile companies, for example, can be overly rigid if they force the reuse of components when they are better off building new technology from scratch to make sure the component is right for the new application. Poor designs or lax quality control will also diffuse defects much more widely than a development strategy focused on building one product at a time. The flip side of these possible tradeoffs, though, is that planned or systematic reuse can ensure that a larger number of projects take advantage of high-quality technologies proven over multiple customer engagements or product design efforts. Customers can also save money from not having to pay for totally new technology and a costly invention process. But designing components for reuse generally requires more time and effort (cost) at least in the first usage compared to building something for one customer in one project. Designers and testers must take into account more potential usage scenarios. Both software and auto companies have found ways to deal effectively with these kinds of challenges, most of the time. It follows that scope economies require more advance planning and managerial skills than ad hoc product development, and this can reduce overall flexibility and creativity. But, as we will see in this chapter as well as in Chapter 6, managers can minimize these tradeoffs as they choose when and when not to pursue scope economies.

The Software Factory: Structuring an Unstructured Technology

In 1984, after having finished the initial study of Nissan and Toyota, I decided to investigate what I thought was the next great industrial challenge for Japan: whether the Japanese were capable of writing world-class software. At the time, Japan made excellent "hard" products (automobiles, consumer electronic devices, semiconductors, and mainframe computers developed in technology tie-up arrangements with American companies). But software was something else, even though it was already becoming ubiquitous for a variety of industrial and consumer products. The Japanese had a reputation for doing little if anything significant in software engineering. Some observers believed the Japanese lacked the creativity or education to be top-notch programmers. Another question was whether it was possible for Japanese firms to transfer to software any of the skills they had mastered in mass production and conventional engineering, where their reputation for productivity and quality was becoming widely appreciated throughout the world.[8]

As I learned more about the Japanese software producers, one thing struck me as different from what I had read about in the United States and Europe: all the major Japanese computer manufacturers were investing heavily in what they called "software factories," with thousands of programmers co-located to build large-scale custom systems for industrial and government applications. The factory approach is similar to what we have seen in India for more than a decade, but in the 1970s and 1980s it was novel and risky. At first blush, the Japanese efforts seemed to be searching for manufacturing-type scale economies and other efficiencies but in tasks that more resembled craft or job-shop operations. As I dug deeper into their efforts, however, I concluded that the underlying principle, at least for applications projects, was really economies of scope. Again, in the late 1980s and 1990s, several Indian entrepreneurs came to the same conclusion—that factory-like approaches were quite suitable for custom software development and producing some types of IT services more efficiently.

In other industries, factory systems had emerged in the 1800s as managers and engineers attempted to move the process of replicating product designs beyond the craft stage. Craft production required highly skilled employees to make multiple copies of guns, agricultural machines, steam engines, and other products—and there were no economies of scale, since each product was essentially unique. Factory production came to mean the use of standardized and interchangeable parts, standardized and simplified production steps, use of less skilled but specialized workers through a division of labor, mechanization and then automation of as many steps as possible, with stricter process, quality, and accounting controls—all of which facilitated economies of scale and learning-curve efficiencies.[9] In software engineering and other technical services work, it was relatively easy to bring large numbers of people together and put them in a single building (though, for small design and engineering tasks, small teams generally worked better than large teams). But what made bringing a large group of people together in such an operation more or less like a "factory" as opposed to merely a large job shop became a central question of my research. There were many implications for practice as well, as the Japanese and later the Indian software companies pushed the software factory concept forward.

A "Factory-ness" Survey

To understand better how companies were tackling large-scale software development, in 1988 I conducted a survey of management practices and emphasis. I collected responses from fifty-two development organizations at more than two dozen companies, divided roughly equally between North America (twenty-seven facilities responding from AT&T Bell Labs, Bell Communications Research, Bell North Research, DEC, IBM, Control Data, Unisys, Data General, Hughes Aircraft, Boeing, Honeywell, Draper Laboratories, Martin Marietta, Arthur Anderson, TRW, EDS, Cullinet, and Computervision) and Japan (twenty-five facilities responding from NTT, Mitsubishi Electric, Fujitsu, NEC, Hitachi, Toshiba, Nippon Systemware, Nippon Business Consultants, and Nippon Electronics Development).

In particular, I wanted to see if there was a spectrum that went from something that resembled craft or job-shop approaches to more factory-like characteristics. I asked eight questions that tried to capture the degree of emphasis on tool and process standardization versus reuse of designs and components.

As seen in Figure 5.1, there was clearly a range of emphasis, with facilities in North America looking more like job shops whilst most of the Japanese appeared more factory-like. The differences in responses relating to process standardization were similar, suggesting that most large firms were adopting standardized approaches to process management and tool support. But the difference in emphasis on reuse was statistically significant (at the 1 percent level). Reported reuse rates were also more than twice as high (35 percent compared to 15 percent) in the Japanese versus American projects.[10] The 1988 survey encouraged me to probe more deeply into how the Japanese were attempting to structure this normally unstructured technology—software development.[11]

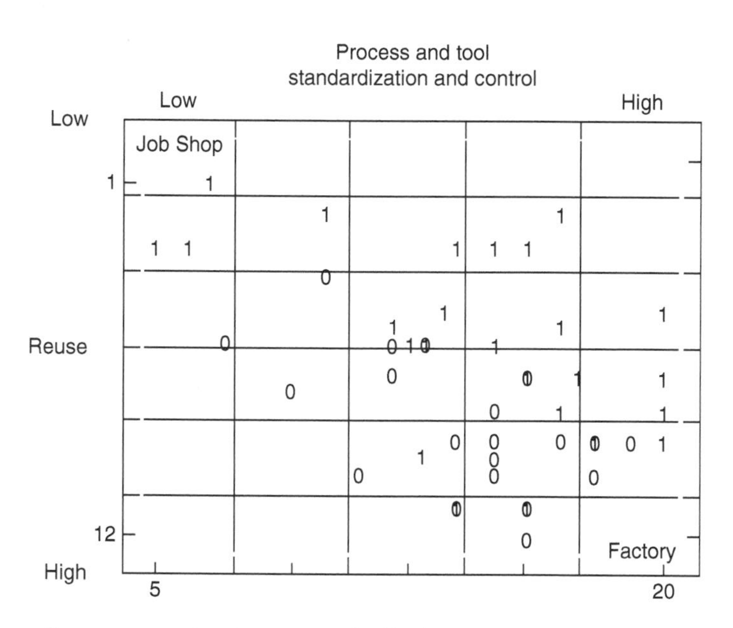

Figure 5.1. Reuse versus tool and process standardization

Notes: $N = 52$(25 Japan, 26 USA, 1 Canada)
Codes: 0 = Japanese facilities
1 = US and Canadian facilities

Source: Cusumano (1991b: 31).

I thought it important to understand the software factory approach in terms of both technology strategy and organization theory. Scholars such as Joan Woodward, Charles Perrow, and Henry Mintzberg had already written about job-shop or craft versus factory production, routine versus non-routine tasks, and bureaucracy versus "adhocracy."[12] Operations management specialists such as Robert Hayes and Steven Wheelright, as well as Roger Schmenner, had proposed "product-process matrices" for manufacturing.[13] I followed their thinking and proposed a simple contingency framework for software product development. For what I called "high-end" custom projects that required unique features or very complex new technology for each customer, a job-shop or craft approach seemed best. On the "low end," where it was possible to build features that many users wanted, I thought that an application-oriented project—such as how Microsoft organized the creation of its software products for mass distribution—was the right approach. In between, it seemed a "flexible design factory" should work—if the product or system features were only partially unique to each customer and the producer could standardize processes and reuse at least some components.[14]

The Japanese mainframe computer companies—Hitachi, Toshiba, NEC, and Fujitsu—largely followed IBM in adopting industrial software engineering practices and therefore pursued very similar approaches (Table 5.1). The Japanese efforts began in the late 1960s or mid-1970s with decisions to create centralized organizations and management control systems for specific software product families. They then standardized around methods and tools (special computer programs that help a programmer write or edit software instructions) tailored to these domains, and gradually provided partially automated support for development and project management. Over time, they refined their processes and tools, increasing their ability to handle different types of projects, though primarily using mainframe computers (Table 5.2).

Factory Concepts in Software Development

There were both technology push and market pull behind the software factory.[15] The push came from the desire to apply manufacturing or at least industrial engineering practices to software development. The pull

Table 5.1. *Japanese software factory organizations, c.1989*

Year est.	Company	Facility/ organization	Products	Estimated employees
1969	Hitachi	Hitachi Software Works	BS	5,000
1976	NEC	Software Strategy Project		
		Fuchu Works	BS	2,500
		Mita Works	RT	2,500
		Mita Works	App	1,500
		Abiko Works	Tel	1,500
		Tamagawa Works	Tel	1,500
1977	Toshiba	Fuchu Software Factory	RT	2,300
1979	Fujitsu	Systems Engineering Group	App	4,000
		(Kamata Software Factory		1,500)
1983	Fujitsu	Numazu Software Division	BS	3,000
1985	Hitachi	Systems Design Works	App	6,000

Notes:
BS = basic systems software;
RT = real-time systems and applications software;
App = business applications software;
Tel = telecommunications software.

Source: Cusumano (1991b: 7).

came from rapidly growing demand for software from the late 1960s throughout the 1970s, as IBM and Japanese mainframe computers became widespread for a variety of commercial, military, and government applications. Most users required custom-built systems. Not surprisingly, a shortage of skilled programmers—called "the software

Table 5.2. *Common elements in the factory approach*

Across a series of similar projects
Objectives
Strategic management and integration
Planned economies of scope
Implementation
Commitment to process improvement
Product-process focus and segmentation
Process-quality analysis and control
Tailored and centralized process R&D
Skills standardization and leverage
Dynamic standardization
Systematic reusability
Computer-aided tool and integration
Incremental product/variety improvement

Source: Cusumano (1991b: 9).

crisis"—became acute in the United States, which had most of the world's computers. This recognition fed many debates on improving performance and quality in software programming and how to make this activity more efficient—debates that continue to this day.

The first public proposal for a software factory came from Robert W. Bemer of General Electric in 1968.[16] His idea was to utilize standardized development support tools and a historical database for project management and budget control.[17] GE's exit from the computer business in 1970 ended the company's commitment to commercial hardware and software production, though interest in software factories continued within and outside GE.[18]

Roughly at the same time, AT&T Bell Labs was beginning to develop large amounts of software for telephone switching systems and billing. One of its key executives was M. Douglas McIlroy, a Ph.D. in mathematics from MIT who was involved in the development of UNIX and related programming languages. He began promoting the idea of systematically reusing components when constructing new programs.[19] In an address at the 1968 NATO Science conference on software engineering, McIlroy argued that the division of software programs into modules offered opportunities for "mass production" methods. He then used the term "factory" in the context of facilities dedicated to producing families of software parts that could serve as building blocks for tailored programs reusable across different computers.[20]

Reception to McIlroy's ideas was mixed at the time. It was especially difficult to create software modules that worked efficiently and reliably for all types of hardware systems and user applications. Nor did anyone know how to catalog program modules so that programmers could easily find and reuse them. Over the next decade, the diffusion of the multi-platform UNIX operating system and C programming language, and reusable component ("class") libraries, brought many of these ideas into practice, initially at Bell Labs. The later introduction of objected-oriented design and modular programming concepts as well as more advanced programming languages such as C++ made reuse even easier and are the basis of sharing in free and open source software. In the 1960s, however, none of these technologies was available. This was also before the era of packaged software—the easiest form of large-scale reuse. Nonetheless, by the late 1960s, the term "factory" had arrived in software and was being associated with computer-aided tools, management control systems, modularization, and the longer-term goal of reusing software components.

Process standardization and improvement, as well as tool support (partial automation), were probably the most important factory-like elements successfully applied to software. Here again we need to look to IBM for the leadership initiatives. This company made several discoveries about how (and how not) to manage software development when it deployed a thousand or more programmers during the mid-1960s to develop the operating systems for the System 360 family of mainframes. A senior IBM manager, Frederick Brooks, shared many of his insights in *The Mythical Man-Month* (1977). This book became famous for the argument that adding people to a late software project generally makes it later: people and months are not interchangeable because of the complexity of communications and management in a large project. On the other hand, IBM made progress in standardizing methods for different phases of software development and introduced a variety of useful tools and management techniques, such as intensive software code and design inspections. IBM's Santa Teresa Laboratory in the mid-1970s also brought together some 2,000 programmers on one site, with specially designed workspaces, in addition to factory-like industrial programming methods.

The First Software Factory: Hitachi

Whilst US companies mainly talked about factory concepts and organizations during the 1960s, some Japanese managers took the idea very seriously—consistent with their focus on building long-term capabilities in engineering and manufacturing. Hitachi became the first company in the world to adopt the term factory (its Japanese equivalent, *kojo*, translated as both "factory" and "works"). The Hitachi Software Works opened in 1969, led by a senior manager from the computer hardware inspection department, Kazuyuki Sakata. A history of independent factories for each major product area had prompted Hitachi to create a separate facility for software when this became a major activity. Managers were trying to improve productivity and reliability through process standardization and set the goal of transforming software from an unstructured "service" to a "product," with a guaranteed level of quality. An economic motivation was the severe shortage of skilled programmers in Japan and numerous complaints from customers regarding bugs in the software Hitachi was providing (most of which, along with the hardware designs, Hitachi was importing and adapting from RCA until 1970).

Hitachi managers concentrated initially on determining factory standards for productivity and costs in all phases of development, based on data they had begun to collect on project management and quality. The Software Works then standardized around structured programming techniques (a top-down design methodology emphasizing a logical hierarchical structure of modules that hide details at the different levels and make a software program easier to understand and modify—and potentially reuse components) and set up training programs for new employees and managers. These efforts emphasized process and reflected an attempt to raise average performance rather than trying to specify every step that programmers needed to perform. After some success, Hitachi then invested extensively in creating automated or semi-automated tools to assist in project management, design support, testing, code generation, and reuse—all for its rapidly growing and highly profitable mainframe business.

At the same time, Hitachi managers underestimated the difficulty of implementing factory concepts such as reusability and process standardization. For example, their attempt in the early 1970s to introduce a "components control system" for reusability failed, as did efforts to introduce one standardized process for both basic software products such as operating systems as well as custom-built business applications. Their desire to distinguish the best methods and tools for different software types led to a separate division for basic software and applications within the Hitachi Software Works, and then to establishment of a second software factory in 1985 dedicated to custom business applications.

Despite the struggles, Hitachi reported remarkable improvements in performance. Company data indicate that they doubled productivity in software development between 1969 and 1970 alone. Productivity leveled as Hitachi focused on refining its tools and methods, but began rising rapidly again in the late 1970s and 1980s, especially after the introduction of reuse support and automated programming tools for business applications development. Hitachi Software Works also reduced late projects from 72 percent in 1970 to 7 percent in 1974, and thereafter averaged about 12 percent late projects a year—remarkable for an industry where most surveys show that up to 85 percent of large projects are late and over budget.[21] This is not to say that Hitachi necessarily got faster or better at software development than other companies (though its reputation for defect-free programs is unmatched in Japan). But internal performance measures improved, and managers at least became more skilled in estimating how long it would take to complete a project. Better quality also reduced unexpected delays. Defects per machine reported from Hitachi customers dropped from an index of 100 in 1978 to 13 in 1984.

The US Pioneer: System Development Corporation

The world's second software factory opened in 1975 at one of the leading US companies in the custom software business, System Development Corporation (SDC), formerly a part of the Rand Corporation and today a Unisys division. The key factory architect was an engineering manager named John Munson. SDC had separated from the Rand Corporation in the 1950s to develop the SAGE missile control system for the US Department of Defense. It later took on other real-time programming tasks as a special government-sponsored corporation before going public in 1970. Top management then wanted more control over software costs and launched a factory initiative in 1972 to tackle five common problems: lack of discipline and repeatability in the development process; little visualization of project progress; poor requirements specification; poor testing tools; and little reuse.[22]

SDC managers outlined a detailed factory process and set up an organization consisting of an integrated set of tools (program library, project databases, online interfaces between tools and databases, and automated support systems for verification, documentation, and so on); standardized procedures and management policies for program design, coding, and testing; and a matrix separating high-level system design (at customer sites) from program development (at the Software Factory). SDC set up the factory in Santa Monica, California, and staffed it with 200 programmers. The company also copyrighted the name "The Software Factory."

Scheduling and budget accuracy improved dramatically for most of the ten projects that went through the SDC factory. Nonetheless, management closed the factory in 1978. Whilst the new process and tools were valuable, SDC's projects were too different to reuse much if any of the components. But perhaps the major problem was that SDC executives did not require project managers to use the factory. In the past, project managers had created their own programming teams to work at individual customer sites, as is typical in IT service projects. Customers generally prefer to work this way too, because they are closer to the programming effort. Munson persuaded several project managers to use the factory, but eventually they reverted to the usual practice of building programming teams at customer sites. The flow of work into the factory stopped. It closed when the last project ended, and the remaining programmers took assignments elsewhere. As Munson mused in our 1987 interview, the SDC Software Factory ended "not with a bang, but a whimper."[23]

In retrospect, organization theorists might argue that SDC executives attempted to impose a factory structure—standardized methods, automated tools, and component reuse goals—on a range of projects whose tasks were too different and better suited to job-shop or craft production. Still, SDC continued to use many of the factory processes and some of the tools. Perhaps most importantly, the SDC factory process heavily influenced the software standards later adopted by the US Department of Defense. SDC had some influence as well as on Hitachi and other factory efforts already underway in Japan, where project volume was larger and project managers did not have the discretion to create their own programming teams.

Acceptance of a factory-like process for software development was not simply a matter of culture, though experienced American project managers resisted the idea. Arthur Anderson (now Accenture), at roughly the same time as SDC, had much more success with a factory process. But, rather than creating a separate facility and providing it with people, processes, and tools, Anderson created a factory-like methodology called Modus and a set of tools referred to as the Design Factory, which made heavy use of form-based tools and COBOL code generators. This firm also recruited new college graduates from different backgrounds, trained them extensively in its particular process, and then put them to work under the guidance of more experienced system engineers and project managers—exactly what the Japanese (and later the Indians) did. The Anderson approach worked well as long as the projects did not encounter too many new requirements or demands from customers for innovative solutions.[24] Of course, as the world moved quickly during the 1980s and 1990s into new platforms—workstations, personal computers, and the Internet—programmers without strong technical backgrounds generally struggled to adapt to the new technology, whether they were Americans, Japanese, Indians, or something else.

Japan's Software Factory "Boom"

Three other Japanese firms—Toshiba, NEC, and Fujitsu—followed Hitachi and SDC to launch their own factory efforts between 1976 and 1979. These firms and Hitachi also expanded and refined their approaches during the 1980s, as software engineering methods and software quality control became major areas of interest and expertise in Japan.[25]

In 1976, Dr Yukio Mizuno began NEC's effort to rationalize software production within his company's computer division, which developed operating systems and other basic software such as compilers and programming support tools. NEC management then set up a research laboratory in 1980 to direct the development of tools, procedures, and design methods for various types of software, and organized a quality-assurance effort to standardize management practices throughout the corporation. Since product divisions did not always accept technology from the laboratory, and there was considerable variety in the systems they built, NEC allowed divisions to modify tools or to add their own design or project-control systems. Therefore, in contrast to SDC, the strategy at NEC was to mix standardized processes and tools with tailored processes as suited the different product domains.

In 1977, Dr Yoshihiro Matsumoto led Toshiba's efforts to create what became Japan's most structured software factory. Toshiba benefited from very focused product lines and was able to create a centralized facility to develop real-time process control software for industrial applications. Toshiba also employed a matrix, with experienced systems engineers assigned to different departments and overseeing the work of programmers that shared the factory infrastructure. Similarities from project to project allowed Toshiba to build "semi-customized" systems that combined reusable design patterns and coded modules with new code. Toshiba's Software Engineering Workbench utilized the UNIX programming environment and included tools for design support, reusable module identification, code generation, documentation and maintenance, testing, and project management support. Particularly important in Toshiba's approach was the design of new program modules (generally limited to fifty lines of code) for reusability, the practice that programmers deposit a certain number of reusable modules in a library each month, rewards for writing frequently reused modules, and the factoring of reuse objectives into project plans and schedules.

Following its main rival Hitachi, Fujitsu established a basic software division within its Numazu computer hardware factory in the mid-1970s to build IBM-compatible operating systems, compilers, and similar programs. Then, in 1979, it formally established the Kamata Software Factory for applications development, though its first mission was to convert Hitachi and IBM programs to run on Fujitsu mainframes. Again like Hitachi, Fujitsu devised automated design-support tools for producing business applications programs.

These similar efforts appeared effective in raising productivity, quality, and process control over what each Japanese firm had earlier experienced. In addition to the improvements noted at Hitachi, Toshiba reported that productivity rose from 1,390 equivalent assembler lines of source code per month (about 460 in Fortran) in 1976 to 3,100 per month (about 1,033 in Fortran) in 1985, including reused code totaling 48 percent of delivered lines. At the same time, quality improved from a range of 7 to 20 faults per 1,000 lines of source code at the end of final test to approximately 2 or 3, with residual faults left after testing averaging between 0.2 and 0.05. NEC claimed improvements in productivity of 26 to 91 percent whilst reducing bugs by about one-third. Fujitsu said it cut bugs for all outstanding operating systems code (newly delivered code plus maintained systems) from 0.19 in 1977 to 0.01 in 1985, whilst productivity in lines of code per month rose by two-thirds between 1975 and 1983.

These numbers are difficult to compare without more careful analysis of exactly what each firm measured. Nonetheless, there is no doubt that the programmer training, management techniques, and tool support introduced at these software factories helped the Japanese develop highly complex software with high degrees of both productivity and reliability. These Japanese firms also carried forward these process skills into the present, though it took them years to learn programming for the Windows and Internet platforms. NEC became known as a leader in software quality assurance. Hitachi became known for zero-defect systems such as real-time reservation and control software for the Shinkansen bullet trains. Toshiba became a world leader in real-time automated process control software for electric and nuclear power plants. These firms also successfully transferred a good deal of their process knowledge to other divisions, such as industrial equipment, consumer electronics, machine tools, and telecommunications gear, all of which use great deals of embedded software that must be largely free of defects.

The Early Software Factories in Retrospect

Despite making significant progress in structuring software development and reducing defects, the early software factory efforts had their limitations. This is common for any major product or process innovation for a complex technology. SDC tried to apply a factory process when its work flow and the preferred work style of its project managers more resembled a job shop. Co-location with the customer helps to reduce misunderstandings or communication gaps in the high-level design process. But, if SDC had used the factory only for similar projects (as Toshiba did), it would probably have been more acceptable for managers to put their programming work through the factory. On the other hand, SDC learned that process is more reusable than actual technology (that is, code modules), and it effectively relied on the factory process even after the factory had closed. Hitachi and NEC seemed to have learned this lesson as well and today emphasize quality control and a repeatable process more than reuse. The Indian IT service companies are another story, though their methodology combines teams located on-site with the customer to define requirements with programming teams back in India for the more labor-intensive work.

The Japanese have also faced other obstacles in software development that they are only gradually overcoming. In particular, Japanese universities were slow to develop strong departments in computer science and information technology. Companies compensated for this lack of skills with software factories. The factories worked well whilst the underlying technology remained relatively stable. But, as the market moved to newer types of computing, Japanese companies (as well as Anderson/Accenture in the United States) had to raise the level of individual skill in their facilities and evolve their processes and tools more rapidly. The factory-like focus on process and tool standardization as well as reuse, and reliance on historical project management and quality data, did not transfer well when new software had to be built for the first time on non-mainframe platforms, especially with employees not trained in computer science or related technical fields. Over time, the Japanese adjusted to the world of client-server computing, the Internet, and mobile systems. But they still have a minimal presence in creating global products or establishing new technology platforms, except perhaps in mobile phone systems for domestic use, such as NTT DoCoMo's i-mode.[26]

Nonetheless, the major Japanese firms today develop all types of software for nearly every computing and communications platform, and they still lead in global surveys of software quality and lines of code productivity.[27] The factory approaches continue to service customers in the local Japanese market, with semi-custom systems. Japanese firms such as Nintendo and Sony are also strong in video-game development, which leverages historical skills in animation, cartoons, and comic books with newer technological capabilities in graphics processing and consumer electronic devices.[28] Except for video games, however, the big Japanese software producers have not focused much attention on creating standardized software products and platforms for global markets. This is very different from the strategies of Microsoft, Oracle, Adobe, and many other American companies or SAP and Business Objects in Europe.

India's IT Services Companies

Indian IT services companies share an intellectual heritage with Japanese software factories and are equally focused on exploiting scope economies in a technical service. But there are important differences. In particular, the Indian entrepreneurs decided not only to hire low-cost engineers and train them in rigorous software development techniques. They also took these organizational resources and made them the foundation of a blockbuster global strategy. The Indian companies can build and maintain nearly any type of business application as well as perform system integration, customize software packages, conduct training, and handle the outsourcing of various back-office operations.

What is more, the Indians did not simply adopt best practices in software engineering from US companies and add some of their own quality control and reusability concepts, as the Japanese did. Rather, in the 1980s and 1990s, they adopted more formally recognized US practices from the Software Engineering Institute (SEI), founded in 1985 at Carnegie Mellon University and initially funded by the US Department of Defense. The SEI practices are based on the five levels associated with the Capabilities Maturity Model or CMM (referred to as CMM-I, for the CMM-Integrated model, since 2002). The SEI practices derive from IBM's experiences through the mid-1980s, so it is no accident that the processes used in India closely resemble those of IBM as well as the Japanese. Most Indian firms deliberately set up their development organizations along the lines of the CMM or CMM-I Level 4 and 5 practices to make sure they would be highly evaluated by potential US clients. What India added to the factory model, however, was something that Japan lacked—a nearly unlimited supply of university-trained programmers and computer scientists who spoke English and worked for relatively low wages.

Of course, the Indian software companies have their own limitations. Not only are wages rising but so is competition from other developing countries or low-wage areas such as Russia and Eastern Europe that have created software outsourcing industries. The Indians also tend to see software primarily as a service and, even more than the Japanese, have lagged in "productizing" whatever original technology they possess and pursuing global product markets. At the same time, the Western product companies have seen their own sales decline. as "free software" and price pressures have become more common. Product firms such as SAP and Oracle are now aggressively pursuing the same IT services business as the Indian firms, which have generated a lot of revenue from installing, customizing, and maintaining the products built by American and European firms. As I noted in Chapter 2, these former "partners" have now become competitors.

Nonetheless, the Indian firms have successfully combined factory-like elements, such as standardized approaches to process and quality control as well as project management and extensive tool support, with more technically skilled people than in Japan, a much lower cost of labor, and excellent English language skills. The Indian companies could not adopt one single development process or toolkit, like some of the Japanese software factories, because they usually conform to the technologies and development approaches of their customers. But they added the structure and assurances that came with SEI certification (by far, most of the world's top-rated firms on the 5-level CMM-I scale are in India). Other books are available that relate the history of India's software business. Here I will briefly sketch out some highlights of how the industry evolved as another illustration of how firms can achieve efficiencies even in a technical service.

The person widely regarded as the pioneer of the Indian IT outsourcing business is Narendra (Naren) Patni.[29] A graduate of the Indian Institute of Technology, Roorkee, Naren came to the United States in the mid-1960s and obtained master's degrees from MIT in electrical engineering (there was no computer science department at the time) and management. In 1972, he and his wife, Poonam, established a company based in Cambridge, Massachusetts, called Data Conversion, Inc., with operations back in Pune, India. They initially took data on paper tapes used in old-style computers and re-input them on magnetic tape for modern computers. Meanwhile, Naren's family established another company back in India to import computer hardware and software. According to Indian laws at the time, a domestic company could import foreign computers and components only if it generated a certain amount of foreign exchange. This law provided an incentive for the Patni family company to take on more export work, including application development for Data General in 1976.[30] Other Indian entrepreneurs followed suit—establishing companies to convert data, and looking for export customers to fund the purchase of computer hardware and software products to resell in India. The Indian government also set up special export zones with tax exemptions to encourage this practice.[31]

I first met Naren Patni around 1988. He came to my office holding a working paper I had just written on the Hitachi Software Factory and deposited in the MIT library. He said something to the effect that, "I can do this in India." Though nowhere near the scale of the Japanese software factories, a few Indian companies were already starting to do this—offering maintenance and some application development services for clients, mainly in the United States and mostly connected to the hardware and software partners they represented in India. The Patni company would go on to distinguish itself mainly by offering low prices and cultivating close relationships with Fortune 500 clients. It would later adopt the CMM practices as well and, following its major Indian competitors, achieved the SEI's CMM-I Level 5 status in 2004.[32]

Although Naren Patni early on had the idea to provide IT development and outsourcing services, the family-owned company took several years to set up an independent software firm. That move finally occurred in 1999, when Patni Computer Systems, Ltd., was spun off and incorporated separately. The new firm went public on the Indian stock exchange in 2003 and then on the New York Stock Exchange in 2004 (NYSE symbol: PTI). For calendar year 2009, Patni had approximately $656 million in revenues and 14,000 employees. (Full disclosure: I became a director of Patni Computer Systems in 2004.)

The original Patni company also spun off the founders of what has become India's premier IT services company—Infosys (NASDAQ: INFY). In 1981, seven engineers, including Narayana Murthy, Nandan Nilekani, and S. "Kris" Gopalakrishnan, decided to leave Patni and establish their own company.[33] They set up a US office in 1987, but the company grew slowly until the later 1990s, when the boom started in Internet technology and Y2K spending. Infosys then established a European office in 1996 as well as launched a series of new IT practices, including e-business (1996), engineering services (1997), enterprise solutions (1998), and IT consulting (1999). The company reached $100 million in revenue in 1999, the same year it received SEI's CMM Level 5 certification, and then kept doubling revenues every few years.[34] For fiscal year 2010, Infosys had approximately $5.9 billion in revenues and 114,000 employees—a remarkable growth story.

The other major Indian players include India's largest IT firm, Tata Consulting Services, Inc. (TCS).[35] This was founded in 1968 to provide IT services to other Tata group companies. F. C. Kohli, who received a master's degree in engineering from MIT in 1950, later became general manager of TCS soon after it was founded. He began the software outsourcing business in 1974. In fiscal 2010, TCS had approximately $6.3 billion in revenues and 160,000 employees. Another more diversified company is Wipro, founded in 1947. This firm moved into the computer hardware business in 1977 under the leadership of the founder's son, Azem Premji, when the Indian government forced IBM to leave the country. Wipro expanded into software development and IT outsourcing in 1980. For fiscal 2010, Wipro (NYSE: WIT) had approximately $4.4 billion in IT revenues and at least 80,000 IT employees, including a large business process outsourcing (BPO) division.[36] In 2009, two other major players with global reputations were Mahindra Satyam (founded in 1987 as Satyam Computer Services and racked by a financial scandal in 2008–9, leading to a merger with Tech Mahindra) and HCL Technologies (founded in 1976).

Infosys provides the best example of how profitable and dynamic these Indian software companies can be. It charges a premium for its services and operates with high levels of efficiency and flexibility, even though Infosys's primary business is custom software development and maintenance. Only about 5 percent of 2010 revenues came from product sales.[37]

The development process closely follows the CMM-I Level 5 guidelines, but projects adapt to either waterfall or iterative development styles, depending on the customer requirements.[38] As in the Japanese factories, there are well-defined repeatable systems and procedures, with excellent tool support and automation for project planning, change control, configuration management, quality assurance, technology standards management, design reviews, code inspections, and testing, as well as specific procedures for process improvement research, statistical measurement, defect prevention, and knowledge sharing across projects. Like the other Indian firms, Infosys has packaged these particular tools and methods into its own well-defined and repeatable "Global Delivery Model."[39]

Infosys's global delivery model emphasizes *distributed teams* to develop software. Projects in the same domain try to follow a common process and deploy the same tool set. They also rely on Infosys's base back in India, where the company located development centers in approximately a dozen cities. At the customer site, Infosys teams generally do system analysis and planning as part of the requirements generation stage, along with some prototyping, high-level design, and user interface design. Back in India, other project members take over detailed design, additional prototyping, coding, unit testing, and documentation, as well as make bug fixes and provide long-term maintenance support. Customers usually receive weekly status reports on their projects by email. Final adjustments and testing occurs back at the customer site. IBM, Accenture, Tata, Wipro, Patni, and other IT services companies in India and elsewhere followed a similar process for global delivery. However, Infosys stands out among the Indian firms for "thought leadership" and the innovation capabilities of its teams. Infosys has also more aggressively set up delivery centers outside India.

Compared to the Japanese software factories, the Indian companies placed less emphasis on component or design reuse. The Japanese, led by Toshiba, did relatively well in this area, probably because their roots were as product manufacturers. They also had large volumes of work and many similar projects channeled into a few large facilities. The Indian software companies, by contrast, achieve economies of scope mainly by reusing methods and frameworks across similar projects for different customers. In addition, most of the Indian firms, as well as other leading IT services firms in the United States, Europe, and Japan, have been increasing their investments in what we might call partial products or design frameworks for particular industries—also referred to as "vertical solutions." These might include a design framework to help an IT services company build a new inventory control system or create a new application that easily integrates with SAP's financial module. The semi-products or design frameworks are not full reuse, but they are a way to improve efficiencies across different projects—again, achieving economies of scope, rather than scale.

From "Lean" to Multi-Project Management

In the software factories of the United States and Japan, as well as in IT services companies in India and elsewhere, there is potentially great tension in the search for scope economies. For one-of-a-kind systems, customers usually like to have a dedicated project team at their site at least part of the time they are building unique technology. Even for mass-market products, project managers and engineers generally like to have the freedom to build whatever technology seems right for the market. On the other hand, to gain efficiencies at the level of the firm or within divisions of the firm, senior managers need to introduce some centralized coordination and reuse of knowledge or technology above the level of the individual project. In other words, firms need to find the appropriate balance between centralization and decentralization, and construct their own contingency frameworks to help mid-level managers decide when to use which approach.

We know from various studies that the most successful product-development projects often take place in relatively small, focused autonomous teams—led by what Kim Clark and Takahiro Fujimoto have called the "heavyweight" project manager.[40] Indeed, Toyota pioneered the use of heavyweight project managers during the early 1950s. This approach helped it leapfrog Nissan and other competitors. Honda also adopted this style of development when it began designing automobiles in the 1960s and continues to use it to great advantage. A number of authors and research programs, including the MIT Lean Advancement Initiative, have focused on ways to combine the heavyweight project manager idea with the general principle that Taiichi Ohno emphasized of eliminating waste through process rationalization and improvement.[41] This "lean" development style offers many advantages compared to the function-oriented style for product development used in American and European automakers (Table 5.3).[42]

Table 5.3. *Lean versus functional product development*

"Functional" management	"Lean" management
Slow model replacement	Rapid model replacement
Infrequent model-line expansion	Frequent model-line expansion
Some radical product improvements	Incremental product improvements
"Lightweight" project coordinators	"Heavyweight" project managers
Sequential, long phases	Overlapping, compressed phases
High levels of in-house engineering	High levels of supplier engineering
Department member continuity	Design team/project-manager continuity
"Walls" between departments	Good communications across functions
Narrow skills in specialized departments	Cross-functional teams

Source: Adapted from Cusumano and Nobeoka (1998: 5).

Traditional "functional" management of product development generally organized engineers into departments with long sequential schedules to replace existing models or add new ones to the portfolio. This structure was good for making relatively "radical" innovations that leveraged the accumulation of narrow but deep functional skills and a focus on in-house engineering rather than outsourcing, such as for new engine technology or new body designs. But these types of projects often had only weak coordination across the different functional areas because functional departments retained most of the power over budgets and personnel. By contrast, lean management of product development at Toyota, Honda, and other Japanese companies, especially since the 1980s and early 1990s, was generally organized in dedicated projects with members assigned from different engineering and manufacturing functions. The "heavyweight" project managers led the teams and had extensive decision-making authority over budgets, personnel, styling, and marketing. This authority allowed the projects to follow short, streamlined schedules, with a number of compressed phases (such as concept development) as well as overlapping phases (such as design and manufacturing preparations). The Japanese also outsourced a lot of design to suppliers who worked in parallel as well. The overall goal was to focus limited R&D resources and quickly to introduce new models and replace existing models, to help them catch up with foreign competitors.

Clark and Fujimoto collected data that showed how the Japanese automakers successfully applied these lean management concepts to product development. The Japanese firms in the late 1980s produced new models in about 45 months with 1.7 million engineering hours, as opposed to 60 months and over 3 million engineering hours at US and European automakers.[43] The gap narrowed in the 1990s, as the Japanese slowed down to spend more time on design and move upscale to build Lexus, Acura, and Infiniti luxury models, as well as to coordinate projects, reuse components, and manage their product portfolios more effectively. By this time, though, Toyota, Honda, Nissan, and other Japanese automakers had caught up if not surpassed most of their competitors in product reliability as well as design quality.

Even in the early 1990s, however, it was clear that the Japanese automakers had run up against some "limits to lean" in product development, as well as in manufacturing, and were shifting to economies of scope strategies. Honda management, for example, realized that it was spending too much money on R&D because the isolated heavyweight teams were not reusing components developed in other projects, such as Honda's innovative engine, steering, and suspension system technologies. Toyota management similarly concluded that lean product development had resulted in too many product variations and too frequent model replacements.[44] Both companies gradually modified their approaches. What emerged in this next stage beyond lean was a more coordinated approach for managing multiple projects simultaneously, with mostly positive and some occasionally negative consequences.[45]

Kentaro Nobeoka and I called this new strategy "multi-project management." It differs from simply using a functional organization or a traditional "lightweight" matrix to make engineering departments design components for different products. The latter scale-oriented approach resulted in the outcome GM experienced in past decades, when most of the products in its half-dozen or so divisions all looked essentially alike. But pre-bankruptcy GM, with the Chevrolet, Buick, Oldsmobile, Pontiac, Cadillac, and GMC product lines, was not alone. We saw most large automakers struggle to balance reuse of designs and components with the desire to have different products with different branding—Ford, with its Ford, Lincoln, and Mercury lines; Chrysler, with its Plymouth, Dodge, and Chrysler lines; Fiat, with its Fiat, Lancia, and Alfa Romeo lines; Honda, with its Honda and lower-end Acura lines; Nissan, with its Nissan and lower-end Infiniti lines; and Volkswagen, with its VW and Audi lines. Even Toyota has based its low-end Lexus products on Toyota Camry/Highlander models—and this was at times too obvious to customers. It is a problem common as well in other engineering and manufacturing industries, such as consumer electronics, where it often is not clear how to distinguish different phone, television, or audio-equipment models.

Multi-project management tackles this problem of balancing reuse versus differentiation with a simple strategy. (1) Create new products that share key components, but design the components upfront to be shared. And (2) organize separate development teams in separate projects with their own managers and personnel, coordinated by a senior

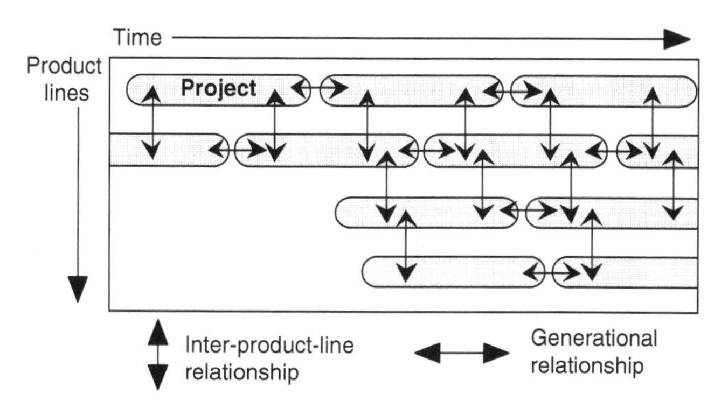

Figure 5.2. A simple model of multi-project product development

Source: Cusumano and Nobeoka (1998: 2).

executive with a portfolio strategy for product development. In this approach, project managers become more "middleweight" than heavy-weight because they report to a senior executive and do not have complete authority over matters such as budgets or which technologies to use. There is some but not total centralization. Figure 5.2 illustrates how we envisioned the multi-product firm as managing a portfolio of chronologically overlapping and interrelated projects that shared key components.

The project separation and multi-project portfolio planning helps ensure that each product will be distinctive and potentially attract different customers, as well as contain a mixture of unique parts and strategically reused components. We also believed it would be easier to manage multiple projects if they overlapped in time because this would facilitate communication and mutual adjustments in the designs. Indeed, we studied nearly every automobile built during a twelve-year period and found that companies following the multi-project approach achieved remarkable improvements in sales and market share because customers appear to respond well to distinctive products with new technology. Companies also experienced dramatic savings in engineering costs. At least during the period of our study, we did not uncover cases of faulty components being reused across multiple projects, though, no doubt, this has occurred on occasion, just as Toyota experienced in 2009–10 with faulty gas pedals and braking software.

The Research and Strategic Framework

To analyze project performance, we first sorted automobile products into four types, based on the age of the underbody platform and chronological overlap between the timing of the initial project and subsequent projects that reused the platform. We measured platform change by analyzing the underbody and floor panel specifications (wheelbase and track) as well as the suspension system and its design. Researchers can use this type of analysis in many different industries to identify reuse of an in-house platform. As shown in Figure 5.3, Type 1, or *new design* projects, developed the platform primarily from scratch. Type 2, or *concurrent technology transfer* projects, reused the platform

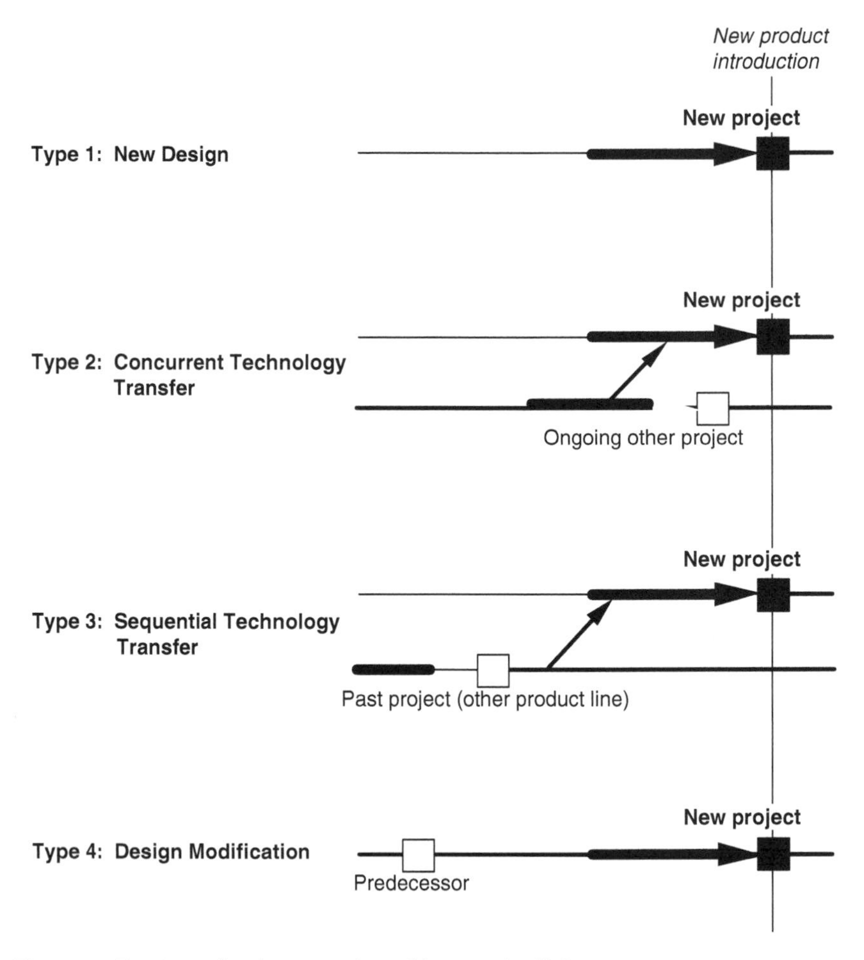

Figure 5.3. Typology of project strategies and inter-project linkages

Source: Nobeoka and Cusumano (1995: 398).

from an earlier project that was still ongoing. Type 3, or *sequential technology transfer* projects, reused their platforms but took them from projects already completed. Type 4, *design modification* projects, carried over platforms from the previous models with minimal changes.[46]

Automakers that relied more often on the concurrent transfer strategy did significantly better than firms relying more heavily on the other strategies. When they developed a new product platform, they quickly leveraged it in another ongoing project. Firms that relied on new designs to refresh their product line-up had to expend more engineering resources to produce new models. Firms that sequentially transferred platforms or relied on design modifications ended up with much older technology in the marketplace.

Table 5.4 summarizes the most important performance results. Most striking was that, when firms relied more on concurrent technology transfer, they showed sales growth over a 3-year period of 37–42 percent more than firms relying on sequential transfer or new designs and 68 percent more than design modification strategies. This finding is especially important because the concurrent transfer projects used about half the engineering hours (actually 33–64 percent less, depending on whether we adjust the data for variables such as project complexity) than projects following one of the other three strategies. Concurrent projects were also faster than the new design projects by 12–17 percent.

To analyze company performance, we collected data on all new products, totaling 210, introduced by the 17 largest automobile manufacturers between 1980 and 1991. For this part of the study, we relied

Table 5.4. *Performance of the concurrent technology transfer strategy*

Metric	Comparison strategy	Performance
Sales Growth (over 3-year period)	New design	42% more
	Sequential technology transfer	37% more
	Design modification	68% more
Engineering hours (per average project)	New design	45–62% less
	Sequential technology transfers	33–64% less
	Design modification	40–63% less
Lead time (per average project)	New design	12–17% less

Note: The ranges for engineering hours and lead time depend on whether one uses raw data or data adjusted for factors such as project complexity.

Source: Cusumano and Nobeoka (1998: 15), based on tables 5.1, 6.2, 6.5, and figure 5.3.

primarily on public data, but we also did interviews with 130 engineers and project managers, primarily from Toyota, Nissan, Honda, Mitsubishi, Mazda, GM, Ford, Chrysler, Fiat, Volkswagen, Renault, and Mercedes. We divided the data into four 3-year time periods—long enough to see the impact of a new product on sales. Model variations designed in a single project, including products with different brand names (such as the Ford Taurus and the Mercury Sable), we counted as only one product. We characterized company strategies by the percentage of products introduced in each 3-year period that followed one of our four strategy types, and calculated product revenues by using average price data. Appendix II, Table II.6, summarizes the data.

Placing firms within a strategic group allowed us to examine market performance as well as some characteristics of their product portfolios during these 3-year time periods. Not surprisingly, when firms relied on new designs, they had the lowest average platform age—0.37 years, in contrast to 5.54 years for design modification strategies and 3.75 years for sequential transfers. Concurrent transfer strategies had an average platform age of 0.91 years. The other important metric is the new product introduction rate, which is the percentage of new products the firm introduced during the 3-year period. Firms relying on new designs replaced the fewest products—only 45 percent—compared to 69 percent when firms followed the concurrent transfer strategy.

This latter finding is significant because there is a strong correlation between the number of new products and sales growth. We can see this in the market-share growth number compared to the previous 3-year period for each firm. New design strategies corresponded to only a 3.4 percent share increase over the previous 3-year period, despite the newness of the products, probably because this strategy was expensive and time-consuming. More importantly, when firms relied primarily on design modifications, they *lost* 15.6 percent in market share from one period to the next. Firms did better when they adopted a scope strategy and leveraged a new platform in more than one project. Doing this concurrently was associated with a remarkable 23.4 percent increase in market share! For example, if a firm had a market share of 10 percent in the prior 3-year period, a 23.4 percent increase would mean that its share rose to 12.34 percent in the subsequent period.

The impact of these different development strategies was even more striking when we consider project performance.[47] We sampled 103 projects at 10 automakers in Japan and the United States, where we were able to get enough cooperation to complete our survey. As expected, the concurrent transfer projects were the most efficient: they had the newest technology as well as consumed the fewest engineering hours because they reused platform technology from a new design project. As seen in Figure 5.4, the lead times for the concurrent, sequential, and design modification projects were similar, but there were substantial differences in lag times. The concurrent projects started an average of 15 months after the base project. This overlapping allowed engineers to make adjustments so that the shared technology suited the new project and was not simply forced into the design or reused by engineers who did not understand the original design, which often occurred with sequential transfers. The average lag time in reusing the platform for sequential projects was nearly 67 months and over 82 months for design modifications.

To analyze engineering hours as well as lead times more precisely, we collected data to adjust for project content, such as cars versus trucks, price, and task complexity (number of body types in each project,

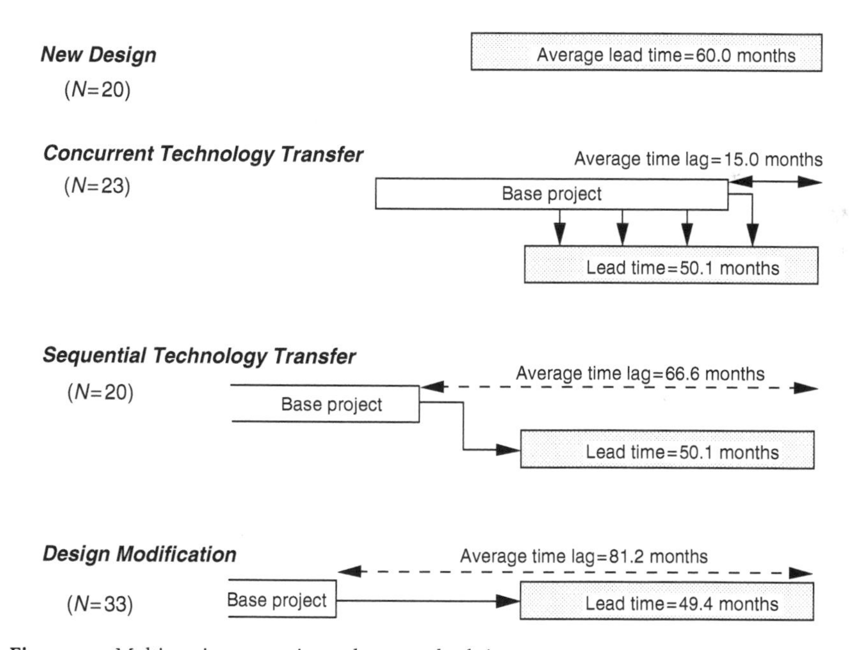

Figure 5.4. Multi-project strategies and average lead times

Source: Nobeoka and Cusumano (1995: 400).

percentage of new design for the engine and transmission as well as the body and the interior, and innovativeness of designs based on manager assessments). Figure 5.5 shows the results. The concurrent projects (which do not include engineering hours allocated to the base project) required 1.2 million engineering hours and had an adjusted lead time of about 51 months. This compared to 2.2 million for new design projects (which had a lead time of 58 months) and 1.8 million for sequential transfers and 2.0 million for design modification projects, both of which had adjusted lead times of about 51 months. The conclusion is clear: given the market data, it makes no sense for firms to introduce design modifications. Managers should plan one or more concurrent transfer projects with *every* new platform design. This shows economies of scope in action!

Organizational Options for Scope Economies

Economies of scope as seen in the concurrent transfer strategies did not just happen; this strategy required considerable planning and

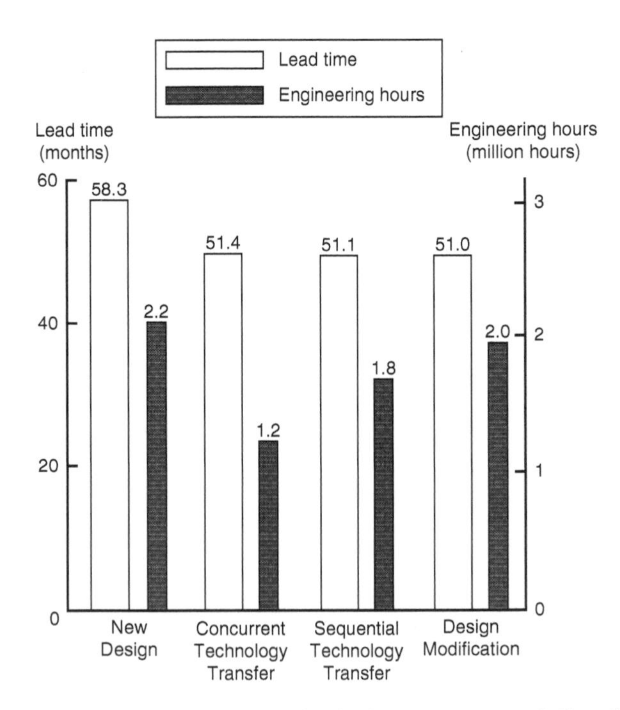

Figure 5.5. Lead time and engineering hours for the four strategy types (adjusted)

Source: Nobeoka and Cusumano (1995: 402).

coordination. The approach also seemed to work better with particular organizational strategies. Advance planning was essential to develop a portfolio of projects and coordinate the design and reuse of components from a base project. It was equally important to overlap the base project and the concurrent transfer projects to facilitate communication across the project teams, make adjustments in designs as necessary, and share the engineering work. Engineers from two overlapping projects—for example, a Toyota Camry sedan and a Lexus or Toyota Highlander SUV—might jointly design a new suspension system or door structural components.

Reusing new technology was also much better than reusing off-the-shelf designs that were five or more years old, as in the sequential and design modification projects. Older technology often required almost as much re-engineering work to reuse as developing new technology from scratch. The reason seems to be that experts on the old technology were often no longer around after five or six years, so it was difficult to consult them when attempting to reuse old technology.

Finally, to make the concurrent transfer strategy work, firms needed to have an executive in charge of the multi-project portfolio and to coordinate the individual project managers. Conflicts could easily arise when, say, a base project had to take extra time to design and test components for reuse in another project. If executives evaluate the manager of the base project simply on budget and schedule for that one project, with no adjustments for helping other products, then project managers will not have the proper incentives to cooperate. This was part of the problem that SDC encountered with its software factory.

Obviously, multi-project management places some constraints on project managers and engineers; they are not free to do anything they wish, as in a heavyweight project management system. There is still ample room, however, for team members to experiment and iterate in joint designs as well as in components or subsystems unique to each product, such as the outer body shell or the vehicle interior—the most visible components to the customer.

All the auto companies we sampled used all four of the product development strategies at one time or another, with the exception of sequential transfer, which the Japanese did not use during the period under study. The US companies also used the concurrent transfer strategy only once. To handle these different projects, automakers followed a variety of organizational approaches, as summarized in Figure 5.6. Most firms (13 out of 20) used a traditional matrix, where engineers in functional departments contributed to projects coordinated by a liaison manager or a project management office. Most of the authority over personnel and budgets in this structure lay with the functional managers. Two companies (Honda and Chrysler, which copied Honda) were organized in product teams, with extensive authority over personnel, budgets, schedules, and content given to a single "heavyweight" project manager. A few companies (GM, Nissan, and Mazda briefly) organized "semi-centers" where they managed some engineering work through functional departments serving all projects, such as for engine or drive-train development. They divided other work into projects (such as body design) under development centers. This hybrid structure made it easier for projects to share engineering work if the firm had enough projects. The fourth option was the pure center organization (Toyota and Ford), where firms broke up their work into

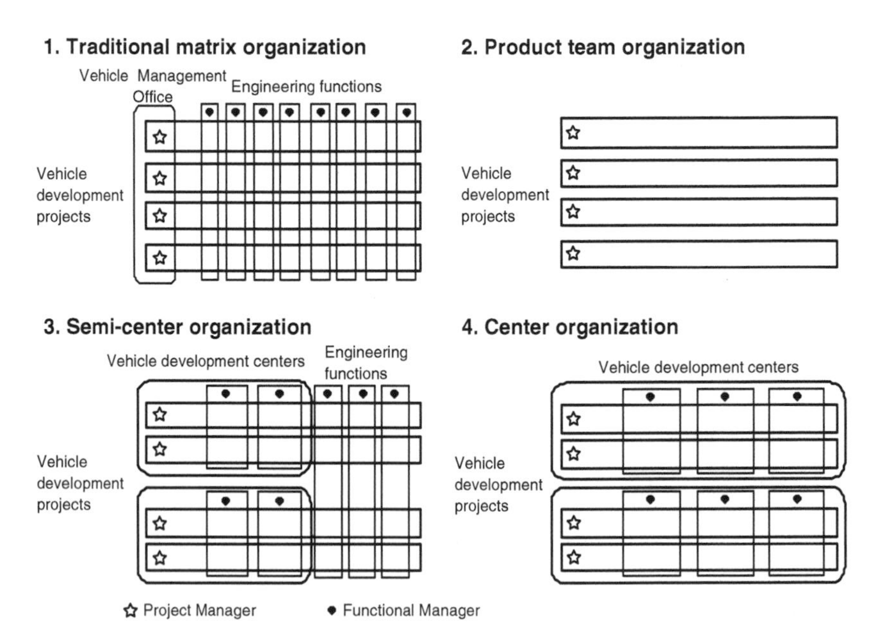

Figure 5.6. Four types of product development organizations

Source: Cusumano and Nobeoka (1998: 53).

clusters of similar projects and duplicated engineering functions as necessary. The center and semi-center organizations were appropriate only for firms with a lot of projects, especially if they were geographically dispersed and found it difficult to coordinate.

Figure 5.7 illustrates a multi-project management organization in the form of what we called a "differentiated matrix." We found structures resembling this concept in firms that used a matrix, such as Mitsubishi, as well as in firms that used a center organization, such as Toyota. Some component teams work on one project (for example, the body design team for Project A or the interior design team for Project C), others work on two (for example, the chassis design team and the interior design team for Projects A and B), and others work on more than two (for example, the body structural components team, as well as the engine and electronics teams, all of which work for Projects A through D). The differentiated matrix brings together the best of both worlds—functional expertise in the functional departments and product teams in a dedicated project that can infuse the design with a distinctive personality and market target.

Companies that had to coordinate many projects found it easier to break out the projects into clusters, and then organize around semi-centers or centers. Toyota, for example, started the heavyweight project

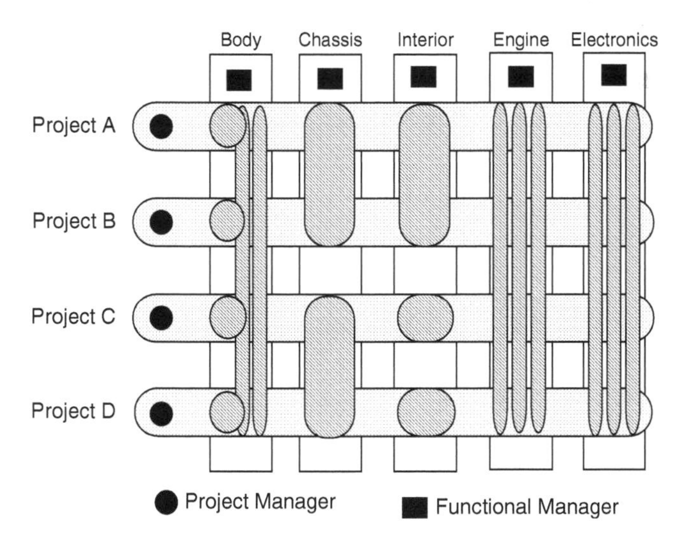

Figure 5.7. Simplified example of a differentiated matrix

Source: Cusumano and Nobeoka (1998: 67).

manager system (it called these managers *shusa*) in 1953 and formalized this structure in 1965.[48] But, by the early 1990s, it had as many as fifteen projects going on simultaneously—too many to coordinate easily. To resolve this problem, in 1992 Toyota introduced a new organization consisting of three product development centers: one for front-wheel drive vehicles (low-cost economy products), one for rear-wheel drive (higher-quality luxury products), and one for trucks and vans (commercial and recreational products). Each of these development centers worked on about five projects simultaneously. Toyota also established a fourth center to create reusable components for the other centers. Overall, Toyota reported that the new structure helped reduce average product development costs by 30 percent and lead times by several months. The improvements mainly came through reducing the need for prototypes by 40 percent as well as increased component sharing, better coordination across the different engineering and testing departments, more simultaneous or joint engineering across projects, and more use of computer-aided engineering (CAE) tools.

Again, I must point out that designing components for reuse can work extremely well in terms of sharing good technology and saving manufacturing and engineering costs. But the quality of the components must be rock solid—otherwise there is a proliferation of flawed parts in many different products. Toyota clearly experienced this potential downside of multi-project management with its massive recalls in 2009–10. Problems with the gas pedal design alone, for example, affected some eight million vehicles across a wide range of car, truck, and SUV products—including Camry, Corrolla, RAV-4, Highlander, Lexus, and Tundra models.

Comparisons between Autos and Software Product Development

When using the differential matrix, managers and engineers have to figure out which components they can fully standardize and reuse unchanged across more than one project and which components or subsystems they need to tailor to each project or perhaps to a cluster of similar projects. This means that multi-project management is more complex than dedicated lean teams because the firm needs to coordinate across functional departments as well as project teams. Figure 5.8 presents a decision-making framework for dealing with this challenge.

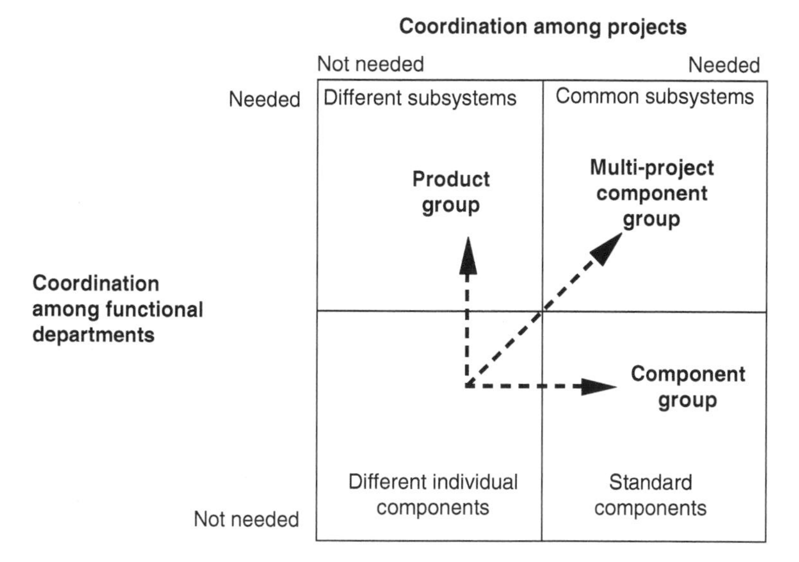

Figure 5.8. Impact of component characteristics on the differential matrix organization
Source: Cusumano and Nobeoka (1998: 170).

The basic idea is that firms should be able to identify four types of component groups or development teams: those where there are different individual components, different subsystems, common subsystems, and standardized components. Again, our research context was automobile product development, but managers can apply this kind of framework to software engineering or other forms of product development. We can even think about service design this way: some service "packages" can share common elements, whilst others need to differ depending on the customer or the type of service contract.

In the case of an automobile, the interior and exterior components generally differ for each product and are often the first things that customers notice. Companies should design these kinds of distinctive subsystems in separate product groups. But standard subsystems and components such as audio equipment, seats, or batteries can be common across multiple products and require relatively little coordination. Separate component groups can easily design these kinds of parts. Other subsystems that are expensive and highly desirable to share, such as underbody platforms, engines, or braking systems, require higher levels of coordination among different engineering groups as well as project teams. In these cases, firms can create multi-project groups. Japanese software factories such as Toshiba used this multi-project approach when they planned the reuse of design frameworks and individual components for multiple customers, even though they were selling "custom-built" systems. In personal computer software, Microsoft also coordinates multiple projects and does concurrent technology transfer when it develops the Office suite.[49]

The standard Office product consists of several subsystems that Microsoft formerly designed, built, and sold as separate products: Word (word-processing), Excel (spreadsheets), PowerPoint (graphical presentations), and, more recently, Outlook (email and calendar client). The original desktop applications (Word and Excel, followed by PowerPoint) since the mid-1990s have been sharing about half their components. But this sharing did not happen until Microsoft deliberately created the Office business unit and formed component groups to build common parts and coordinate requirements across the separate product teams. For example, even before the creation of the Office suite, the Word, Excel, and PowerPoint products all did text processing, printing, cut-and-paste, file management, and some graphics. The individual product groups built these shared components. Because the Word group had the most experience with text processing, file management, and printing, in the mid-1990s Microsoft management (in a sharing effort initiated by Bill Gates) asked this team to build these modules for the other Office product groups as well as other application product units. The Excel group had the most experience with graphics, so it built this module for itself and the other groups. Microsoft also found that, if too many groups were involved in deciding the design requirements for the common component, then the design task became too complex to manage. Therefore, one group always took the lead. The lead group then became responsible for consulting the other groups. If there were disagreements on the requirements, then the three most important groups—Word, Excel, and Office—each cast a vote to decide.

When it came to components that other product groups could reuse with no change in the design, such as a standardized printing module or file management module, Microsoft simply used component groups (see Figures 5.7 and 5.8) that needed little or no coordination with other functional (component) departments. If the components required special features or design changes, then Microsoft used multi-project component groups. For example, users who do text processing in Excel and PowerPoint do not need all the features available in Word, so the Word group had to consult with the other product groups as well as design the standard text-processing module so that different product groups were able to cut features according to their needs. Each product in the Office suite also had some unique components exhibiting major interdependencies with other subsystems from different groups, like the presentation wizard in PowerPoint. The PowerPoint product team could build this feature on its own but had to make sure this technology worked with other features imported from the other groups. Design reviews, daily builds, and integration and system testing were all useful ways to coordinate requirements and synchronize component development. Overlapping projects, as in concurrent technology transfer, was also useful in software development to coordinate changes, make adjustments in real time, and share the engineering work.

Lessons for Managers

First, economies of scope can be extremely valuable to firms for operations with no obvious economies of scale but lots of opportunities to share technology or specialized knowledge. This is really the best way, other than automation, to create an efficient and effective organization for product development or professional services. Second, managers must be up to the task of managing a more sophisticated organization that revolves around sharing technology and expertise, as well as finding the middle ground when conflicts or tradeoffs arise. If your company organizes by business units that compete for profits and often fall into turf battles over how to allocate revenue and costs, then you are unlikely to have much success with economies of scope. This, of course, leaves such firms wide open to competition from companies that master a scope strategy, overcome the organizational or technical tradeoffs, and more effectively use company resources and capabilities.

For example, project teams give up some autonomy and flexibility when senior managers ask them to share components or standardize processes. Taken to the extreme, customers will probably complain as well if the resulting products do not meet their needs or if the scope strategy proliferates a poorly designed or unsafe component. But uncoordinated projects also do not leverage good technology or good ideas across multiple projects, for multiple customers. It follows that something in between an efficiency-oriented structure, like functional engineering departments, and a creativity-oriented structure, like a heavyweight project team, may be the best alternative most of the time for firms in software, automobiles, and many other industries. Software factories and multi-project management in automobiles both rely on a differentiated matrix, with senior executives overseeing clusters of related projects and component sharing. They represent intermediate approaches and work well when executed well, particularly when the technology is relatively stable.

It also appears that very new technologies, or technologies undergoing disruption, are not well suited to scope economies (that is, to software factories or multi-project management) until the technology reaches some level of stability and predictable performance. This seems to be one reason why Japan had difficulty transitioning its mainframe software factories and an emphasis on reusable components to PC and Internet software. By contrast, the Indian IT firms have been successful for a wider variety of projects by focusing on reusing knowledge or domain expertise rather than components. It may also explain why multi-project management seems to have worked well in the automobile industry with relatively mature components and not so well with new technologies, such as use of sensors and software to replace mechanical parts. Reuse has also worked well in relatively stable segments of PC software, such as features in desktop applications (Microsoft Office).[50]

But it clearly seems a waste of engineering resources and a missed market opportunity not to share new technology across as many products as possible and whilst the technology is fresh and customers are more likely to respond positively. In automobiles, the concurrent transfer strategy and the differential matrix are effective ways to do this as long as quality is under control. There were no such benefits when firms pursued only new designs, and limited benefits when firms reused older technology, as in sequential transfers and design modifications.

When delivering professional services, several aspects of the software factory and multi-project management also seem broadly applicable.

Domain focus is important because this raises the likelihood that a firm will encounter similar customer problems and see opportunities to repeat tasks, standardize processes, reuse knowledge, or even introduce automated or semi-automated ways of interacting with customers, along the lines of Google, e-Bay, Amazon, and other Internet service companies. Clayton Christenson and his co-authors in their 2009 book *The Innovator's Prescription* have made very similar recommendations for rationalizing healthcare services and the hospital system in the United States.[51]

Finally, the practical question: if scope economies have such obvious benefits, why do not *all* managers and firms embrace this principle? First, a company needs to be large enough to have multiple products that can share components or reuse other forms of knowledge and technology. Many firms fit this requirement. But then managers need to go another step. They need to rethink their product or service portfolios as well as their implementation strategies. They will probably also have to restructure their organizations and incentive schemes. As with the cultivation of distinctive capabilities, not all managers will want to make this effort and undertake the risks of extensive sharing, unless the competition forces their hand. Senior managers also have to preside over a more complex organization that deliberately creates a tension between centralization and decentralization, and then votes for both—such as by imposing multi-project managers over project managers and adopting a "middleweight" project manager structure. Many managers dislike matrix structures because people then have multiple bosses and need to share authority and make joint decisions. This is true. But the Japanese (with their software factories and automobile companies) and Microsoft (with the Office product groups) have shown that it is possible to set up a process where there is primary and secondary authority within a matrix, so it is clear who is in charge of what.

In a more strategic sense, managing economies of scope well is really about achieving balance—between centralization and decentralization, structure and creativity, efficiency and flexibility. I have relied on technical examples in this chapter because they are how I discovered the importance of scope. But managers need to see this principle in their own context. Most important is to think about what is best for the whole organization over the long term versus what is best for a particular project or an individual product. Firms, if they are to use their resources effectively, need to manage portfolios of projects and learn how to share and reuse technology and other forms of knowledge, and maintain proper levels of design and production quality. Flexibility, rather than just efficiency, is very much a part of scope economies, though flexibility is also a much broader idea. That is why it is the subject of the next chapter.

Notes

1. The classic study on related versus unrelated diversification at the corporate level is Rumelt (1974). Other key studies on corporate diversification include Chandler (1962), Bettis (1981), and Ravenscroft and Sherer (1987). For a literature review, see Ramanujam and Varadarjan (1989).
2. See Abernathy and Wayne (1974) and Abernathy (1978).
3. Chandler (1990: 24).
4. See, e.g., Teece (1980) and Panzar and Willig (1981).
5. For an econometric analysis of knowledge spillovers in pharmaceutical R&D, see Henderson and Cockburn (1996).
6. See, e.g., Helfat and Eisenhardt (2004). For economies of scope in services, see Leggette and Killingsworth (1983) and Hallowell (1999).
7. My thanks to Bettina Hein for reminding me of open source software as an example. For consulting firms as an example of scope in services, see Hansen, Nohria, and Tierney (1999).
8. The following sections are based mainly on Cusumano (1989b), and more detailed treatments in Cusumano (1991b), as well as two articles: Cusumano (1991a, 1992).
9. See Hounschell (1985). This list is from Cusumano (1992: 453).
10. I first reported on this survey in Cusumano (1989b: 27–8). For a detailed discussion of the methodology and results, see Cusumano (1991b: 447–55, appendix A).
11. In another survey, in 1990, Chris Kemerer and I found similar differences in reuse rates between Japanese and US projects. See Cusumano and Kemerer (1990).
12. Woodward (1965); Perrow (1967); Mintzberg (1979).
13. Hayes and Wheelright (1984); Schmenner (1990).
14. Cusumano (1991b: 30; 1992: 462).
15. This section also follows the discussions in Cusumano (1989b), as well as (1991a, b, 1992).
16. For some direct comments, see Bemer (undated).
17 Bemer (1969).
18. I fondly recall being invited to give a talk at GE in 1988 on my software factory research by Don McNamara from Corporate Information Technology. Don later put me in touch with Bob Bemer.

19. McIlroy's biography can be found at www.cs.dartmouth.edu/doug/biography (accessed Mar. 28, 2009).
20. McIlroy (1969).
21. See a collection of survey summaries in www.it-cortex.com/Stat_Failure_Rate.htm (accessed Mar. 28, 2009).
22. Accounts of the SDC factory are based on Bratman and Court (1975) and Bratman and Court (1977). Also, interview with John B. Munson, Oct. 4 and 5, 1987. There is also a self-published history of the company. See Baum (1981).
23. Cusumano (1991b: 137).
24. Cusumano (1991b: 102–7).
25. Some of the articles in English discussing the practices of these Japanese firms include the following: Kim (1983); Mizuno (1983); Matsumoto (1984, 1987); Tajima and Matsubara (1984); Kishida et al. (1987).
26. For a discussion of NTT DoCoMo, see Gawer and Cusumano (2002: 214–29).
27. See, e.g., Cusumano et al. (2003).
28. Aoyama and Izushi (2003).
29. For the most detailed account of the story of Patni and other Indian software companies, see Sharma (2009), esp. pp. 257–64 for the Patni story. Also Karki (2008: 168–77). See also Weisman (2004).
30. Interview with Narendra Patni, Dec. 18, 2008, Cambridge, MA.
31. An excellent discussion of the various laws promoting development of the Indian software industry is Athreye (2005).
32. Patni Computer Systems, Ltd. (2004).
33. See Sharma (2009: 264–71).
34. Infosys Technologies, Ltd. (2003).
35. See Sharma (2009: 271–9).
36. A useful book on Wipo is Hamm (2007). For the latest financial results see www.wipro.com.
37. Infosys Technologies, Ltd. (2010).
38. Cusumano (2004: 188–90).
39. See www.infosys.com/global-sourcing/global-delivery-model/default.asp (accessed May 27, 2009) for more details.
40. See Clark and Fujimoto (1991).
41. See http://lean.mit.edu and Walton (1999).
42. Cusumano and Nobeoka (1998: 5). This table is also reproduced in Walton (1999: 15).
43. Cusumano and Nobeoka (1998: 7), from Clark and Fujimoto (1991).
44. Cusumano (1994: 28–9).
45. See the discussion in Cusumano and Nobeoka (1998: ch. 1).
46. This section on firm-level performance relies on Cusumano and Nobeoka (1998: 137–56), which largely follows Nobeoka and Cusumano (1997).
47. This section is based on Cusumano and Nobeoka (1998: 115–36), which relies primarily on Nobeoka and Cusumano (1995).

48. For a detailed discussion of the Toyota case, see Cusumano and Nobeoka (1998: 19–49).
49. This example of Office is from Cusumano and Nobeoka (1998: 171–2). For a more complete discussion, see Cusumano and Selby (1995: 384–97), as well as Staudenmayer (1997) and Staudenmayer and Cusumano (1998).
50. My thanks to Michael Bikard for his comments on the potential relationship between the maturity of the technology and the feasibility of achieving scope economies such as in software factories or automobile product development.
51. Christensen, Grossman, and Hwang (2008).

6

Flexibility, Not Just Efficiency

The Principle

Managers should place as much emphasis on flexibility as on efficiency in manufacturing, product development, and other operations as well as in strategic decision making and organizational evolution. Their objectives should be to pursue their own company goals whilst quickly adapting to changes in market demand, competition, and technology. Firms also need to be ready to exploit opportunities for product or process innovation and new business development whenever they appear. Moreover, rather than always requiring tradeoffs, flexible systems and processes can reinforce efficiency and quality, or overall effectiveness, as well as facilitate innovation.

Introductory

In the last several chapters, I introduced three distinct but related principles around capabilities, pull concepts, and scope economies. All promote agility in one form or another. Agility seems essential to staying power—surviving and thriving over years and decades, and despite the ups and downs of markets and other unfortunate events. At the same time, to evolve capabilities and implement pull and scope concepts, firms require a high degree of flexibility—the subject of this chapter. Firms also need to be efficient at what they do, though balancing efficiency and flexibility is what really makes operations effective for particular products and markets at specific points in time. Achieving an effective balance should be the real goal of managers, not flexibility or efficiency by themselves.

I first recognized the importance of flexibility and the need for balance when studying Toyota in the early 1980s. We saw in Chapter 3 how indirect technology transfer led to a slower start in mass production at Toyota compared to Nissan. But the focus on cultivating internal capabilities put Toyota in a better position to understand how to rethink the process of mass production as well as how to design its own products. In manufacturing, we saw how the pull system, supported by other techniques, enabled Toyota to achieve a remarkable degree of product mix flexibility and efficiency. More specifically, Toyota learned how to vary the sequence of models assembled on its final production lines—with no apparent penalty in productivity or manufacturing quality—rather than making the same model over and over again in very large batches.

Indeed, Toyota managers generally found that varying the mix in final assembly kept employees alert and allowed the company to match assembly work with the orders coming in from dealers on a weekly and daily basis. Despite the company's recent quality problems, Toyota achieved flexibility in a number of ways: rapid set-up or changeover times for equipment; multi-skilled workers who could handle the variety in tasks and assignments; suppliers able to adjust their deliveries quickly; components and subsystems that were easy to assemble; and liberal use of worker overtime. Several of these shop-floor techniques required consent from the Toyota company union, which became very cooperative with management after 1950. This relationship gave Toyota an important advantage over Nissan (which had a more independent and much less cooperative company union) and Western auto companies (which organized very independent unions at the industry level).[1]

In product development, we saw that Sony and JVC were highly flexible in R&D because of their orientation toward trial-and-error learning. We also saw that PC and Internet software development focuses on daily builds and accommodating late design changes, and as such is a highly flexible process (and sometimes too flexible). We can also interpret the software factory and multi-project management as organizational structures that, to be effective, must balance the flexibility requirements of individual projects with the potential constraints and efficiencies of reusing designs, components, tools, and other technical knowledge across different projects or customer service engagements.

There is no easy way for managers to achieve the appropriate balance between flexibility and efficiency, or between their counterparts in creativity and structure. Scholars of technological innovation such as my MIT Sloan colleague Thomas Allen have argued in favor of using different organizational structures and processes (such as multi-functional teams and projects rather than functional departments) as well as physically co-locating engineers, scientists, and other key project personnel to improve communication across traditional function-oriented boundaries.[2] But much of the organization theory literature poses flexibility and efficiency, as well as creativity and structure, as opposites. The claim is usually that efficiency requires structure, formalization, bureaucracy, standardization, hierarchy, and automation. By contrast, flexibility generally requires the opposite because organizations need to make frequent and rapid ad hoc adjustments to change or unanticipated conditions. But empirical research on manufacturing and product development has not fully supported the argument that tradeoffs must exist in practice. Some studies find tradeoffs, whilst other studies do not.[3] Different techniques and management philosophies, such as the pull system, or multi-project management, help firms achieve an appropriate balance. My experiences with Toyota and other firms also suggest that flexibility and efficiency (as well as manufacturing quality, though not necessarily design quality) often go together, and that flexible processes facilitate product and process innovation. A brief review of some of the research illustrates what we really know about flexibility.

In the organizational literature, several scholars have identified the need for some sort of contingency framework to understand where and why flexibility might be more useful than focusing on efficiency. I noted in the previous chapter that I read some of these works when trying to understand how theorists might view the software factory. Several concepts and frameworks from Joan Woodward, Charles Perrow, and Henry Mintzberg were particularly helpful (Table 6.1).[4] For example, a "machine bureaucracy," like a highly automated mass-production factory, can operate with minimal flexibility because it rarely runs into exceptions. However, if customer requirements or a factory's production plans change unexpectedly, then managers may conclude

Table 6.1. *Organizational structure and technology*

Structure	Technology	Tasks and problems	Characteristics
Machine bureaucracy	Routine, mass production	Few exceptions, well defined	Standardized and de-skilled work, centralization, divisions of labor, high formalization of rules and procedures
Professional bureaucracy	Engineering	Many exceptions, well defined	Standardized and specialized skills, decentralization, low formalization
Adhocracy	Non-routine	Many exceptions, ill defined	Specialized skills but few or no organization standards, decentralization, low formalization
Simple structure	Unit or craft	Few exceptions, ill defined	Few standardized specialized skills, centralized authority but low formalization

Sources: M. Cusumano (1991b: 40, table 1.3), derived from Woodward (1965), Perrow (1967), and Mintzberg (1979).

this is the wrong structure. They need a more flexible factory. This could still be very structured, but it needs to handle exceptions, albeit within a specified range. It follows that software factories and similar highly structured engineering or service organizations would have to break down non-routine tasks so that they appear more routine and predictable to managers and workers. In this case, a factory-like engineering process could handle exceptions, such as unique customer requirements, or even some innovations like the Internet, but only if most of the potential variations are reasonably well understood in advance.

There is also a large literature on strategic flexibility and the relationship to organizational capabilities. Most of the research cited in Chapter 3 on dynamic capabilities is really about how organizations can adapt to change. Related concepts include the idea of "post-modern management" talked about by Mel Horwitch in the late 1980s as well as the "ambidextrous organization" from Michael Tushman and Charles

O'Reilly. They argued that managers can pursue different objectives within the same firm by creating different kinds of organizations internally, with different objectives, cultures, people, and processes, integrated by senior management.[5] For example, firms can establish an R&D lab with highly creative people and special incentives to pursue new ideas whilst managing mass-production factories using less creative people and conventional metrics. This kind of ambidexterity does not necessarily help firms adapt quickly to unanticipated change, however. Instead, firms need to reinvent themselves periodically. Highly relevant to this argument is Richard D'Aveni's concept of "hypercompetition"—the idea that there is no such thing as long-term competitive advantage. Firms that succeed over long periods of time are actually changing the basis of their competitive advantage as the environment changes.[6]

Kathleen Eisenhardt and Shona Brown in the latter 1990s studied this issue by looking at how firms can balance structure with the ability to endure transitions in fast-paced environments. They cited the usefulness of introducing major strategic and technological changes in predictable time intervals, along with shorter planning cycles and modularity in product designs so that all of a firm's technologies do not have to change at once.[7] More recent theoretical work has shown as well that unpredictable or fast-changing environments require less structure so that firms can pursue unanticipated opportunities and respond quickly to change, whereas firms perform better with more structure in stable environments.[8] These are merely a few examples of a much broader stream of research on flexibility in organizations, technology, and strategic management.[9]

Other important studies relevant to this chapter are in operations management, where flexibility became a popular topic in the 1980s and early 1990s. The diffusion of programmable control systems, robots, and machine tools as well as computer-aided engineering and design tools made it possible for factories to make a greater variety of parts and final products with less impact on productivity and quality than using traditional manual methods. In 1984, Joel Goldhar and Mariann Jelinek even published an article in the *Harvard Business Review* under a title that would surely fit my previous chapter—"Plan for Economies of *Scope*." The authors touted the flexibility benefits of computers in manufacturing and encouraged managers to connect this capability more directly to product strategy and other operations. Another closely connected stream of thought at the time was the concept of "mass customization," such as practiced by Dell to configure mass-produced products to meet varying customer specifications, within limits. Joseph Pine, a former MIT Management of Technology master's degree student, popularized the term with a 1992 book.[10]

Another influential thinker in operations management was the late Ramchandran Jaikumar. In a 1986 *Harvard Business Review* article, Jaikumar examined why Japanese flexible manufacturing systems (FMS) in the machine tool industry, led by companies such as Yamazaki, Hitachi Seiki, and Mori Seiki, proved to be far superior to American counterparts in the number of different parts they could produce and the ability to vary volumes and introduce new parts—whilst achieving higher utilization rates. (Full disclosure: Jaikumar was one of my supervisors at the Harvard Business School in 1984–6 and I accompanied him on the initial visits to these Japanese plants.) His conclusion was that the Japanese firms chose to invest in a series of projects that challenged their R&D and manufacturing engineers to design in high levels of flexibility as well as achieve untended (100 percent automated) operations as much as possible. Not all FMS operations turned out to be efficient and profitable; they were expensive to develop and fell out of favor in Japan when the country went into recession from 1989. Nonetheless, Jaikumar made an important argument: firms need to invest in "intellectual assets" by challenging manufacturing engineers with specific types of projects that, over time, push their knowledge and technological capabilities to the limits. Jaikumar referred to this philosophy as "managing above the line," in contrast to optimizing short-term payback schemes on a per project basis or relying on conventional notions of efficiency and profitability.[11]

Other authors have contributed to this topic with useful taxonomies that sometimes measured and tested different types of flexibility in manufacturing, such as product mix, production volume fluctuations, new-product introduction, and delivery or lead times. Some research has focused more on modeling the potential tradeoffs between different types of flexibility and efficiency (cost, productivity, inventory levels) as well as quality. Most of these studies argue for a linkage among environmental uncertainty, plant-level flexibility, and higher performance, though these variables are difficult to compare precisely in different contexts.[12]

We also have a lot of recent research on how firms can use different product development strategies such as modularity, common components, and in-house product platforms to enhance flexibility in the sense of product variety.[13] Again, these approaches aim at scope rather than scale economies in engineering, and do place some constraints on projects, such as limiting the components they can use or develop independently. But we can also interpret modular or platform-based products as enhancing the kinds of flexibility that take advantage of scope in manufacturing, such as for product mix or rapid new product introduction.

Other research has looked at how firms can use specific engineering processes to achieve a better balance of structure and creativity. Durwood Sobek, Jeffrey Liker, and Alan Ward studied this issue at Toyota. They described the use of "rigid" engineering standards for particular activities, analogous to the high-level "meta-routines" that guided the highly disciplined but flexible practices followed in Toyota's factories. These engineering practices, like Toyota's manufacturing practices, were apparently all subject to careful and continuous scrutiny, experimentation, and modification according to the needs of individual projects, with the standards and modifications reinforced by professional supervision and training. So, whilst they described a process that was highly formalized, repeatable, and efficient, it was also flexible and open to continuous improvement. We shall see how such processes help Toyota overcome the quality problems of 2009–10.[14]

Overall, the research suggests that flexibility takes many forms and can serve different strategic purposes. Managers can rely on different types of flexibility both as a hedge against uncertainty and as a potential competitive weapon against less nimble competitors. Flexibility is also more complicated than what we see in particular rules, processes, or techniques. Rebecca Henderson and Kim Clark, for example, have shown that established firms have great difficulty adapting mentally to new technology architectures and system concepts.[15] Clay Christensen has argued that innovations from firms with initially inferior technology (such as small disk drives and personal computers compared to large-scale storage systems and mainframe computers) can dramatically disrupt both the technology visions and the business models of entrenched firms.[16] To deal with threats such as these requires a kind of intellectual and strategic flexibility that is related but different from what we see at the technology and operations level. In any case, flexibility in different forms is essential to deal with rapidly changing, unpredictable markets.

As noted in the Introduction, I illustrate the various dimensions of flexibility by relying on three very different examples. First is flexibility in manufacturing, based on a large-sample study of printed-circuit board (PCB) production. Second is flexibility in product development, returning to the example of PC software and agile or iterative development at Microsoft and other companies. Third is flexibility in strategy and entrepreneurship based on my experience studying Netscape in the 1990s and its competition with Microsoft in Internet software.

Flexibility in Manufacturing

By the early 1990s, my colleagues in the International Motor Vehicle Program at MIT were already studying flexibility in automobile assembly, whilst Jaikumar and others were looking at flexibility in machine tool production. The research, however, was largely theoretical or based on small-sample studies, and left several key managerial questions unanswered. My own work on the automobile industry was also limited to firm-level measures, such as vehicle productivity, investment, or inventory turnover. Accordingly, Fernando Suarez (for his doctoral thesis) and I, inspired by what we understood about Toyota, began new research to examine flexibility and effectiveness at the factory level. We were particularly concerned with questions such as what are the different types of flexibility, how do they affect company performance, and what type of tradeoffs exist, if any?

The Research Setting

We decided to look at the printed circuit board (PCB) industry because manufacturers were already studying flexibility issues and had collected relevant data. A number of companies affiliated with MIT were willing to share their data with us on a confidential basis. And, since PCBs were used in various applications, we were able to create a sample consisting of numerous companies with diverse plants that probably had to balance or trade off flexibility and efficiency. To study the tradeoffs, we collected data from 31 PCB plants belonging to fourteen electronics firms in the United States, Japan, and Europe (Table 6.2). These plants produced boards for forty-one different applications, ranging from automotive and consumer electronics to computers, measuring instruments, medical equipment, telecommunications, and business machines.[17] This was not a random sample; all but one of the companies were members of MIT corporate programs or research centers. We visited more than half the plants in person and collected data on the others through a written survey and telephone interviews, and published the results in 1995 and 1996.[18] Overall, we found significant relationships among different types of flexibility but no specific tradeoffs with quality or costs.

In the mid-1990s, the PCB industry split into two major groups of players: captive plants (which we studied) that produced for downstream factories or divisions of the same company, and independent factories or contract manufacturers that sold their assembly services to different companies. Assembling PCBs consisted primarily of placing different components on a wired "raw" board by machine or by hand. Components varied greatly in shape, technological sophistication, and process requirements; some were simple resistors and others complex microprocessors. At this time, most plants used two basic technologies to place components on a board: through-hole placement and surface-mount

Table 6.2. *PCB plants and final applications by region*

Final application	USA	Europe	Japan	Totals
Automotive	4	2	1	7
Consumer electronics	0	5	5	10
Computers	7	3	3	13
Measuring/medical instruments	4	0	0	4
Telecommunications	2	2	2	6
Business equipment	1	0	0	1
Totals	18	12	11	41

Source: Suarez, Cusumano, and Fine (1996: 223).

technology. The latter was a newer technology that would soon become the industry standard.

Several characteristics of our research sites are worth noting. First, most of the plants were highly automated, which means that human factors were less relevant than in more labor-intensive industries, such as automobile assembly. Second, the production process was highly standardized, so there was less variation than in other assembly industries. Third, PCB assembly is an intermediate industry that supplies boards for other factories to use in final applications, so the "customers" are manufacturing firms rather than end users. Finally, our study covered only captive plants, which tend to be more insulated from market pressures than plants competing in the open market.

Types of Manufacturing Flexibility and Implementation Strategies

When we reviewed what other people had written about manufacturing flexibility, it immediately became apparent that there were many different kinds of flexibility, with overlapping definitions. Eventually, we adopted a framework with only four basic types of flexibility: *product mix*, *new-product introduction*, *volume*, and *delivery time*, as defined below. We concluded that other types of flexibility were all variants of these four. For reasons of data availability, we concentrated on the first three.

MIX FLEXIBILITY. We used four variables to measure mix flexibility in the case of PCB producers: (1) the number of different board models assembled by each plant; (2) the number of different board sizes used during assembly by each plant; (3) the range of board density handled by each plant (in components per square centimeter); and (4) the number of product categories (for example, video cassette recorders, televisions, and stereos) in which the boards were used. The literature equated a highly flexible product mix with a broad product line, which in turn correlated with larger market share and higher profits.[19] But other researchers typically measured this type of flexibility by the number of products that a factory produced at any point in time. We believed this measure was too simple. For example, a plant that produces two very different products (such as a personal computer and a laptop) should have greater mix flexibility than a plant that produces two similar products (such as two personal computers that differ only in speed and RAM characteristics). Therefore, we selected the four variables noted above.

NEW-PRODUCT FLEXIBILITY. We measured new-product flexibility by determining the number of months from the earliest stage of design to the first production run of a batch of salable products. Fewer months meant greater new-product flexibility.[20]

VOLUME FLEXIBILITY. We defined this as the ability to vary production with no detrimental effect on efficiency and quality. That is, volume flexibility is not the same as volume fluctuation, which may be associated with higher costs and lower quality levels. Instead of simply measuring fluctuations in volume over a certain period, we used a formula that took into account volume fluctuation, cost per placement (efficiency), and the fraction of PCBs that required repair (quality).

DELIVERY-TIME FLEXIBILITY. This refers to the ability to alter the time required to deliver components or finished products to the customer. Firms generally have a usual lead time for handling orders and can respond more quickly to rush orders or special cases, such as to replace a defective shipment. Since we examined an intermediate component, most of the plants in our sample were captives and delivered their boards to other factories or assembly stations within the company or the same plant. Only a few plants tracked delivery times, so we were unable to measure this variable.

When it came to implementing flexibility, the research on Toyota and lean production suggested that firms used several strategies, rather than one. These included production technology, production management techniques, relationships with suppliers, human resource management, product development strategies, and accounting and information systems. We were able to test the effects of these different implementation strategies on the three types of flexibility described above except for accounting and information systems (the firms were reluctant to provide data here).

PRODUCTION TECHNOLOGY. We expected companies using automated programmable technologies to have greater mix and new-product flexibility. The PCB assembly equipment introduced by the early 1990s was easily programmable and therefore could be used, theoretically, to make a variety of board designs without much cost or quality penalty. New equipment could also be linked to other parts of the factory, such as design and procurement, which should make it easier for a plant to handle a greater mix and bring out new products more rapidly. To our surprise, we found that newer, more automated processes tended to be associated with less mix and new-product flexibility! It appeared that the companies in our sample were using programmable equipment to run their largest production batches rather than to increase flexibility. Jaikumar had also flagged this problem as typical of American FMS plants, which did not exploit the flexibility potential of their technology. Nonetheless, we did find that more automated production technology correlated with greater volume flexibility.

PRODUCTION MANAGEMENT TECHNIQUES. We expected Toyota-style production techniques to affect mix and new-product flexibility. My study of Toyota in 1985, as well as earlier and later research, suggested that specific techniques could reduce machine setup times and improve worker involvement in the production process—characteristics that should facilitate a more complex product mix and introduction of new products more quickly and without harming quality.[21] These management techniques did, indeed, correlate with mix and new-product flexibility. For instance, firms in which a high percentage of workers were involved in formal problem-solving group activities, such as quality circles, had greater mix and new-product flexibility.

SUPPLIER RELATIONS. Our study suggested that a close relationship with suppliers positively affected all three types of flexibility. A plant that can subcontract orders for which it has no adequate in-house capability should be able to increase product variety, speed up prototyping, or increase volume without much cost or quality penalty. Closeness to suppliers should help a plant procure the right components when needed for assembly or prototyping. Otherwise, procurement may become problematic as product variety increases or if factories increase volumes too quickly (as Toyota seems to have done in the 2000s).

HUMAN RESOURCE MANAGEMENT. We expected human resource management to be strongly related to volume flexibility. Toyota, for example, relied heavily on worker overtime as well as the use of temporary workers to accommodate unplanned or seasonal increases in demand. The PCB assembly industry was relatively automated, making people less important than in the automobile industry. But the theoretical link between this factor and volume flexibility was so clear that we expected it to have a significant impact anyway. Plants that tended to use temporary workers should have an advantage in adjusting the work force to volume changes. Plants with wage structures linked to plant or division performance should also have an advantage in adapting to changing volumes: In periods of low sales, the payroll burden will automatically fall. We found support for the latter notion: Plants with flexible wage schemes did have a cost advantage, in terms of volume flexibility, over plants with fixed wages.

PRODUCT DEVELOPMENT PROCESS. Factories that benefited from design for manufacturability principles (that is, had policies to reuse components across board models to minimize equipment changes) had greater mix and new-product flexibility. Higher component reusability seemed to allow these plants to handle a greater variety of both existing and new products. We thought this finding was especially important for companies that needed to deal with increasing product variety or rapid product development. As in software factories, managers can promote component reuse by specific policies, such as mandating the use of menus or lists of preferred components and incentive schemes.

To summarize, we found the following relationships between implementation strategies and flexibility types. (1) More automated plants tended to be less flexible in terms of product mix or new product introduction, despite the programmable nature of most equipment used in the industry. (2) Nontechnology factors—high worker involvement, close relationships with suppliers, and flexible wage schemes—appeared to increase mix, volume, and new-product flexibility. And (3) component reusability appeared to raise both mix and new-product flexibility.

Tradeoffs and Relationships among the Flexibility Types

We realized when doing this study that managers would not want to emphasize flexibility if this increased costs or decreased quality. Accordingly, we wanted to analyze the precise relationships among cost and quality, and mix and new-product flexibility. (Our measure of volume flexibility already took cost and quality into account.) We measured quality in two ways: the number of non-repairable boards per million at the post-assembly check and the percentage of boards that underwent some repair through the assembly process. Both measures were necessary because some plants had few post-assembly defects but did extensive board repair during assembly (that is, they inspected and "repaired-in" quality, rather than built it in). Quality figures varied substantially. For example, defects per million at the post-assembly check varied from zero to 14,000.

We found no relationship between quality and flexibility. That is, plants with greater mix or new-product flexibility did not have more defects than plants with less flexibility. Likewise, we found no relationship between cost and flexibility. Companies were unwilling to give us detailed accounts of each plant's cost structure, so we had to rely on a common industry measure: cost per component placed. These figures varied widely, from less than \$.01 to roughly \$.40. Nonetheless, none of the correlations between cost and flexibility was statistically significant. Nor did we detect any significant tradeoffs between mix or new-product flexibility and either cost or quality. This finding fit with studies of automobiles (by International Motor Vehicle Program researchers) and air conditioners (by David Garvin), which suggested that high quality, instead of being costly, is often associated with low costs or high levels of productivity, and that improvements in mix flexibility do not seem to increase costs or worsen quality levels to any significant extent.[22]

The factories in our sample also achieved different types of flexibility through different emphases on production technology, production management techniques, supplier relationships, human resource management, and product development strategies. We believed that, if the different flexibility types required different configurations on all these factors, then it would be very difficult to achieve them all at once. A truly flexible plant may be impossible to achieve, even though the term "flexible factory" had already come into use in the 1990s, particularly in the machine tool industry. But, whilst one of our objectives was to see if there were tradeoffs between two or more types of flexibility, during the data analysis we began to see *relatedness instead of tradeoffs*. Flexibility types seemed to reinforce each other rather than work against each other. Moreover, these interrelationships had major implications for manufacturing strategy as well as for understanding the concept of flexibility more deeply.

MIX FLEXIBILITY AND VOLUME FLUCTUATION. The research to date had not carefully distinguished between volume fluctuation and volume flexibility. Therefore, we thought it important to consider the relationship of volume fluctuation to mix flexibility. The plants in our sample with the most mix flexibility had the lowest volume fluctuations and seemed to enjoy the benefits of a more stable production flow. This was mainly the result of the "cushion" or buffer effect provided by the broader product mix. For instance, plants that could switch among PCBs for many final applications (such as, computers, consumer electronics, and medical instruments) were not so adversely affected if the demand for one product shrank unexpectedly. By extrapolation, all the factors that increased mix flexibility should increase production volume stability. Closeness to suppliers and subcontractors, for example, should have a stabilizing effect on production volume. Not only will such relationships increase mix flexibility and therefore give the plant a cushion, but the subcontractors themselves may be a source of cushion as firms increase or decrease subcontracting in response to demand fluctuations (a common practice in Japan).

MIX FLEXIBILITY AND VOLUME FLEXIBILITY. When we analyzed the relationship between mix and volume flexibility, we found none. This was not surprising given that the factors associated with mix flexibility (for example, Toyota-style production techniques) were different from those associated with volume flexibility (for example, human resource policies).

VOLUME FLEXIBILITY AND NEW-PRODUCT FLEXIBILITY. Our results showed a weak or insignificant correlation between these two types of flexibility.

MIX FLEXIBILITY AND NEW-PRODUCT FLEXIBILITY. These two types of flexibility appeared to go hand in hand, and sometimes reinforced each other. Several plants in the sample had great flexibility in both mix and new-product introduction; other plants seemed to become increasingly less flexible on both dimensions. The reasons seemed clear: the same factors that correlated highly with one correlated highly with the other. For example, component reusability increased mix flexibility and shortened design cycle times. Similarly, worker involvement in problem-solving group activities was important for both types.

We also considered the implications of mutually reinforcing relationships. Some scholars had championed the idea of a "focused factory," one that has trimmed down its product variety to specialize in a narrower product line.[23] In the language of my principles, this is a vote for efficiency over flexibility. But our results showed that this policy may have consequences not only for mix flexibility (which drops when a plant gets more focused), but also for new-product flexibility. Managers who decided to produce only a few products may have implicitly sacrificed new-product flexibility and left themselves open to volume fluctuations. In the long term, this can jeopardize a factory's ability to maintain high levels of capacity utilization and thus to operate efficiently. Conversely, plants that stress rapid new-product introduction will naturally tend to increase their mix flexibility over time, assuming the rate of product obsolescence is not too high. This dynamic, in turn, will tend to smooth production volume fluctuations because of the cushion effect of mix flexibility on volume fluctuations. Thus the relationships amongst these three flexibility types seemed to have potentially powerful consequences for long-term factory operations and company performance.

We can see the connections among these different flexibility types as well as the different ways of achieving them in Figure 6.1. This is a causal loop diagram commonly used in system dynamics models. The + sign indicates a positive relationship, 0 no correlation, and – a negative relationship. This is useful in particular to visualize the positive relationship or self-reinforcing effects between mix flexibility and new-product flexibility. We can also see that mix flexibility reduced volume fluctuations, shown by the negative sign, and that worker

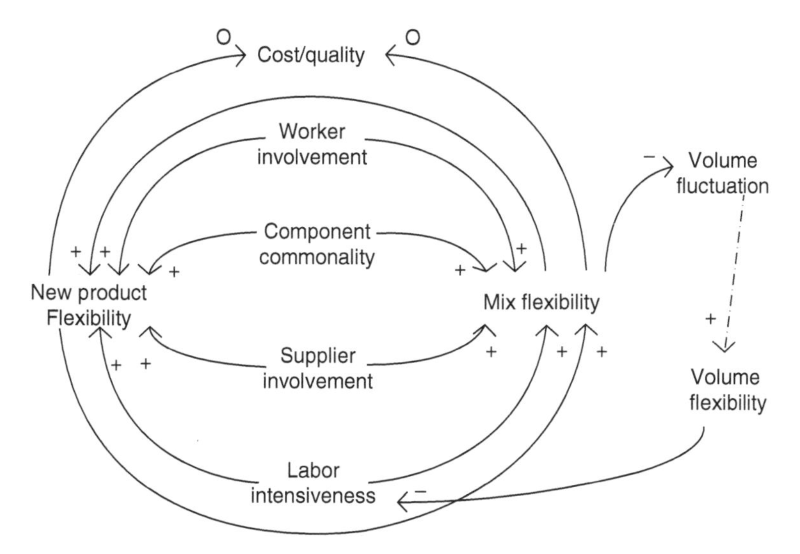

Figure 6.1. Causal loop diagram of printed circuit board study findings

Source: Suarez, Cusumano, and Fine (1996: 239).

involvement, component commonality (reuse), and supplier involvement enhanced new-product and mix flexibility. The o around cost and quality indicates no penalties on these dimensions for new-product and mix flexibility.

Managerial Implications

Flexibility by itself is not useful. Managers need to seek the right level of flexibility, as well as efficiency and quality, for their particular markets and competitive strategies. What is the right level, in turn, depends on customer needs, competitor capabilities, and other factors, such as market uncertainty or government regulation (for example, in healthcare or telecommunications). But it is clear that managers have several levers they can use to balance flexibility with efficiency and quality. They can rely on technology, such as programmable automation. Potentially more useful, though, are specific production management techniques and policies for managing suppliers, workers, and product development, such as the degree of modularity or use of in-house platforms.

It was striking that flexibility (at least in our sample) had much more to do with non-technology factors than with technology itself. Worker involvement in problem-solving groups, particular ways of relating to suppliers, and component reuse in product design seemed far more important in determining flexibility. In fact, plants with the most computers (programmable automation) were the least flexible. This finding continues to have important implications for the way an organization uses its resources and theoretical capabilities, and the type of investments necessary to improve flexibility. Before buying the latest technology, for example, managers would be better off maximizing the flexibility in their existing factories, equipment, and suppliers.

It is also instructive to recall why Taiichi Ohno never liked too much automation or robots in Toyota factories. Whilst he was still in charge of production management during the 1960s and early 1970s, the technology was not as flexible in practice as it was in theory. That did change, and Toyota eventually became a leader in introducing robots and other forms of automation and computers. But Toyota has always done this with an eye to maintaining flexibility. Ohno taught us that programmable robots and automation do not by themselves make a factory flexible. Management policies and practices, as well as product design, achieve this aim—together.

The PCB study also showed that an organization could be very flexible in some ways and less flexible in others; thus it was not necessarily so useful to aim for a "flexible factory" or a "flexible production system." Moreover, our results suggested that some types of flexibility tend to move together, such as mix and new-product flexibility, whereas volume flexibility responds to a different dynamic. This again has important implications: managers can choose the dimensions along which they want their factories (and their organizations) to be flexible. Depending on the environment, a firm may require more of certain types of flexibility And, of course, flexibility should never come at the expense of quality or customer safety.

Flexibility in Product Development

Just as we can analyze different types of flexibility in manufacturing, we can analyze flexibility and tradeoffs in product development. The best example from my research again takes us back to Microsoft's agile or iterative style of software development for the PC and Internet markets—in contrast to the slower-paced and more structured world of software engineering for mainframe computers. For PC and Internet software, the managerial challenge quickly became how to add enough structure to avoid chaos in projects but still encourage individual creativity and product innovation, speed to market, and adaptations to unpredictable changes in technology and customer needs.

As described in Chapter 4, the "old" way to structure software engineering and many other kinds of product development was to adopt a "waterfall" style or even a less rigid "stage-gate" process, and enforce rules based on this kind of sequential process flow. For example, managers in a waterfall-ish project would normally require a detailed requirements document and then a detailed design document. They would then try to "freeze" the design, and then programmers would write code and create test cases based on the original requirements and the detailed design, with minimal allowance for change. Of course, a rigid waterfall or stage-gate process could be more efficient in the sense that projects might minimize rework and retesting by not allowing design changes after a certain point. But, in fast-paced markets when the challenge is to invent and innovate, or in settings where development is best done together with the customer providing frequent feedback, then the waterfall and stage-gate approaches become too inflexible. They often result in an inefficient process to the extent that project members will still have to go back and change their designs or throw out their work if it does not meet the customer's expectations or evolving market needs.

In fast-paced markets like PC and Internet software during the 1980s and 1990s, Bill Gates and other Microsoft managers tried to retain the nimbleness and flexibility of small teams even as product size and complexity increased dramatically. This occurred as the PC industry moved from character-based DOS and simple DOS-based applications to much larger graphical programs for the Apple Macintosh from 1984 and then for Windows PCs from 1990. Microsoft has not always succeeded in its attempt to "make large teams work like small teams," especially when team size reached into the thousands for Windows Vista. But Microsoft has succeeded often enough, especially in applications projects like Word and Excel (now part of Office), which it first introduced for the Macintosh and later for the PC. The company still follows the same agile or iterative principles and they again proved useful in the very successful Windows 7 project. Microsoft completed this in fall 2009, under the leadership of veterans Steven Sinofsky and Jon DeVaan. Both had previously led the Office group and came over to run the Windows team and "fix" the problems in Vista.[24] The principles they helped refine in the 1990s and that they adapted for Windows 7 are the focus of the next discussion.

Making Large Teams Work Like Small Teams

We can perhaps see the best expression of this need to combine flexibility and structure in one of my favorite quotations from *Microsoft Secrets* (1995).[25] Dave Maritz, Microsoft's former test manager for Windows 95, explained how he used his background as an Israeli tank commander to impose a simple but rigid rule to keep the highly individualistic software developers coordinated:

> In the military, when I was in tank warfare and I was actually fighting in tanks, there was nothing more soothing than people constantly hearing their commander's voice come across the airwaves. Somebody's in charge, even though all shit is breaking loose... When you don't hear [the commander's voice] for more than fifteen minutes to half an hour, what's happened? Has he been shot? Has he gone out of control? Does he know what's going on? You worry. And this is what Microsoft is. These little offices, hidden away with the doors closed. And unless you have this constant voice of authority going across the e-mail the whole time, it doesn't work. *Everything that I do here I learned in the military... You can't do anything that's complex unless you have structure... And what you have to do is make that structure as unseen as possible and build up this image for all these prima donnas to think that they can do what they like.* Who cares if a guy walks around without shoes all day? Who cares if the guy has got his teddy bear in his office? I don't care. I just want to know... [if] somebody hasn't checked in his code by five o'clock. Then that guy knows that I am going to get into his office. [Emphasis added][26]

The picture Maritz describes resembles a highly flexible but synchronized army: divisions of Microsoft programmers working in individual offices (akin to individual tanks), with coordination from email and a common process. The comment about five o'clock refers to the deadline the Windows 95 group had for checking in code to the daily build, which I explained in Chapter 4. Again, the build creates a working version of the product so that engineers can test newly added functions to make sure existing features are still working. The daily build is one of those few "rigid" structure points that force everybody to come together on a project.

Again, most Microsoft managers and programmers avoided the higher level of structure and bureaucracy that we saw in Japanese software factories or advanced software engineering facilities in India following Software Engineering Institute practices (CMM-I Level 4 and 5). Rather, as discussed in Chapter 4, they generally tried to "scale up" a more informal small-team style of software development, relying on daily builds and continuous testing to integrate the work of several relatively small feature teams. These teams normally consist of 5–8 developers, parallel "buddy" testers, and a program manager (see Figure 4.6). The teams follow a few high-level principles (which Adler et al. would call "meta-routines"). Table 6.3 provides a list of seven principles that Richard Selby and I identified at Microsoft in the 1990s. These are now common practice in most software companies around the world that follow an agile process.

One obvious way to retain the benefits of small teams is to keep projects small, such as by placing limits on their size and scope. For example, rather than do one large project that covers, say, four years,

Table 6.3. *Making large teams work like small teams*

- *Project size and scope limits* (clear and limited product vision; personnel and time limits)
- *Divisible product architectures* (modularization by features, subsystems, and objects)
- *Divisible project architectures* (feature teams and clusters, milestone subprojects)
- *Small-team structure and management* (many small multi-functional groups, with high autonomy and responsibility)
- *A few rigid rules to "force" coordination and synchronization* (daily product builds, immediate error detection and correction, milestone stabilizations)
- *Good communications within and across functions and teams* (shared responsibilities, one site, common language, non-bureaucratic culture)
- *Product-process flexibility to accommodate the unknown* (evolving product specifications, project buffer time, recognition of an evolving process)

Sources: Cusumano and Selby (1995: 410); also Cusumano (1997: 15).

it is better to cut the objectives and teams in half and schedule two 18-month or 24-month projects. This kind of scheduling became more difficult in the later 1990s and early 2000s, as Microsoft wanted to catch up to Netscape fast and added hundreds and then thousands of programmers to the Windows team. The first goal was to add a browser to Windows, but then Microsoft needed a media player, a search engine, and various other technologies as well as better security features to compete with Google, Linux, Sun Microsystems (Java), and other firms—simultaneously. Nonetheless, the idea is right: managers need to set boundaries on what each project will attempt to accomplish. In the automobile industry, design changes after a certain point require expensive changes in manufacturing equipment such as metal stamping dies and plastic molds, and physically affect adjacent components. Auto companies also deal with hundreds of suppliers for each new model. Consequently, there is usually much more discipline regarding change control in automobile design compared to software development, except for programming applications that are truly mission-critical, like space shuttle software, which SDC worked on, or nuclear power plant control software, in which Toshiba specialized.

Auto companies generally limit design changes as much as possible to the modeling and prototyping phases, and increase flexibility by using computers and other techniques that make it easy to produce many prototypes and test models. Automobile designers also make heavy use of computer-aided tools, simulation programs, and other techniques that allow projects to delay some final decisions to as late as possible in the schedule. For example, engineers can lock in certain specifications and interfaces, such as how big a car door will be and how it must fit into the automobile frame, and allow some manufacturing preparations and die cutting to proceed in parallel on those assumptions. But they can delay final decisions on, say, the precise contour of the exterior surface of the door panels until the last possible moment.

It is also useful to set limits on the number of people who will work on any one project. Microsoft was much more successful at this in application projects such as Word, Excel, and then the Office Suite, compared to Windows. The applications historically have had much clearer and simpler goals than the operating system group. In addition, managers can focus projects by setting time limits, such as 12 or 18 months for a new version of a software product, or four years for a new model for an automobile. The time limit, especially when combined with a limit on personnel, forces teams to make decisions on what they can do with the resources allocated. This promotes efficiency and forces creativity. Without setting priorities and time limits, a project team will often try to do too much and not do anything very well.

In the Microsoft case, once more, the applications groups have generally been much more adept at setting time and personnel limits than the Windows group, which has often had vague targets and no practical limits on time or manpower. The application features are also usually less technically coupled with other functions than in the operating system. So it is generally easier simply to stop work after a period of time and ship an applications product. The team can postpone the unfinished features until the next release, which usually comes much more frequently than an operating system release.

The different degrees of flexibility between applications and operating-systems development at Microsoft also reflect different underlying product architectures as well as different competitive conditions. Windows evolved in an ad hoc manner on top of the DOS code base, resulting in more complex, interdependent functions. The Windows group has little competition, generally plans projects that last many years, and has learned to place much more emphasis on reliability, compatibility with prior applications and hardware, and security. These attributes now require years of testing because of the program's enormous size and overly complicated architecture. Applications groups have placed more emphasis on modularity and usability, and generally organize smaller, shorter, and more focused projects.

The Microsoft examples illustrate that, even more than limitations on project resources such as time and people, product architecture can play a crucial role in enhancing flexibility as well as enabling large teams to work more like small teams. The key technique, used widely in many industries, is modularization that breaks down large, complex systems into a set of smaller and more self-contained features, functions, or subsystems. It follows that managers can then subdivide large projects into smaller subprojects and teams that logically mirror the subdivided structure of the products. Microsoft, for instance, routinely divides its applications products and projects into relatively small features and feature teams. This kind of architecture also makes it possible to prioritize features and group them into clusters, and then build the most important or technically interdependent ones first, and then go to the next set, and so on.

Another useful technique is to divide one large project into subprojects, which Microsoft referred to as milestones. The feature teams go through a full cycle of development (design and coding), testing, and stabilization for each subset of features or components, before moving on to the next milestone. The objective is to avoid building and then trying to integrate too many features at once. Project managers also want to avoid targets that are too far into the future and allow for too much change to creep in. Again, at least before Windows 7, Microsoft was much more successful in limiting change and following a tight milestone schedule in applications. With products divided into smaller features, and even features divided into subfeatures, it becomes possible for projects to break down engineering assignments into pieces or "chunks" that a handful of people can usually accomplish in small amounts of time, such as half a day. The smaller groups and individuals also feel more direct responsibility for their work and have the knowledge and autonomy to make adjustments or modifications in their part of the product.

Now we come to the crux of the process: the few rigid rules that force individuals and small groups to work as a team (that is, efficiency) but do not overly inhibit individual creativity or the ability to make changes in the product design (that is, flexibility). When we observed Microsoft in the 1990s, Selby and I saw three practices that accomplished this goal. (1) Every project had to create a daily build. (2) No individual check-in of new code was allowed to break the build, and so programmers had to fix any problems immediately or back out their changes. And (3) projects had to subdivide themselves into milestone subprojects to reduce the complexity of what they were doing and focus the team more narrowly.

We were struck by the lack of other rules. For example, developers were free to come into work whenever they pleased, as well as to contribute to the product build as frequently or as rarely as they liked. However, on the day when developers checked in their work, they had to do so by a particular time, and they were not allowed to check in code that caused the evolving product to stop working. It required considerable investment to do this continual testing—buddy testers assigned to each developer, and a host of automated testing tools and test cases.

No team-based process is likely to be effective without good personal communication and formal or informal practices that facilitate the appropriate exchange of information. We saw several practices that worked in Microsoft. For example, despite some functional specialization and division of labor, people shared most tasks and responsibilities for those tasks. As discussed in Chapter 4, program managers and product managers generally took joint responsibility for writing product vision statements. Program managers and developers together defined the product features. Developers and testers both tested the code. Program managers, developers, and testers all helped answer customer support phone lines after the release of new products. In addition, Microsoft tried to do as much development as possible for the same product in a single location, and used common programming languages and other tools. Most teams also avoided bureaucracy as much as possible, though this became more difficult as the company and teams grew in size.

Finally, we saw how it was possible to design flexibility into a process in order to accommodate unforeseen change. In particular, Microsoft assumed that the product specification would continue to evolve until some very late point in a project in order to incorporate feedback from in-house users of prototypes, usability labs that tried out new features with typical customers, and beta users. The vision statement guided the overall objectives of a project, but this was not the same as a waterfall specification document. In the mid-1990s at least, managers expected the final feature lists for Microsoft products to change and grow by 20–30 percent compared to original estimates when a project started. Also, to accommodate change, projects included unscheduled "buffer" time (about 20 percent in application projects and as much as 50 percent for operating systems projects and totally new products). The practice of holding postmortem meetings and writing reports after each milestone and the project's conclusion also encouraged people to evolve the development process and improve as the teams gained experience.

None of these practices is peculiar to software development. In nearly any industry, it makes sense to restrict time and people because these limits force teams to focus their efforts and deliver something to the customer before the competition. It is good engineering practice to divide complex systems into modules or subsystems, as well as to reuse components and design some components for reuse—as in the software factory or multi-project management. It is good management practice to create project teams that map to the product architecture—to the components or subsystems. Teams should also be more effective and innovative if they have the autonomy to experiment and improve their designs or processes as they go along.

At the same time, projects need a few strict rules to force individuals to communicate frequently and to coordinate their work. Good communication mechanisms, a single location (or perhaps state-of-the-art communications technology), shared tools and techniques, and a minimum of bureaucratic rules and regulations encourage effective coordination. Common sense suggests as well that project managers should allow product specifications to evolve, schedule in buffer time, and include real-time process improvements. Managers need to accommodate many unknowns as well as the useful learning that comes with short development cycles in fast-paced industries. Of course, firms have to hire individuals with the ability to function with a minimum number of rules and exercise the discretion to be efficient when that is the best course of action and to be flexible on other occasions.

The Hewlett-Packard Study

In Chapter 4, I noted that the kinds of techniques Microsoft used were now relatively common around the world, at least as indicated in the global survey. Prior to launching this study, my co-authors and I did a more detailed analysis of these kinds of techniques at Hewlett-Packard (and Agilent, which was spun off during the research). Our objective was to compare the performance of projects that emphasized iterative or agile versus waterfall-ish techniques. We ended up with data on twenty-nine projects for the analysis and published our results in a fall 2003 issue of *IEEE Software*.[27]

The most important finding was that, at least within a single development culture (HP and Agilent), particular techniques appeared to function as a coherent set of practices. When used together, they eliminated tradeoffs between efficiency and flexibility. In the case of HP and Agilent, software projects were able to be more flexible in the sense of accommodating late design changes with a minimal impact on quality when they used several techniques together, rather than just selecting one or two. More specifically, when they used beta releases of early prototypes to get user feedback, conducted design reviews, and ran regression or integration tests on each build (that is, after each change or addition of new functionality to the working product), then the correlation between having an incomplete design when coding started and "bugginess" disappeared.

Statistical analysis (multivariate regression) also provided some very specific insights. Three factors explained approximately three-quarters (74 percent) of the differences in defects among the projects: how early customers saw prototypes, whether or not the projects did design reviews, and whether or not the projects conducted regression or integration testing on each build. We also found some striking results with regard to quality and productivity levels compared to the median project:

- Releasing a prototype earlier (that is, with 20 percent of the functionality done compared to 40 percent, the median level) was associated with a 27 percent reduction in the defect rate and a 35 percent increase in lines-of-code productivity.
- Conducting integration or regression testing at each code check-in was associated with a 36 percent reduction in the defect rate, again compared to the median project.
- Holding design reviews was associated with a 55 percent reduction in the defect rate.
- Conducting daily builds was associated with a near doubling (93 percent increase) in lines-of-code productivity.

Of course, there are different ways to measure productivity and quality; in particular, lines of code are a very crude metric, and defects per line of code say nothing about customer satisfaction or the economic value of the product. (We tried to measure economic productivity by asking for data on sales and profits, but the firms preferred to keep this information confidential.) Nonetheless, we have some concrete results from a relatively large-sample study that confirm observations found in other research. In particular, the finding that certain practices, when used together, seem to eliminate tradeoffs among productivity, quality, and flexibility is remarkably similar to the PCB study. This is also what many researchers have observed in other studies of manufacturing and engineering operations.

Flexibility in Strategy and Entrepreneurship

I cannot end this chapter without another example to emphasize that flexibility is not only important in technology and operations management: it is an equally powerful principle for strategy and entrepreneurship, and helps ensure that firms can remain competitive under changing conditions. I made this argument in *The Business of Software* (2004) when presenting a list of eight key factors to consider when evaluating a high-tech start-up (see Table 6.4).[28] For me to invest in a start-up or recommend that others invest, first, I want to see a management team with technical and marketing skills as well as experience (successes and failures). The firm should be in an attractive market (large enough, not too much competition, and so on) with a compelling new product or service. There should be evidence of serious customer interest as well as a plan to overcome the "credibility gap" many customers have in buying products or services from a new company that might cease to exist in a year or two. The business model should show an early path to larger operations and a way to

Table 6.4. *What to look for in a start-up company*

1. A strong and balanced management team
2. An attractive market
3. A "compelling" new product, solution, or service
4. Strong evidence of customer interest
5. A plan to overcome the "credibility gap"
6. A business model emphasizing early profits or scalability
7. Flexibility in strategy, technology, and product offerings
8. The potential for a large investor payoff

Source: Cusumano (2004: ch. 5).

make money—scalability and profitability. Investors should be able to see a large payoff, which depends on how all these factors come together with the size and attractiveness of the market. Of equal importance, though, the management team and the technology need to be flexible enough to evolve—to go through a process of trial, error, and experimentation.

My interest in how flexibility extends to strategy and entrepreneurship dates back to my early study of Toyota and Nissan as well as other new engineering companies formed in pre-war Japan, including Ricoh and other firms spun off from Japan's Institute of Physical and Chemical Research.[29] But the importance of flexibility remains somewhat controversial in the entrepreneurship literature. Researchers have found that strategic focus is critical to success to the extent that start-up companies usually do not have enough resources to do more than one thing well.[30] My point is somewhat different from this finding. I agree that new firms should focus on one product concept or business strategy at a time. But start-ups often need several tries to get their strategy, business model, and product line right. If the mindset of the founders is too narrow, or if the product technology or production process is too rigid—such as a technology that can be used for only one thing, or a manufacturing system that is extremely expensive and time-consuming to change—then firms cannot easily evolve their strategies and technologies to find a formula that works.

There are many examples where strategic and organizational flexibility have been important, but here I wish to focus on Netscape and its battle with Microsoft over Internet software. The Internet has been an enormously disruptive technology that challenged managers and entire organizations in nearly all industries to rethink the way they plan, compete, and develop new products and services. Netscape adopted nearly identical product development techniques to those used at Microsoft, but with more emphasis on being fast and flexible—even "slightly out of control," to use the words of one engineering manager. Their goals were to maximize innovation, creativity, and speed to market, but still deliver reliable products to consumers and enterprise customers.[31]

Netscape was founded in May 1994 in Mountain View, California, and quickly became the fastest-growing software company in history to that time. Its main product, the Navigator browser, captured more than

60 percent of the market in less than 2 months after release in December 1994. Netscape eventually built an installed base of more than 38 million users, making Navigator the world's most popular personal-computer application in the 1990s. The company also generated $80 million in sales for its first full year and reached an annual sales rate of more than $500 million within 3 years By contrast, it took Microsoft almost *14 years* to reach comparable revenues. Shareholders also made out well, even though Netscape ultimately lost the browser war to Microsoft. AOL purchased the company for a value that ended up at $10.4 billion when the merger was consummated in spring 1999.

Many pressures compelled Netscape, and other companies of that era, to be unusually fast and flexible. For example, as David Yoffie and I described in *Competing on Internet Time* (1998), in getting the company to ramp up so quickly, Netscape hired both an extraordinary number of superbly skilled programmers, who created much of the Internet infrastructure we use today, and many experienced managers from the software, computer hardware, and telecommunications industries. The internal capabilities they created, as well as prescient investments in enterprise-class infrastructure suitable for billion-dollar companies, helped Netscape grow with Internet usage and get big fast. Company executives also had a clearer idea earlier than most competitors of the Internet's potential and devised a flexible product road map to capitalize on that vision. Of course, Netscape would have done better had Microsoft not been nearly as flexible and much more powerful. Microsoft had the Windows platform to distribute its own browser and server technology, and a lot more people and money. Microsoft's more extreme actions also violated antitrust law, though, by the end of the 1990s, it mattered little. Microsoft had taken over the browser market, whilst Netscape became a part of AOL and was no longer a leading Internet software company. But there remains much to learn about flexibility in how Netscape responded to the battle with Microsoft, and how Microsoft was able to react.

From Vision to Products

Jim Clark and Marc Andreessen created Netscape based on a powerful but simple idea. They realized that the Internet was a new form of networking and communications technology that could fundamentally change how people lived and worked. Clark and Andreessen did not know exactly what forms these changes would take, and they dealt with this uncertainty by remaining flexible in strategy. But they were determined to create a business that would put them at the heart of the new "networked" world.[32]

Andreessen had led the team that developed Mosaic, the first popular Web browser, whilst he was a college student and working for the National Center for Supercomputing Applications at the University of Illinois. Then he met Clark, a distinguished computer scientist and professor at Stanford University as well as the founder of Silicon Graphics, the high-performance work-station company. Andreessen convinced him that the Internet would someday become like the phone system. He also argued that the browser could become a "universal interface" for the Internet and work on different types of devices: personal computers, touch-screen kiosks, cell phones, electronic organizers, and the like. Andreessen intended to base the browser on what were being called "open standards" like HTML (Hypertext Markup Language), rather than proprietary standards (such as Windows or UNIX). Both men agreed that the proprietary model of competition was likely to fail with the Internet. As Andreessen later explained, "In a networked world, we're headed toward everything being interconnected, and in that world, anything that doesn't talk to everything else [gets killed]."[33]

Neither Clark nor Andreessen was sure how to make money from this remarkable new technology. Neither was John Doerr, one of Silicon Valley's premier venture capitalists at the firm Kleiner Perkins, and an early Netscape backer and company director. In our 1997 interview, he recalled his first meeting with Andreessen a few months after the company had been formed. Doerr was unsure how to make money, but he immediately saw the opportunity of a lifetime:

> I vividly remember Marc sitting in his chair [in the summer of 1994]. Twenty-three years old and he said, "This software is going to change the world." And there was an alignment of the planets, as far as I was concerned... Bill Joy [the co-founder of Sun Microsystems] once told me, "John, someday you guys are going to back an eighteen-year-old

> kid who'll write software that will change the world." And so here's Andreessen, just five years older than eighteen, and I'd seen Mosaic, the UNIX version of it, running on a Sun Web Explorer in January of that year. Marc earned three dollars and sixty-five cents an hour, or whatever the University of Illinois had paid him, and he posted this thing on the Web, and two million people were using it. You would have to be as dumb as a door post not to realize that there's a business opportunity here.[34]

Four principles drove Netscape in these founding years. Three—the power of networks, the promise of a universal interface, and the need for open standards—would serve as a source of stability for the company, providing a basis for strategy and technology development. The fourth principle—which Yoffie and I described as "everything else was open to change"—directly addresses the importance of flexibility in a dynamic, uncertain world. Senior managers in the company expected the distinctions between different networks (proprietary, open, and so on) to blur over time. Clark and Andreessen also worried about detailed business plans becoming obsolete overnight, so they deliberately tried to be broad in the corporate vision. They intended to develop software for Windows and other platforms as well as follow open standards. They decided not to focus the company on a particular niche but instead be flexible in choosing which markets to enter. Indeed, Netscape would go on to introduce a surprising variety of products. However, as put by one senior executive, Roberta Katz, the big picture vision of the Internet and its potential remained fixed: "There's always kind of this overarching vision there, and then the question is, 'Okay, in today's marketplace, since we have to make money, what's the best way to do it?' "[35]

For example, in 1994–5, Netscape focused simply on basic products to take advantage of the World Wide Web, the graphical version of the older Internet. Publishers, retailers, and Internet service providers were starting to enter the market, but they still needed products and infrastructure technology. About two dozen companies were planning to release browsers, most of them based on the Illinois technology, which the supercomputer center was licensing to Microsoft and other firms. In response, Andreessen pushed his team to develop a browser from scratch (but modeled on Mosaic) to avoid any licensing or copying issues, and released a beta version in October 1994. Netscape then released two servers by the end of the year to take advantage of the network effect between the browser and the server. One server supported publishing documents on the Web and the other supported electronic commerce, such as selling books. At this time, Netscape also designed a set of "integrated applications" to facilitate Internet commerce and other functions. Within a year or so, Netscape had introduced nearly a dozen specialized servers supporting everything from online storefronts and Internet publishing to chat rooms, Internet-based email, and groupware.

By late 1995, however, Andreessen was becoming concerned that use of the Internet and Internet commerce was growing too slowly. This was something of a chicken-and-egg problem common in platform markets: not enough people with browsers and Internet access to encourage companies or individuals to buy servers and build websites—or not enough websites to compel people to use browsers and get Internet access. On the other hand, Netscape managers noticed that most of their sales were coming from companies that wanted to use browsers and servers for internal networks such as email and collaboration. Eureka! Andreessen pushed his engineers to use their knowledge of the open Internet protocols and the consumer Internet—the entire focus of Netscape's original business plan—to create internal corporate networks using Internet technology behind "firewalls," which the company decided to call "intranets." Netscape quickly added support for messaging and collaboration in its second version of Navigator, released at the end of 1995, and hired hundreds of salespeople to sell its technology to corporations. By mid-1996, Netscape held 80 percent of the new intranet market.

When other companies began creating intranets in late 1996 and 1997, Netscape moved quickly again. This time, Andreessen's team came up with the concept of an "extranet"—using private leased lines or secure Internet connections to extend the intranet to select outside parties, such as a company's supplier network, or premium customers and partners. Chrysler, for example, was an early adopter and deployed a Netscape extranet to connect its Supply Partner Information Network to 12,000 business partners. Netscape also acquired a high-end application server company to boost its technology skills.

By 1998, Internet commerce was growing strongly. Netscape had distributed nearly forty million browsers to corporations, who paid, and individual users, most of whom did not pay. Individuals generally received the software on a trial basis and were supposed to pay but did not have to—Netscape's "free, but not free" strategy.[36] And, whilst Netscape largely used open standards, several of the ways it implemented Internet technology were unique and tended to resemble proprietary solutions—the strategy of "open, but not open."[37] Netscape by this time had also developed an enterprise software business of nearly half a billion dollars a year, fueled by these different products.

With Internet usage finally picking up momentum, and tens of millions of Navigator users defaulting to the Netscape website when they started up their browsers, the opportunity was finally ripe for Netscape to become a major player in Internet commerce and online services, including content distribution. This was true, even though Netscape (like Microsoft) viewed itself primarily as a software product company and had a very difficult time deciding to diversify into services. But Netscape's website, named Netcenter.com, had quickly become the most popular destination site on the Web. It was hard for Andreessen to deny that Netscape was no longer a pure product company and probably would have a better future offering a combination of products and services through its website portal. In reality, the value of Netcenter.com and its forty million or so "eyeballs" eventually persuaded AOL to acquire Netscape for such an exorbitant price.

Trial, Error, and Tactical Successes

Netscape created billions of dollars in value for its shareholders, but it also made several strategic mistakes. These errors helped Microsoft catch up in browser technology and prevented Netscape from fully exploiting what had become its greatest asset. For example, Netscape rolled out browser versions and new features so quickly in evolving Navigator into the Communicator suite that the system architecture became very difficult to modify. This situation provided an opening for Microsoft to come in with neater, more modularized technology and improve Internet Explorer (IE) more quickly. Internet time also slowed down, at least measured by the time between releases of key browser and server product versions. For example, Netscape released four new versions of Navigator every 6 or 7 months between December 1994 and August 1996. Release intervals then slowed down to 10, 14, and 18 months. Microsoft released its first version of Internet Explorer in August 1995 and then the second in November, merely 3 months later. IE version 3 came 9 months later, in August 1996. By this time, Microsoft had matched Netscape in technology, and went ahead with IE 4 (September 1997), thirteen months later.[38] Moreover, Microsoft was giving away IE for "free" to everyone by bundling it with Windows 95, which users paid for directly or indirectly—another version of "free, but not free."

As we now realize, Microsoft violated antitrust law by pressuring PC manufacturers not to bundle Navigator, among other illegal actions. But Netscape compounded its problems by trying to use the new Java language before it was ready to rewrite Navigator and wasting more than a year of development time, if not more. After the sale to AOL in 1999, the open source community organized through the Mozilla Foundation had to work with Netscape engineers for several years to fix the problems and produce a new, streamlined browser. This technology is today at the heart of the successful FireFox browser. (In 2009, according to various sources, Firefox had 25–30 percent of the global market in browser usage, compared to about 60 percent for Internet Explorer, which had over 90 percent of the market as recently as 2002.)

Andreessen felt his biggest mistake was not to see the value of the website earlier. He admitted in a 1998 interview: "I thought [using our website] was a distraction. It's kind of funny to think about how many people have had the opportunity to make billion-dollar mistakes. I absolutely thought we were a software company—we build software and put it in boxes, and we sell it. Oops. Wrong."[39] But Internet time was moving so fast that it is hard to fault Andreessen and other executives at Netscape. CEO James Barksdale was also very savvy, but was late to see the value of the website portal. Netscape started out as a software products company, not a content and services company. Though it evolved, this change was dramatic and the opportunity much larger than anyone could have imagined. Netscape recovered quickly enough to build up Netcenter.com and sell itself to AOL, the online services company that later merged with TimeWarner. AOL was spun off in 2009 and continues to struggle with adapting its original private network and dial-up service to the open Internet.

Netscape managers, with all their challenges, were superb at experimenting and making quick, tactical decisions. It is possible that they could have used more balance—more strategic planning as well as frequent reviews of those plans and implementation efforts, not to lock themselves in but to give the company more stability at least for some longer periods of time. But, as another senior manager, Rick Schell, told us, "When the market is going to change in six months to a year, you can't spend two months worrying about a strategy. So you have to compress the time...In general, our time horizons are six months to a year."[40] Microsoft, by contrast, forced Internet time to slow down by "embracing and integrating" the Internet with Windows technology, and taking more time to roll out a strategy and new Internet-enabled products. Microsoft continued to work on its traditional 3-year planning cycle. With the pressure of Internet time, however, Bill Gates put in six-month reviews to increase Microsoft's ability to change or adapt decisions in shorter increments.

Judo Strategy

After observing Netscape closely for more than a year, Yoffie and I came away with another impression: the inherent flexibility in its vision, planning, and product development capabilities, combined with its life-and-death struggle with a much larger and powerful opponent (Microsoft), encouraged managers to think and act much like judo players. We called this approach "judo strategy" and distinguished it from the "sumo-strategy" style of Microsoft, which tended to rely on sheer market power and overwhelming resources to win battles. The term judo strategy was not completely novel; Yoffie came up with the term inspired by prior academic research on "judo economics."[41] But the Netscape case inspired this concept, which Yoffie and a co-author later expanded into an entire book.[42]

Judo is the "art of hand-to-hand fighting in which the weight and efforts of the opponent are used to bring about his defeat." The purpose of strategy for a company behaving like a judo player, similarly, is to turn an opponent's strength into weakness, rather than confront the strength of the opponent directly—as sumo wrestlers usually do.[43] The "judo approach" to competition emphasized "the use of *movement* and *flexibility* to avoid unwinnable confrontations and the use of *leverage* to undermine competitors' strength by turning their historical advantages (installed base, high existing prices, established distribution channels, and so on) against them."[44] Four principles seemed to capture the fundamentals of judo strategy as we observed in the competition between Netscape and Microsoft:

- Move rapidly to uncontested ground in order to avoid head-to-head combat.
- Be flexible and give way when attacked directly by superior force.
- Exploit leverage that uses the weight and strategy of opponents against them.
- Avoid sumo competitions, unless you have the strength to overpower your opponent.

Netscape brilliantly executed the core principles of judo strategy in its early days. When it was a start-up, we can see the concept of *movement* in how Netscape managed to avoid head-to-head struggles with more than two dozen competitors in the browser market, including several that had a head start. It sold the browser only over the Web and did not try to bundle it with different services, such as Yahoo! did. When Netscape did choose to go against larger competitors, it focused on pioneering nascent market segments such as Internet email, Internet groupware, servers, intranets, extranets, and applications where opponents were relatively weak and where it had a technology advantage. Netscape was also highly *flexible*, shifting focus to different products and even the website portal as competition changed. The company also embraced Microsoft technology and made sure the Navigator browser worked as a complement to the Windows platform. Perhaps most important, we can see Netscape's use of *leverage* in how it attacked competitors so that it was difficult for them to retaliate. In particular, Netscape publicly criticized Microsoft's proprietary technology strategy and presented itself as the champion of "cross-platform" technology and "open standards." It designed Navigator to work with Windows as well as UNIX, OS/2, Apple, and Java. Microsoft was so committed to Windows that it would never follow down this path.

Despite its many accomplishments, though, Netscape's browser and the Internet more generally represented a new industry platform for computing and communications that did not require Windows. This technology directly challenged the very essence of Microsoft's existence—as long as Bill Gates and his lieutenants were running the company. Ultimately, the judo match became a sumo match, with the smaller player competing with the monopoly power of the much larger company and the Windows franchise. Netscape's flexibility got it only so far, and help from the antitrust authorities came too late. Microsoft and Bill Gates were also flexible enough to embrace and extend the Internet, and quick enough to recover lost ground. They demonstrated true staying power. In fact, Microsoft's platform leadership position with Windows enabled it to reset the pace of Internet competition to the slower Windows release cycle. Microsoft shipped hundreds of millions of copies of Windows 95 and then later versions of Windows, with a free browser technically equivalent or superior to Navigator by 1997. At this point, there was little for Netscape to do except sell itself to a larger company, which might have a better chance at competing with Microsoft. This was not to be the case for AOL. Nonetheless, as we saw in Chapter 1, the Internet platform has challenged Microsoft to its core and brought another powerful company to the forefront of the Internet software and services business—Google.

Lessons for Managers

First, as with the pull concept but even more broadly, to emphasize flexibility over efficiency is a general philosophy of management. It can extend to everything from operations to strategy and entrepreneurship. It follows that flexible organizational or technical capabilities are useful only if managers exploit them. Second, we can show that there are minimal tradeoffs with cost and quality if managers implement flexibility properly and fully utilize the capabilities of technology, people, and design concepts such as modularity. Therefore, the strategic benefits of being able to adjust manufacturing systems easily, change product designs late in a project, or alter a competitive strategy in short intervals would seem to far outweigh any negatives, such as greater uncertainty in operations or planning. Third, flexibility in all its forms is an enabler of innovation and agility, and that has to be good for staying power. In sum, there is no good reason, in most cases, for managers to vote for efficiency over flexibility, given what we know.

The potential relationship between flexibility and innovation is especially significant. Flexibility in manufacturing or operations more broadly can help firms rethink their most fundamental assumptions and leap to new levels of performance and customer service. Firms should be able to adapt more easily to changes in market demand or to introduce new products more quickly—again, with minimal impact on productivity and quality. Flexibility in product or service development can help firms respond quickly to new information on customer needs or competition, as well as introduce new products and features or modify existing products quickly and efficiently—again, with minimal tradeoffs, albeit with some potential constraints. Flexibility in strategy and entrepreneurship can help start-ups and established firms seize new business opportunities and adapt, especially in unpredictable environments and difficult economic conditions.[45]

Flexibility in different functional areas of the firm can also be highly complementary. For example, changes in strategy are likely to require changes in products and services, which the firm has to be able to design, produce, deliver, and support quickly and efficiently. Most important seem to be organizational capabilities and specific techniques or mechanisms that help an organization adapt quickly, exploit unforeseen opportunities for innovation, and counter competitive threats. Some types of flexibility can enhance quality in the sense of identifying defects more quickly or bringing to the customer new features or responding to feedback more effectively.

It is also important to recognize that strategic or operational flexibility alone may not lead to organizational effectiveness and any particular competitive advantage, just as efficiency has little value by itself if the company is reusing flawed technology or making bad products that no one wants to buy or use. Flexibility appears especially important in certain kinds of market conditions. But companies must still be able to develop the right products or services at the right time, and do so better than the competition. This skill may benefit from flexibility but may depend more directly on superior capabilities or better timing or a better initial strategy. How flexible or efficient a firm needs to be thus depends on a variety of contextual factors. In some cases, firms may need to be equally flexible in technology management as in strategy. As we can see in the case of Toyota, however, an organization that can routinely *combine* world-class levels of efficiency with flexibility and quality—in manufacturing, product development, and other operations—is generally very effective and hard to beat. It will still have its ups and downs, and may even suffer uncharacteristic product disasters, such as Toyota experienced in 2009–10 with unintended acceleration, and as Ford experienced a decade earlier with Firestone tires on its Explorer SUV models. But a firm with true staying power must be able to recover from adversity or lapses in quality as well as adapt to new conditions and competitive threats.

Then there is the practical question: if focusing on flexibility rather than efficiency has such obvious benefits, why do not *all* managers and firms embrace this principle? As with the other five enduring principles, there is no simple answer to this question. Surely, we can assume again that managers face constraints on time and resources, as well as conflicting short-term versus long-term incentives. For example, creating a flexible factory generally requires more upfront investment and employee training. It usually requires changes in the relationship with suppliers as well as unions and workers. There are many benefits, as we have seen—the ability to make more new products or handle variations in demand better, or even to introduce a pull system and eliminate unnecessary inventory whilst improving productivity, quality, and learning cycles. But there are short-term costs before a firm reaps these long-term benefits.

Similarly, iterative or agile product development is more complex to manage and will nearly always take more time to perfect than a traditional waterfall process—even though we can show there need not be tradeoffs with quality or cost if managed properly. Moreover, iterative or agile development can support experimentation with new products and features as well as dramatically reduce rework by incorporating early feedback from customers, catching defects early, and making other adjustments to changes in market requirements.

When it comes to strategy and entrepreneurship, most managers and entrepreneurs should want to be flexible enough to adapt to market changes or uncertain competitive conditions. But organizations—and individuals—generally need some stability to get things done and may be slow to respond to the suble signs of change or negative information. At the opposite extreme, managers who embrace too much strategic flexibility create a tension with consistency and persistence. (As Henry Mintzberg once wrote, "Organizations that reassess their strategies continuously are like individuals who reassess their jobs or their marriages continuously—in both cases, people will drive themselves crazy or else reduce themselves to inaction."[46]) Be all this as it may, knowing when to change a strategy or an organization's behavior before it is too late is clearly important and essential for staying power. I believe managers can sharpen their judgment skills by thinking carefully about the principles and examples described in this book.

Notes

1. The labor story is extremely important to understand what happened at Toyota as well as Nissan. For a detailed discussion, see Cusumano (1985: ch. 3).
2. Allen (1984). See also Allen and Henn (2006).
3. For an excellent review of the theoretical and empirical literature on the flexibility–efficiency tradeoff, see Adler, Goldoftas, and Levine (1999).
4. For a more recent review of the literature as related to factory-like concepts in software development, including software factories and CMM, see Adler (2005).
5. See Horwitch (1987), Tushman and O'Reilly III (1996), and O'Reilly III and Tushman (2004).
6. See D'Aveni (1994).
7. Brown and Eisenhardt (1997, 1998); Eisenhardt and Brown (1998).
8. Davis, Eisenhardt, and Bingham (2009).
9. See, e.g., Lawrence and Lorsch (1967), Aaker and Mascarenhas (1984), Harrigan (1984), Sanchez (1993, 1995), Volberda (1996, 1999), Ghemawat and Del Sol (1998). A useful collection of papers is Birkinshaw and Hagström (2002).
10. Pine made this argument in his MIT Sloan master's thesis as well as Pine (1992).
11. Jaikumar (1986).
12. For reviews of the operations flexibility literature, see Sethi and Sethi (1990), Gerwin (1993), and Upton (1994). See also Pagell and Krause (2004). This last paper attempted to replicate prior results from the machine tool industry and did not find, in a multi-industry study, significant correlations among uncertainty, flexibility, and performance.
13. See the references for product platforms in Chapter 5.
14. Sobek, Liker, and Ward (1998).
15. Henderson and Clark (1990).
16. Christensen (1997).

17. Details of the sample, methodology, and the statistical analysis can be found in Suarez, Cusumano, and Fine (1996).
18. The references are Suarez, Cusumano, and Fine (1996) and also Suarez, Cusumano, and Fine (1995). The discussion here adapts the 1995 text.
19. Kekre and Srinivasan (1990).
20. We largely followed Clark and Fujimoto (1991) on this metric.
21. In addition to Krafcik (1988a, b) and Womack, Jones, and Roos (1990), see MacDuffie (1995) and MacDuffie, Sethuraman, and Fisher (1996).
22. One can find this argument, and some data, in Garvin (1988), as well as Krafcik (1988a, b).
23. See Skinner (1974).
24. See Microsoft Corporation (2009) and Pogue (2009).
25. This section closely follows Cusumano (1997), which is based on material in Cusumano and Selby (1995: 409–17). This topic is also discussed in Cusumano (2004: 144–59).
26. Interview with Dave Maritz, then Test Manager, MS-DOS/Windows, Microsoft Corporation, Apr. 15, 1993; originally quoted in Cusumano and Selby (1995: 18–19). See also Cusumano (2004: 146).
27. The median project had 170,000 lines of code, 9 people, and lasted 14 months. The median project also had 40% of the functionality complete when the project released the first prototype, 35.6 defects per million lines of code (0.04 per 1,000) as reported by customers in the 12 months after release, and a productivity rate of 18 lines of code per person day (360 per month). See MacCormack et al. (2003). Also Cusumano et al. (2009).
28. Cusumano (2004: 210–11).
29. See Cusumano (1989a).
30. See the discussion of success factors in Roberts (1991a); also Roberts (1991b).
31. Cusumano and Yoffie (1998: 222).
32. Cusumano and Yoffie (1998: 17–88).
33. Cusumano and Yoffie (1998: 25).
34. Cusumano and Yoffie (1998: 22).
35. Cusumano and Yoffie (1998: 27).
36. Cusumano and Yoffie (1998: 97–100).
37. Cusumano and Yoffie (1998: 133–8).
38. For a table on the release intervals, see Cusumano and Yoffie (1998: 233).
39. Cusumano and Yoffie (1998: 38).
40. Cusumano and Yoffie (1998: 67–8).
41. Gelman and Salop (1983).
42. Yoffie and Kwak (2001).
43. Definitions and sources discussed in Cusumano and Yoffie (1998: 89).
44. Yoffie and Cusumano (1999).
45. Again, my thanks to Michael Bikard for pushing me to clarify these distinctions.
46. Mintzberg (1987: 72).

Conclusion: The Power of Ideas and Research

I have argued in this book that six principles have enduring value for managers and should create staying power for the firm. The underlying ideas are not short-term management fads but have had a lasting and growing impact on management strategy and practice over the past several decades. The examples illustrate how these principles have helped firms in a variety of industries adapt to change and remain competitive even in unpredictable markets or bad economic conditions. Again, the six principles emerged from my reflections on some three decades of research on firms ranging from Microsoft and Toyota to Intel, Apple, Google, and many others. The principles also have substantial support in academic research crossing several disciplines. But simply knowing the principles is not enough. Managers need to figure out how to apply the basic ideas to their specific contexts, and do so better than the competition.

How to Understand and Apply the Six Principles

To recap. The first two principles (platforms and services) represent broader ways of thinking about strategy and business models, as well as innovation more generally, for product firms or service companies delivering mostly standardized services. The next four principles (capabilities, pull, scope, and flexibility) are all about *agility*—how to be flexible and quick, in operations as well as strategic change. Some practices associated with agility, such as dynamic capabilities, just-in-time production, iterative or prototype-driven product development, flexible manufacturing, modular architectures, and component reuse, are now commonly regarded as essential best practices in innovation and engineering management. Perhaps most important, though, is for managers to recognize the strategic value of agility. To a large extent, agility enables an organization not only to adapt to change but also to *create* variety, planned and unplanned. Another term for planned and unplanned variety may in fact be *innovation*—in products, production processes, strategy, and business models.

We also saw in this book how platforms and services involve new kinds of capabilities and broaden corporate boundaries, such as when managers extend pull and scope concepts to suppliers and other ecosystem partners. In addition, industry-wide platforms and services place new demands on strategic and technological flexibility, especially in a world where information and digital technologies have become ubiquitous. Platforms and services impact on agility as well: they can play a huge role in helping managers prepare for and influence change within their markets, whether the change is commoditization or technological disruption. We can summarize the key managerial applications of the six principles as follows:

- Managers, whenever possible, should adopt a *platform* strategy (or a *complements* strategy for someone else's platform) and build special technical and marketing capabilities around this approach. The goal is to take advantage of network externalities and a broader ecosystem for innovation that turns suppliers and even competitors into complementors and partners. Platform thinking encourages a new type of scope economies in the sense of leveraging the capabilities and resources of other firms to help produce complementary innovations. Strong network externalities also limit how agile the competition can be: customers and complementors cannot or will not easily switch to another platform.
- Managers of product firms, whenever possible, should pursue *services* as a strategic lever to help sell products, create new products, de-commoditize old products, build deeper relationships with customers, and generate another source of profits and revenues, especially in mature markets or difficult economic times. Professional services such as technical support, consulting, customization, system integration, and training also represent a

direct pull effect from the market and require a different set of capabilities from the ones product firms commonly emphasize. At the same time, product firms that master services should find these to be another important source of differentiation and competitive advantage.

- Managers should base their firms on distinctive organizational *capabilities* and focus on evolving them over time rather than be overly preoccupied with short-term strategies or detailed strategic planning. Of course, all managers want to produce superior products and services of superior value to customers. But extraordinary capabilities also seem to require extraordinary efforts to acquire a deep knowledge of industry dynamics, the technology, and customer needs, as well as how these elements are changing. Managers, therefore, need to be on the lookout to avoid complacency and view all capabilities as dynamic: nurturing them is an ongoing process, particularly with rapidly changing technologies and in uncertain markets, where lapses can more easily occur. These kinds of situations are also where the most important threats or opportunities for innovation and new business development are largely unknown.
- Managers should incorporate *pull-style concepts and systems* wherever possible to link their production and service operations as well as product development efforts directly to the market in as close to real time as possible. Sometimes firms know exactly what to produce and in what quantity, and precisely how operations are proceeding and how products will perform in the marketplace. But especially when lacking this certainty, pull systems can generate real-time information and are much more responsive to change than push systems relying on rigid, sequential processes that incorporate feedback or new information slowly or not at all, and are difficult to alter. Again, however, a pull system in manufacturing does not correct flawed product designs, and can provide useful input for non-manufacturing operations only to the extent that managers devise the proper feedback mechanisms and pay attention to new information.
- Managers should constantly look for opportunities to maximize *economies of scope*, which involve sharing—across different departments, projects, and activities—a common base of knowledge,

information, product inputs, and technology. The goal is to achieve efficiencies in areas of the firm not usually subject to scale economies. Scope economies may be more difficult to identify and exploit than scale economies. Scope strategies such as reusing components in many different products also demand a higher degree of quality control in production and design than simpler operations focused on single-product economies. But this greater risk in implementation makes scope economies a potentially powerful source of competitive differentiation for firms that successfully make the investment.

- Managers should pursue *flexibility* in most if not all areas of operations and decision making, and organizational evolution, while understanding that they need not trade off efficiency or quality. To the contrary, flexible thinking, processes, and technology are often essential to creativity and can enhance quality and productivity. They enable the firm to capitalize on unexpected opportunities for innovation and new business development. Most of all, flexibility in manufacturing, product development, services, research, and skills development, as well as in strategy and entrepreneurship, can help the firm respond quickly to adversity and adapt to changes in the marketplace and competition.

I have emphasized that these six principles should transcend specific company examples. They should be timeless, even when best understood or implemented as part of an "if this, then do that" kind of contingency framework. For example, capabilities are such a broad notion, and so deeply attached to strategy, that they will always be important for competitive advantage. But pull systems, scope economies, and flexibility, as well as platforms and services for product firms, will be more or less important under different conditions. Moreover, all firms encounter variations in their businesses and life cycles. We need to take these ups and downs as well as other factors into consideration when thinking about what is enduring and what is not. We should not be misled by short-term trends, unusually good or bad performance, or even the occasional corporate disaster. I have relied on examples from several firms to illustrate the principles, and the performance of these organizations has varied over time. But, unlike prior authors writing about best practices, I believe that we can understand what makes for staying power only by observing how firms respond to a variety of conditions. Depending on when we view any particular firm, some will be doing better or worse than some competitors or what they have done in the past. But to be overly influenced by these natural variations or human error detracts from the importance of the underlying principles.

For example, we all know that Microsoft has found it impossible to sustain its early growth rates. When you have been expanding sales at 40–50 percent a year and attain a 90 or 95 percent share of your market, there usually is only one place to go—down. Even so, the company has done extraordinarily well financially. It remains a powerful platform leader and has been able to adapt, sometimes begrudgingly, to technological change: character-based computing to graphical user interfaces, languages to operating systems to applications, consumers to enterprises, and the desktop to the Internet and mobile systems. At the same time, Microsoft has hesitated to move much beyond the desktop Windows and Office platforms, strategically and technologically. It remains remarkably weak in the fastest growing segment of computing—mobile software.[1] Microsoft does not fare well when we compare it to Apple in terms of deeper capabilities for innovation, though "new-to-the-world" type of innovation is not what Microsoft has been about. So there are minuses to the DOS–Windows legacy (novelty remains constrained by the technology) as well as pluses (the enormous desktop revenue and profit streams continue).

Not surprisingly, Microsoft experienced its first year-over-year decline in revenues in fiscal 2009, at least in part resulting from the worldwide financial collapse and recession in 2008–9. This one-year drop reflects a longer-term maturation of the personal computer business and Microsoft's limited diversification. But it does not mean that the principle of "platforms, not just products," is any less important now compared to when Microsoft was growing sales at 40 or 50 percent a year. And we continue to see the six principles at work in Microsoft today, in varying degrees: persistent attempts to expand the platform strategy to video-game, mobile, and digital media markets; gradual preparation for a world dominated by automated online services that enhance, extend, or even substitute for traditional software products;

distinctive technical and marketing capabilities cultivated for the PC market; a pull style of product development that has helped generate an unending flow of products and features, all refined incrementally; and scope economies as well as flexibility in strategy and engineering management. To become a truly outstanding performer in the future, however, Microsoft needs at least periodically to reinvent its technology platforms as well as its competitive strategy.

The other company that has occupied a lot of my time—Toyota—recently became the world's largest as well as the most scrutinized automobile producer. Toyota also did poorly in the recession compared to its previous performance. It recorded a loss—another first—in fiscal 2009. To be fair, no automaker foresaw the massive and rapid collapse in consumer demand. And, unlike most of its competitors, Toyota chose not to close factories (with the exception of NUMMI in California, jointly owned with General Motors) or cut back much on employment. At the same time, however, Toyota managers in recent years became overconfident in the quality of their products and in their ability to expand continuously. More disturbing, they appear to have hastily introduced new technologies and components, especially from non-Japanese suppliers, to boost overseas production more quickly as well as to keep down expenses. I have briefly discussed how Toyota introduced into mass production some technologies, such as new types of pedals and braking controls, that it had not yet perfected. One devastating outcome is that Toyota vehicles were involved in several fatalities that were due to sudden acceleration, apparently caused by sticking gas pedals and possibly faulty sensors and braking software. As noted earlier, the company recalled more than 9 million vehicles in 2009–10 and encountered other uncharacteristic quality problems that received less publicity. This is not the Toyota that millions of loyal customers have come to know.

Other companies discussed in this book have experienced ups and downs as well. JVC, under financial distress for many years, finally left the Matsushita group and merged with another Japanese consumer electronics producer. Sony has struggled for years with declining profits and sales. It has fallen behind Samsung in brand value, and at times seemed overly dependent on the vicissitudes of the video-game console market. IBM struggled with the transition from mainframes to smaller computers, though it has excelled in the transition from hardware to software and services. Intel's sales and profits rise and fall as PC sales and semiconductor demand wax and wane. Intel is also gradually diversifying beyond the PC market, but this has been difficult and slow. Most of Cisco's sales and profits depend on the investment cycles of the telecommunications equipment makers and service providers, though it is adding more consumer products, such as Linksys wireless routers and Flip cameras. Even Google and Qualcomm—our latest monopolists—have been unable to sustain the remarkable growth and profit rates of their early years as consumer demand slows and advertising dollars shrink in the global recession. By contrast, Apple is finally doing well now that it has adopted a broad platform strategy that links together its various products and automated services.

Countries have their peaks and valleys, too. Japan has experienced mostly stagnation, deflation, and occasional modest economic growth since 1989. In large part this is because of political gridlock and massive subsidization or protection of inefficient sectors of the economy (see Appendix I). But long-term investments in distinctive manufacturing and engineering capabilities generally have stood up well. Japan remains a very wealthy country and the second largest economy in the world (with China very close behind) largely because of products that competitors still find hard to match in terms of technology, quality, and price. In automobiles, of course, Toyota, Honda, and Nissan remain the leading firms, and I fully expect Toyota managers to rededicate themselves to quality and safety appropriate for their advanced technologies. Sony, Sharp, Sanyo, and Panasonic are still world-class consumer electronics firms, though not all have had equally stellar financial results every year. Canon, Nikon, Ricoh, and Olympus remain leaders in cameras, copiers, scanners, and other optical products. Fujitsu, Hitachi, NEC, and Toshiba have had their troubles with flat demand, too many lifetime employees, and too many businesses. But they all have deep and broad capabilities as producers of computer hardware, software, and many other consumer and industrial goods. NTT, NTT Docomo, and NTT Data continue to form a powerful block of companies in the telecommunications sector, albeit with most of their sales in Japan.

Beyond these and other global Japanese giants, there exist many medium-size Japanese technology companies with near-monopoly shares in critical businesses. Nearly all rely on the same kinds of design, precision engineering, and manufacturing skills that Sony and JVC honed before they won the VCR competition and that Toyota has displayed for most of its history. As recently noted in *The Economist*, for example, Japan Steel Works is the only company in the world able to forge 600-ton ingots for nuclear reactors. Shimano sells as much as 70 percent of the world's bicycle gears and brakes. YKK produces half the world's zippers. Nidec makes 75 percent of the hard-disk drives in computers. Mabuchi manufactures 90 percent of the micro-motors in automobiles for adjusting rear-view mirrors. TEL accounts for 80 percent of LCD panel-etching machines. Covalent makes 60 percent of the containers used to hold silicon wafers and 70 percent of the carbon brushes in electric motors. Shin-Etsu produces 50 percent of the photo-mask substrates in semiconductor manufacturing. Murata has 40 percent of the global market for capacitors—tiny electronic devices sold in the millions that hold electricity in everything from cell phones to computers. Mitsubishi Chemical makes nearly all the red phosphorescent materials used in white-light LED bulbs. Kyocera dominates many segments of the integrated-circuit component business.[2]

The bottom line is that we can expect vacillations in the performance even of outstanding firms and highly advanced economies replete with companies following the best management practices. But managers who grasp the principles described in this book—all of which have withstood the tests of time and geography, as well as rigorous academic scrutiny—should create firms that stay ahead of the competition most of the time and adapt quickly to unpredictable change as well as adversity.

Agility and Staying Power

It follows that agility, including all the principles and practices that contribute to this, should be essential for firms to survive and thrive in good times and bad. Change—in technology, consumer preferences, market demand, and global politics—is the rule in the modern world, and often it takes managers by surprise. In the previous century alone, the telephone, radio, television, automobile, airplane, transistor, and computer, as well as the Internet and modern cell phone, have all appeared somewhat out of the blue to most observers. All have had a revolutionary impact on society, individuals, businesses, and economic development.[3]

But we also know that major innovations and even revolutions—including all the technologies mentioned above—usually give some warning signs and unfold relatively slowly. There are accidental inventions put quickly to use, like the X-ray. But most technologies find commercial applications only after years of incremental innovation and experimentation. This process of necessity includes trial and error—much like the way Thomas Edison invented or the way the Internet evolved. No one knows exactly when commercially successful products will appear. But, for most technological and business revolutions, there exist at least subtle hints that the future will be different from the past. Some exceptionally aware or lucky people will interpret these signals correctly. Others will work to make a particular vision of the future happen. Henry Ford did this with the mass-market automobile. Bill Gates and Steve Jobs did this with the personal computer. Marc Andreessen and his colleagues at Netscape did this with the Internet.

Revolutions may start with a bang but unfold slowly, or start slowly and then unfold with a bang. Either way, unlike completely unpredictable catastrophes or accidental blockbuster inventions, technological revolutions usually give off some early signals and take decades to reach their full impact. The speed of change in markets and society in general is not so fast when things get in the way like "sticky" information, huge sunk costs, and high customer switching costs. But the fact that change is usually not so abrupt provides entrepreneurs, engineers, and managers—and their *firms*—with time to act, if they are sufficiently agile and prepared. Predictable and unpredictable catastrophic events will occur, such as the collapse of the Internet investment bubble in 2000, or the great recession of 2008–9, or the product flaws that Ford and Toyota experienced between 2000 and 2010. But then the most agile and capable firms should still be in a position to survive and to thrive again some other day.

The computer and the Internet are great examples of how technological revolutions—as with Internet time itself—unfold more slowly than we might think. The electronic computer was invented in the mid-1940s, after decades of evolution in electro-mechanical calculators and telephone relay switching equipment a well as electronic circuits, beginning with the vacuum tube. But not until the early 1980s did the computer—now the *personal* computer—really begin to transform society and individual lives, and not only large organizations, in a revolutionary way. The Internet also has its roots in the Arpanet network created by the US Department of Defense in the 1960s. Not until the mid-1990s—with the invention and combination of hyperlinks and the World Wide Web with a graphical browser running on a personal computer and accessing servers with Web content—did the Internet begin to transform society and individual lives, as well as business, in a truly revolutionary way. So too with other technologies, like the automobile with an internal combustion engine, or radio and television, or seemingly simple products like the video cassette recorder, a forerunner of modern media storage and recording devices, and the smart phone: all took decades to become mass-market products or services and transform our personal and business lives.

For firms to have true staying power and perform well across generations of change in products, technologies, customers, and competitors, almost by definition they must be agile and just quick enough to enter the windows of opportunity before they close. Alfred Chandler, in *Scale and Scope* (1990), found that sheer size seemed to contribute to staying power. But we also know that large organizations, because of bureaucratic forces and other mechanisms, find it difficult to change quickly and can also lose their focus and attention to detail. So the cause of what Chandler observed could be the other way around: exceptional qualities more correctly associated with agility may have enabled some firms to evolve more effectively than others. They stay around long enough to become big enough to evolve their capabilities and exploit scale or scope opportunities. Which is cause and which is effect? My vote is for scope rather than scale, and for agility in general.

Other Implications for Strategy and Theories of the Firm

Agility is useful for the *practice* of management, especially for surviving change or adversity. But what are the implications of agility or any of the six principles for deeper views of how managers or entrepreneurs should organize their companies? For example, managers and entrepreneurs need to decide what to do inside versus outside the firm, with partners, suppliers, or even customers. These boundary questions go to the heart of what academics call "the theory of the firm." This is an old topic associated mainly with the economists Ronald Coase (a Nobel laureate) and Edith Penrose, but the big questions remain fundamental to any ongoing quest to understand strategy and implementation as well as how best-practice thinking has changed over time.

For example, economists and other academics generally want to know if companies have any advantage over loose associations of individuals contracting with each other in an open market—like a network of suppliers organized through a Web marketplace. To justify creating a company, or to perform particular activities internally, the firm should be able to do things that the market (or the government) cannot. Or firms should be able to do things more efficiently and effectively, such as by reducing what Coase called "transaction costs."[4]

Transaction costs have long been part of the everyday language of MBAs and practicing managers. They come in the form of extra time, risks, or money spent trying to engage individual workers or other independent entities to do things like make components, assemble products, or provide services needed in the course of production and other operations. It is difficult to find and train workers, especially in rapidly growing economies. It is more difficult to write employment contracts that cover all different kinds of work environments and contingencies. It is even more challenging to develop people with critical knowledge and retain them. Moreover, firms usually invest in what Oliver Williamson (co-recipient of the 2009 Nobel prize in economics) called "specific assets"—equipment, knowledge, technologies, and the like that may be good only for producing what the owners or managers of a firm want to make but not much else.[5] So one theory of the firm argues that this organizational form is a sort of a repository and mechanism to house specific assets or knowledge and avoid the transaction costs of trying to do the same thing with market forces and external contracts. Now we know how Williamson would explain GM's decision to make a majority of its parts in-house—pursuing what economists, consultants, and strategy professors would call a high level of "vertical integration."[6]

The problem we have seen in the automobile industry, and which Toyota and other Japanese firms ruthlessly exploited, is that vertical integration also can make a firm relatively *inflexible*. In part this is because of sheer size and the bureaucracy required to do so many things with so many people in one firm. GM employed over 600,000 people at its peak in the late 1970s—practically a small nation. Toyota in the same period had only about 50,000 employees in the parent firm and maybe double this number in its major subsidiaries. Even in 2009, including consolidated subsidiaries, Toyota employed only 300,000 workers. These numbers are important because we know from decades of research that agility generally declines with organization size. Different structures, such as use of subsidiaries or semi-autonomous decentralized business units and teams, can reduce this problem. But they create others, such as more difficulties in coordination, knowledge management, and quality control (as Toyota also experienced).

We also know that organizations find change difficult because people get used to doing the same things. In automobile manufacturing, Toyota, Nissan, Honda, and other Japanese firms and their suppliers benefited not only from less vertical integration but also from less restrictive union regulations and lower wage scales, or no unions at all (like many, but not all, of the Japanese auto transplants in the United States). Looking at the bigger picture, though, it should not surprise Americans that an external market for auto components became cheaper and more innovative than the internal divisions at GM and Ford once the overall size of the auto industry grew.[7] The negatives of vertical integration seem especially apparent when we think about how modern information technology and the Internet greatly facilitate the functioning of global marketplaces and global supply chains.[8] But there is no one right approach for all situations; what to do depends on the circumstances. For example, when it comes to the complex electronic components and software Toyota uses in 2010, it may be better to hire more software engineers and bring more of the design, testing, and production skills in-house; or to rely more on overseas R&D centers in the United States and Europe, which may have easier access to software skills. If Toyota cannot create in-house the new capabilities it needs, then perhaps the time has come simply to change the way it procures and tests components so that similar flaws never recur.

There is much to think about when imagining what firms of the future should look like if they fully embrace the six principles proposed in this book. Of course, I expect future firms to focus on both platforms and services, not just products. They will understand that distinctive capabilities are far more useful than planning systems or short-cut strategies. But they should also be preoccupied more with organizing development in small groups, building more prototypes, and leveraging better direct pull-style linkages to customers. Economies of scope and flexibility broadly conceived should become commonplace as well, just like transaction costs, economies of scale, and efficiency have been in the past.

I suppose I am arguing that successful firms of the future need to be more sophisticated in their motivations as well as in strategy and management. I have not even touched upon the issues of the environment or sustainable business models, or the challenges of overcoming poverty and religious strife, or extending Western notions of entrepreneurship and economic development to less fortunate parts of the world. But surely managers must consider all these issues and more as they think about staying power.

To be fair, economists and organization researchers concerned with theories of the firm have long argued that firms are complex organizations and usually do more than simply pursue efficiency. Penrose argued that firms bring together unique resources and capabilities necessary for growth, even though shortages of managerial talent are equally likely to restrain growth.[9] Joseph Schumpeter (at least in his early writings) as well as Israel Kirzner argued that entrepreneurs establish firms to exploit economic or speculative opportunities, and thereby introduce innovation into the economy.[10] We know that some entrepreneurs create firms in order to change their lives or to change the world, while many simply want to own a market or do well enough to remain independent. And managers do not always function as profit-maximizing "agents" of the major stockholders; their decisions often reflect incomplete information, standard operating procedures, and conflicting motivations such as financial compensation, job stability, and even job satisfaction.[11]

But the six principles do provide a framework for thinking about what firms of the future should look like, in any market. They also suggest how industry structure (such as Michael Porter's five forces—power of buyers and suppliers, entry barriers, extent of rivalries, and substitutes) and generic competitive strategies (Porter's low cost, differentiation, or focus positions) relate to staying power. If we combine Porter's ideas with the concept of platform leadership, we see how managers can use the firm to *influence and leverage* market structure and industry dynamics for their own ends. Then entrepreneurs and the managers they hire can gamble with a higher probability that the future will move in a particular direction such that emphasizing cost or differentiation or focusing on a particular niche or investing in new platform technology will be profitable. DeBeers did this by orchestrating a tight global monopoly on the diamond market for 100 years before production in countries outside its control reduced its influence.[12] (Diamonds may be forever, but monopolies are more fragile, though still hard to displace.) Can we say that JVC and Matsushita/Panasonic significantly influenced the market for VCRs, just as Microsoft and Intel did for the personal computer? What about Google for Internet search, eBay for Internet auctions, Qualcomm for wireless technology, Apple for digital media, Adobe for editing tools, and Wal-Mart for mass-market retailing? Yes, these and many other firms have used platform strategies to shape their markets or supply chains, and that is a major reason why they usually persist and out-perform the competition.

An agility-based theory of the firm suggests that a distinguishing feature of the very best companies should be to anticipate and exploit change. But, while inefficient, inflexible firms may have no obvious theoretical reason to persist, the practical reality is somewhat different. Capitalism takes different forms in different countries. Certainly, we have seen these alternatives appear in Japan, the United Kingdom, and Western Europe, as well as Korea, China, and India—as just some examples.[13] In 2008–9, we even saw a more "protective" form of capitalism become prominent in the United States as government agencies from two vastly different presidential administrations prevented or delayed bankrupt firms from failing in financial services and automobiles. These kinds of actions are not unprecedented in the United States, where the federal government (and some state governments) have long assisted distressed firms and industries. But government intervention affects which firms endure and which do not.

These recent actions in the United States and abroad suggest that, in practice, another purpose of the firm is *social welfare*—a mechanism to *preserve* wealth and public harmony by providing employment. Here again, though, agility-related principles such as flexibility and capabilities are still very important. Governments can prop up failing enterprises such as General Motors or Citibank for only so long. Eventually, the burden on taxpayers and national treasuries becomes too great. Even subsidized entities must adapt or they will disappear or shrink to irrelevancy. Again, firms that deeply embrace the six principles talked about in this book are unlikely to become candidates for social welfare. They should be able to survive adversity as well as economic and environmental catastrophes.[14]

The concept of a platform leader with its own ecosystem also suggests an alternative to the traditional boundaries of the firm. It has some connection to notions of the "virtual" organization and low levels of vertical integration.[15] It has some affinity to "open innovation" and "democratized innovation."[16] It is in part related to arguments that the future of work is all about distributed organizations and distributed innovation.[17] But platform leadership differs from these other important ideas: it suggests there is a special role for *the firm as a leader* in markets driven by network externalities and where complementary innovations are more likely to occur if some entity provides direction, facilitation, and coordination. It is also probable that large ecosystems improve the survivability of the platform leader and its complementors, as long as these firms can overcome catastrophic events and evolve their technology, products, and other capabilities from one generation to the next.

Toyota and Microsoft—Strange Bedfellows, or Not?

It is impossible to conclude this book without coming back to the comparison of Toyota and Microsoft, two firms not usually talked about in the same sentence. It may be chance that they have occupied such a large portion of my research over the past three decades. Both have been extraordinary firms in their markets and their dominance has persisted over multiple generations of products, technologies, and customers. Today, both firms also face imposing challenges and are not performing as well as they once did. Still, no one should deny that both have demonstrated remarkable staying power and should continue to be extraordinary firms for a long time to come. But what specifically do Toyota and Microsoft have in common? Managers must also be asking how their companies can become more like these two firms, minus their negative features.

There are some obvious and non-obvious similarities. Both companies are centered in semi-rural locations, away from the mainstream cities where most of their competitors and customers exist. Both, perhaps as a result, have evolved distinctive, in-bred cultures. The corporate cultures capture the essential values of their founders and early managers (called "meta-rules" by some observers and "principles" by others, including this author), such as the unrelenting drive for market power and at least continuous incremental improvement. But they can mobilize enormous resources when threatened or challenged. Both tend to hire young people and nurture them in the distinctive company culture and values. Both have preferred to develop new products and process technology independently, though Microsoft has been making more acquisitions as time goes on, with limited results. Both remain far more renowned for their process skills than for product innovation or invention skills. Both have benefited from enormous ongoing revenue streams (Toyota's are due to its superior productivity and premium pricing; Microsoft's are due to its dual platforms of Windows and Office). These superior financial resources have made the two companies far wealthier than their competitors and able to do more things or invest more heavily in preserving or expanding distinctive capabilities. Both have deceptively strong marketing organizations, which complement their technology skills and leverage their superior financial resources. But both companies are also relatively conservative when it comes to anticipating and embracing change.

Both Toyota and Microsoft, at least in their earlier days, relied heavily on a kind of "paranoia," in the sense of Intel's Andy Grove: Toyota used to pursue continuous improvement almost like a corporate religion; and Microsoft, at least when Bill Gates ran the company, used to behave as if some upstart company was sure to come along any day and steal a good share of its business. Of course, Microsoft once did this to IBM, and Netscape almost did it to Microsoft. Internet start-ups (such as Google) have been doing this to Microsoft for the past decade. Apple has also leapt way ahead of Microsoft in innovation by linking computing to consumer electronics and digital media. For its part, Toyota has to deal with a constantly struggling local economy as well as a resurgent Nissan, an always innovative Honda, resilient European automakers such as Volkswagen, and new threats from Hyundai as well as Chinese and Indian automakers. Toyota in 2010, as we have seen, also faced the greatest threat in its history: quality and safety problems that surfaced as the company overtook GM in sales. The historical legacy of Toyota as a conservative but fundamentally agile, learning organization is now being put to its most serious test.

Toyota and Microsoft both inspired my enduring principles—in varying degrees. The last four clearly have been essential to both firms. Toyota and Microsoft are very much centered on distinctive capabilities cultivated largely in-house. Both organize around pull concepts, Toyota more so than Microsoft. Both exploit vast economies of scope in product development as well as enormous economies of scale in production (or product replication). Both are highly efficient and structured but also relatively flexible. They have been agile enough to adapt to significant technological changes, though this is an ongoing story.

Microsoft (along with JVC and Intel) provided most of the inspiration for the first principle—industry platforms. Toyota seems to have benefited less from this idea, since the automobile market is not one with obvious network externalities. At the broader level, though, this is certainly not true. An automobile is not so useful without complementary innovations such as modern roads, fueling stations, insurance companies, and repair shops, as well as the many component and material suppliers. Mass-market automobiles are relatively standardized company platforms with well-understood interfaces to critical components and services. Moreover, it is impossible for one firm to provide all the parts and services necessary to make, sell, and maintain automobiles in volume. So there are, in reality, lots of network effects surrounding the modern automobile and the internal combustion engine. This is why shifting technologies will be difficult and slow. Toyota has had the most commercial success among automakers in nudging the traditional platform away from pure gasoline to a hybrid gas-electric technology, best seen in the Prius. But Toyota's hybrid technology is also a new platform and involves complex interdependencies with a variety of suppliers and partners. Toyota still has more to do to perfect the technology in this new platform. Moreover, a more challenging revolution in automobile technologies is yet to come. It will not occur until the world shifts to a truly different platform, such as electric motors powered by hydrogen fuel cells.

In terms of the supply chain, and despite recent quality problems, Toyota has created the most efficient global base for production in the world auto industry. Toyota will probably continue to rely on its network of subsidiaries and first-tier suppliers for about half of all components and assembly operations. To become a part of this different kind of platform involves a lot of overhead for suppliers, who must link their production systems and R&D processes to Toyota's. Therefore, to join Toyota's ecosystem and remain a viable member involves some fairly strong indirect network effects, and probably some direct effects as well, because of the specific R&D and manufacturing investments the suppliers make for Toyota products.

As for the role of services, especially automated services, these had already become a major topic of discussion at Microsoft when I first began to study the company in the early 1990s. MSN began as a private AOL-like online network, which Bill Gates and other Microsoft people then called "the Information Highway." We can see evidence of Microsoft's strategic and technological flexibility in how quickly the company changed course and turned MSN into an Internet service while adapting Windows and Office to what Bill Gates in May 1995 called "the internet tidal wave."[18] MSN has since generated billions of dollars in losses for Microsoft. But MSN has also served as a learning laboratory for delivering online software products, content, updates, patches, advertisements, and other automated services to customers. More importantly, with more than a decade of experience with online services and advertising, Microsoft now has the technical capabilities to deal better with the new world of digital services, software as a service, and cloud computing—or whatever we call the "servitized" versions of what once were software products.

Nonetheless, the future is unlikely to be as good for Microsoft as the past has been. The golden days of expensive personal computers seem on hold for the present, especially since PCs now share the computing market with hundreds of millions of relatively cheap mobile phones, netbooks, and other digital devices such as Apple's iPod, iPhone, and iPad. We are in a much more diverse technology world today than in any time in the past. Microsoft will do well to follow IBM's lead and remain a powerful and profitable player in a market with several powerful and profitable companies. A lot of market power has also shifted to users through Internet technology (which can make or break mass markets overnight) and movements such as open-source software and social networking. But there are still billions of people who have yet to purchase personal computers, game consoles, handheld computers, or smart phones. Microsoft will probably never get 90 percent of these new markets. But many people will still become Microsoft customers in the future, especially as Microsoft learns how to play more effectively in a world of multiple competing platforms.

Toyota also has much to do in the future, with products to fix, a reputation to rebuild, and the marketing challenge of reaching billions of people yet to purchase an automobile. It needs to get back—quickly—to making the most reliable cars money can buy, at a reasonable price. Toyota's manufacturing skills are still intact. Senior managers put too much emphasis on overtaking GM in sales and meeting demand for the Prius, when they should have been paying more attention to customer feedback on their basic models and the risks of new technology and their aggressive overseas expansion. But, as the Japanese, American, and European automobile markets recover from the global recession, and as overseas markets grow, Toyota should once again become the pride of Japan and remain the world's largest automobile manufacturer. But, first, Toyota managers need to learn or relearn a more general lesson: except in some very specific instances, managers should never see scale as more important than scope. And scale should never become more important than quality, safety, and corporate integrity.

How to Learn

In conclusion, I want to come back to the issue of how—or how should—managers (and researchers) learn. Where do ideas and examples that help firms fix the mistakes of the past and inspire enduring principles usually come from? Where *should* they come from—academics or practitioners? My experience is that, at least in strategy and innovation, the key ideas and practices generally come from *practitioners*—from firms at the leading edge of what they are doing. It is there in the trenches, so to speak, and not in the "ivory tower," where *firms* conduct real-time experiments and generate most of our useful knowledge about management, both positive and negative. Once researchers identify effective practices, researchers study them more deeply and teach about them in business schools. Consultants catch on and build practices around them. Then more managers and students of management adopt the ideas as standard best practices. This process is similar to the emergence of a "dominant design" in product life-cycle theory.[19] But, if everyone knows what everyone else knows, then differentiation of the firm must shift from common best practices to other dimensions, such as implementation skills or use of new technology, or to ideas that will lead the next generation of thinking about best practices.

For example, business schools generally have made distinctions in how they teach about product firms as opposed to services firms. But we have seen that the traditional distinctions between what is a product and what is a service—never so clear to begin with—has become increasingly vague and even irrelevant in some industries. What is most important is how effectively firms solve compelling customer problems. In many industries, we see product firms trying to take better advantage of services to help their product businesses. We also see firms trying to turn products into services or services into products. These transformations remain understudied in business schools. But managers and entrepreneurs are way ahead of the academics here. They have been "servitizing products" and "productizing services" for many years, especially since the arrival of the World Wide Web in the mid-1990s.

Managers seem at their best when examining what is right before them, not necessarily what is behind them or far in front. This is where academics can play an enormously significant role in helping everyone learn more about what we know and do not know. Knowledge requires some certainty of cause and effect. It requires case studies and broader sets of data, as well as careful analysis. It is when firms and universities collaborate to explore fundamentally new phenomena—like the inner workings as well as the strengths and weaknesses of the Toyota production system, the implementation of agile product development, or the true dimensions and tradeoffs of focusing on flexibility versus efficiency—that we get the best understanding of what makes for excellence in management and true staying power for the firm.

When considering what academics believe versus what managers know, I often think back to Toyota's Taiichi Ohno and the period when Toyota was a struggling automaker on the verge of bankruptcy. I told the story in Chapter 4 how, just after the Second World War, Ohno followed the lead of American supermarkets and aircraft manufacturers to experiment with a pull system for production management. He kept refining and extending the concept as he saw positive results. Toyota also had the skills in-house to exploit opportunities that managers did not foresee. In this case, it was important that Ohno came to automobiles from another manufacturing business and brought with him no set assumptions about the mass production of automobiles. He was able to "think outside the box." But Ohno's process innovations also took advantage of the more flexible equipment and incremental learning philosophy that Kiichiro Toyoda had already instilled in the company over the prior decade.

Eventually, this one principle—pull, don't just push—transformed manufacturing at Toyota and then many other companies around the world, in many industries. But not until the 1980s did academics systematically study and mathematically model what pull systems actually did. Only then did "we academics" formally acknowledge that pull was superior to push in most instances, albeit with some limitations, and understand why Toyota achieved such dramatic improvements in productivity and quality, as well as flexibility, all simultaneously. And, as Toyota overcomes problems with its latest technologies and overseas factories, managers and researchers will have a unique opportunity to reflect on the limitations of the pull system as well as on how better to apply pull-style feedback mechanisms to product design, testing, and technological innovation more broadly.

We have another similar example in software development. Today we can easily explain why firms such as Microsoft and Netscape deviated from conventional good practices in software engineering during the 1990s and adopted iterative or agile techniques that emphasized speed, flexibility, and change. They needed to experiment with new product designs and features to find out what customers wanted, get to market more quickly than mainframe companies used to do, and leverage rapidly evolving technologies. Most firms in 2010 consider agile, iterative, or prototype-driven development as standard best practice in software engineering and product innovation more generally. Most firms also understand the importance of establishing themselves as platform leaders.

The best managers usually figure out what to do even in unprecedented situations long before academics devise a theory or mathematical model to explain what is going on. Even Alfred Chandler's observation in 1962 that structure should follow strategy was not really a theory. It was a powerful idea based on detailed research into how huge corporations successfully diversified and managed scale and complexity.[20] This kind of thinking is what academics do best. Theorists and practitioners need to understand why something works and why something else does not, as well as what happens by chance and what is deliberate or inevitable. We should all be concerned with success as well as failure, and we need to understand causality well enough to correct mistakes and prepare better for the challenges of the future. To improve management, we need both the power of ideas as well as the knowledge that comes with empirical research. To me, at least, it follows that the search for enduring principles should begin with practice, both good and bad, but that is never where it should end.

Notes

1. Peers (2009).
2. *The Economist* (2009).
3. Taleb (2007) makes the point that rare events happen more often then we think. This leaves the future essentially impossible to predict, although we know that change is inevitable and catastrophes will happen, eventually.
4. Coase (1937).
5. See Williamson (1975, 1985).
6. Monteverde and Teece (1982).

7. Stigler (1951) makes this point that markets will deintegrate as they grow and new entrepreneurs will enter, making components and providing various services.
8. The book that has gotten the most attention on this theme is Friedman (2005).
9. Penrose (1959).
10. Schumpeter (1934). Also, Kirzner (1937, 1985).
11. See Cyert and March (1963) and Eisenhardt (1989a).
12. Gemawat and Lenk (1998).
13. On how ideas have spread regarding different types of "capitalism," see, e.g., Guillen (1994) and Huang (2008).
14. Because of my prior work on the automobile industry and role as co-director of the International Motor Program until recently, I was asked by media and government representatives whether I favored bankruptcy or a bailout for the American automakers. Early on I favored a structured bankruptcy that eliminated the top managers and replaced the boards of directors at GM and Chrysler. See MIT News (2008).
15. Jacobides and Billinger (2006).
16. Chesbrough (2003) and von Hippel (2005).
17. Malone (2004).
18. Gates (1995).
19. Abernathy and Utterback (1978) and Abernathy and Clark (1985).
20. Chandler, Jr. (1962).

APPENDIX I

THE RESEARCH CHALLENGE: IN SEARCH OF "BEST PRACTICE"

It is important to say something more about the research approach that led me to the six principles talked about in this book and why I believe they are enduring. Academics and consultants, as well as many successful managers, have written extensively about "best practices" that they believe lead to competitive advantage. Much of this research, however, raises as many questions as it answers.

One major problem is that what seems to work for a company in one time period, industry, or national setting often does not work for other companies in different circumstances or even for the same company in another time period or a different industry. We must, therefore, face the possibility that enduring principles or "best practice" in a given area of management may not exist in any absolute sense—that is, for all firms, industries, and environmental or institutional conditions over multiple generations. For these and other reasons, to evaluate the relevance of management research, managers and students of management need to be able to make judgments on their own. They need to form their own assessments as to which principles are potentially enduring and which are simply management fads or too easy to imitate.

Another problem is that many different styles of research exist. These variations can lead to both different insights and different conclusions. Each style of research has its advantages but usually produces an incomplete picture of a given phenomenon. Sometimes the academic lens of one methodological discipline—such as that of the economist, sociologist, psychologist, operations researcher, or business historian—acts more like a "silo" and obscures a broader view of what is really happening (not unlike the story of blind men touching different parts of an elephant and not realizing what is before them). The obvious conclusion here is that no one approach to management research is likely to tell the whole story about what leads to competitive advantage. We need a multidisciplinary, eclectic combination of methodologies, including qualitative and quantitative approaches to research, and different levels of analysis and abstraction (products, teams, firms, industries, nations, regions, etc.) to understand how firms perform and why.

Academics must also come to terms with the argument that "practice" (that is, firms), and not "theory" (that is, academia), is the source of many if not most of the enduring ideas in management. Theorists and empirical researchers may well be the first people to explain, document, or mathematically model a great idea. There probably are some areas, such as financial modeling or combinatorial optimization, and perhaps human resource management, where theory has led practice. But my experience studying firms such as Toyota, Microsoft, Apple, Intel, IBM, and Google suggests that academics generally trail leading-edge firms in introducing the practices that end up winning markets—including all the practices associated with my six principles.

Case Studies vs. Large-Sample Research

Most academics in business schools at some time confront the *case study*, at least for teaching purposes and often for research as well. Harvard Business School, founded in 1908, pioneered business school education and research largely by adopting the case method for teaching. A business school teaching case is usually an in-depth descriptive look at a particular organization or managerial situation to illustrate one or more pedagogical lessons, though without giving the reader a complete answer. In fact, there often is no one right answer to the problem highlighted in a case study. The professor guides students through a "Socratic" discussion, with the goal of improving their analytical and judgment capabilities.

The opportunity to probe an interesting situation or organization in depth is the great strength of the case study, especially when extended to book length. But the obvious weakness is that we cannot generalize from one or even a few examples. The cases may be too unusual. Or non-obvious factors or random chance may have influenced what we see. Case studies can have great value to generate ideas, if selected carefully. But, ultimately, they are only exploratory and illustrative.[1] Cases or small-sample studies do not bring certainty—at least, not *statistical* certainty—about what might represent an enduring principle or a best practice in management.

To overcome these limitations, scholars in fields ranging from medicine and physics to economics and management have turned to *large-sample research*. Here the goal is to study a phenomenon by analyzing a bigger percentage of the relevant "population," hopefully with a sample that is random or at least not obviously biased in a particular direction. The goal is also to analyze the data statistically so that we can attach probabilities to propositions and add "controls" to see if what we think is happening really is happening. How large the sample and how sophisticated the statistical methods usually depend on the questions asked and the available data.

My undergraduate and graduate school training in history prepared me to appreciate and conduct detailed case studies, including my doctoral thesis work for *The Japanese Automobile Industry*. After arriving at MIT Sloan in 1986, however, I decided to combine detailed case research with large-sample statistical analysis for my new project on software factories and their engineering practices. In particular, I wanted to know if we could measure "factory-ness" in software development and compare the effectiveness of different approaches in Japanese and US firms. Collecting survey data on a relatively large number of projects seemed the only way to do this. Though I had taken an introductory course in statistics in college, I quickly learned that I needed to use more advanced methods. In addition, I struggled with problems such as how to define robust "variables" as well as increase the reliability and comparability of the data. Then there was the difficulty of separating out what is cause and what is effect, and interpreting results from the "50,000-feet-high view." With limited information, we often make assumptions about how an organization might have made decisions or behaved, and this can produce wrong conclusions about underlying causes. The most famous case taught in business schools illustrating the use and misuse of statistics versus the insights available only through a detailed case study centers on Honda and why it succeeded in the motorcycle industry. The Boston Consulting Group, in a detailed analysis, argued that it was scale economies and learning curves. A deeper look revealed that Honda's success was due more to Soichiro Honda's love of engines and the distinctive capabilities in engine development that his company acquired as a result, as well as some marketing insights and good fortune.[2]

Then we have some best-selling books that appear more rigorous than they really are because of problems in their samples and the questions asked, or in the lack of statistical controls. Two examples come immediately to mind. One is from 1982, *In Search of Excellence*, by the consultants Tom Peters and Robert Waterman.[3] They were relatively ambitious: Peters and Waterman's team reviewed sixty-two of McKinsey's best-performing clients, used some financial and subjective criteria to weed out some firms, and then focused on forty-three companies that illustrated a small set of principles seemingly fundamental to success. Their list: *a bias toward action, closeness to the customer, a spirit of autonomy and entrepreneurship, a focus on productivity through people, hands-on management, a strategic focus on what the company is good at, simple and lean staffing, and a simultaneous combination of tight centralized values with looseness or decentralization at the workplace level.* Whilst it is hard to argue with these principles, soon a number of the highlighted firms ran into major problems or even disappeared (Atari, Data General, DEC, IBM, NCR, Wang, Xerox). What was enduring excellence and what was luck—or bad luck—seem hard to distinguish, in retrospect. We might also wonder exactly why Peters and Waterman choose these companies and not others. How do we really know that the factors they talked about, and not other factors, were responsible for the performance, good and bad, of these firms?

A more recent best-seller by Jim Collins, *From Good to Great* (2001), suffers from this same lack of controls.[4] It is harder to detect because Collins did use a lot of numbers and does appear methodologically rigorous. As with *In Search of Excellence*, there are also many useful ideas and anecdotes about management in Collins's book, which builds on his previous best-seller, *Built to Last* (2004). But, again, because of the structure of the study, it is not possible to determine which concepts represent enduring best practices and which do not. At least in part to remedy this problem, Collins recently published another study on why great firms fail, though this latest work has its own flaws and does not pay much attention to prior findings on the subject.[5]

For *From Good to Great*, Collins and his team of MBA researchers at Stanford Business School looked at nearly 1,500 of the largest companies to find the best performers in terms of stock-market value. Then they analyzed in some depth the best eleven of those companies: Abbott Laboratories, Circuit City, Fannie Mae, Gillette, Kimberly-Clark, Kroger, Nucor, Philip Morris, Pitney Bowes, Walgreens, and Wells Fargo. The sample they chose through statistics: only firms that had a particular stock-market performance before and after a certain date made the cut. Choosing a sample this way is fine, and the book ends up with a complete population of firms with a particular level of stock-market performance. But then Collins and his research team, relying on interviews with managers and reading secondary materials, turned to subjective methods and extracted a small set of principles that seemed to describe why these companies had done so well. Their list: *a particular leadership style of mostly internally promoted CEOs, a focus on talented people, clear understanding of internal strengths, simple fact-based performance goals, a disciplined culture centered on commitments, a reinforcing use of technology, and momentum built from early successes.*

Once more, it is hard to argue with the relevance of these qualities. Collins also cites the opinion of one friendly statistician that his results are very unlikely to be random. But, unfortunately, like Peters and Waterman, Collins made no attempt to measure and test these attributes, such as to correlate one or more of them with stock performance or any other measures. As a result, we have no way to determine *statistical* significance, or prioritize the factors, or control for other companies that might have exhibited the same or most of the same factors but did not have comparably good or great stock performance. It is also a potential problem when a large group of researchers collaborate on subjective analyses; it is very hard for them to be consistent.

Moreover, as with *In Search of Excellence*, several of the companies Collins highlighted did not do so well after publication of the book—again, making us wonder what are and are not enduring practices that give firms staying power and the ability to maintain superior performance. Abbott struggled with fraud lawsuits and problems with its drug Oxycontin. Circuit City encountered several years of dismal performance and dramatic losses in its stock value (even before the crash of October 2008). Its new CEO resigned before the company declared bankruptcy and closed down early in 2009. Fannie Mae went bankrupt in the 2008 subprime mortgage crisis and had to submit to a government takeover. Gillette did reasonably well but not well enough to avoid a takeover by Proctor & Gamble in 2005. Even Walgreens stumbled financially and saw a major decline in its stock price, again, before the crash of October 2008 and a strong revival afterwards.

Some of Collins's points resemble what we saw in Peters and Waterman (Table I.1)—understanding what the firm is really good at (internal strengths) and a focus on (talented) people. But other points seem completely different in the two books. My list of enduring principles also has some similarities but is mostly different from these two books.

One common characteristic of all three books is that the lists are subjective, even though Peters and Waterman as well as Collins went through a specific process to get to their group of firms. My list also differs in that it is primarily a set of themes found in my research. I do not claim to have compiled *THE* list of anything except recurring, powerful ideas fundamental to strategy and innovation management. In addition, I have chosen principles with a particular goal in mind: each contrasts with a more rigid approach that does *not* seem to represent best practice as we

Table I.1. *Comparison of best practice ideas*

In search of excellence (Peters and Waterman 1982)	Good to great (Collins 2001)	Staying Power
1. Bias toward action	1. Internally promoted CEOs	1. Platforms, not just products (when possible)
2. Closeness to the customer	2. Focus on talented people	2. Services, not just products (or platforms)
3. Autonomy & entrepreneurship	3. Understanding of strengths	3. Capabilities, not just strategy
4. Productivity through people	4. Fact-based performance goals	4. Pull, don't just push
5. Hands-on management	5. Momentum from successes	5. Economies of scope, not just scale
6. Focus on what firm is good at	6. Reinforcing use of technology	6. Flexibility, not just efficiency
7. Simple, lean staffing	7. Momentum from early successes	
8. Simultaneously loose–tight or centralized–decentralized		

Sources: Peters and Waterman (1982); Collins (2001).

currently understand it. Moreover, there is considerable theoretical and empirical research, some by myself and much more by other researchers, that supports the importance of these principles. In short, compiling definitive or objective lists of enduring principles or best practices in management is very hard to do.

Hybrid Methods: Qualitative and Quantitative

Of course, it is possible to go beyond the best-sellers and use more rigorous methods. We might begin with a qualitative study to improve our basic understanding of a problem, devise some metrics and collect data to analyze statistically, test some hypotheses based on theory or careful observation, and then drill down through detailed case studies and intensive fieldwork to probe the phenomenon in more depth. Intuitively, this approach is very appealing and, at MIT Sloan, we often encourage our doctoral students to follow this path. I have also used this hybrid approach in most of my work since 1986. But there are drawbacks. This type of research is generally very time-consuming, and the researchers have to master two very different skill sets: qualitative and quantitative.

One academic study I found inspiring followed a similar methodology to that of Collins but primarily used historical analysis to compare genres of capitalism at the national level. This is the book *Scale and Scope: The Dynamics of Industrial Capitalism* (1990), written by the late (and Pulitzer prize-winning) business historian Alfred Chandler, who began his career at MIT and finished it at the Harvard Business School.[6] This represented an enormous project: analyzing financial data on the top 200 corporations in the USA, Britain, and Germany (600 in total) between the late 1880s and the 1940s. Chandler concluded that size brought specific advantages in terms of both scale and scope economies, in operations ranging from production to marketing, distribution, and professional administration. This was especially true in the United States, where firms were bigger. The nature of capitalism also varied in these countries, apparently driven very much by the average size of the dominant firms.

Chandler used simple descriptive numbers rather than statistical analysis to rank firms by assets and market value. But the idea was bold: measure an entire population, identify quantitatively which were the top organizations, and then describe those firms in more detail qualitatively. Because it focuses on national comparisons, *Scale and Scope* does not go into the detail of Chandler's other writings, such as the classic *Strategy and Structure* (1962).[7] This earlier book explained how DuPont, General Motors, Standard Oil (Exxon-Mobil), and Sears moved to a multi-divisional organizational structure—which we now see as an organizational best practice—after adopting business diversification and globalization strategies. In *Scale and Scope*, the primary idea is that size matters. Firms can take advantage of size both through conventional scale economies, where costs decline with volume production or distribution of a single product, and through scope economies, as in the production, marketing, and distribution of multiple products using the same facilities or channels.

Two other books stand out for effectively combining large-sample analysis with detailed fieldwork. Both resulted from multi-year research programs on the automobile industry at Harvard Business School and MIT.

The more scholarly of the two is by Kim Clark and Takahiro Fujimoto, *Product Development Performance: Strategy, Organization, and Management in the World Auto Industry* (1991). This study is based primarily on Fujimoto's 1989 doctoral thesis, supervised by Clark, the former dean of the Harvard Business School.[8] They documented the advantage of Japanese automakers in product development productivity (lead time and engineering hours) as well as total quality through a meticulous study of twenty-nine projects at twenty-two automakers in the USA, Japan, and Europe. Confidentiality agreements prevented Clark and Fujimoto from identifying specific firms, so they focused on regional comparisons. But, even so, they illuminated some striking differences. Most importantly, the best firms and projects, which were mainly in Japan, utilized what they called "heavyweight" project managers and project teams. The firms also relied extensively on overlapping activities, short concept development phases, tight coordination across engineering and manufacturing functions, and "black-box" designs and components from suppliers. Several other related techniques facilitated fast product development times and very efficient use of engineering resources. The great strength of the book is the detail with which the authors analyzed development processes at the component and project level, the linkages to manufacturing capabilities and competitive strategies, and the rigor with which they collected and analyzed both qualitative and quantitative data.

The more managerial of the two books is by James Womack, Daniel Jones, and Daniel Roos, *The Machine that Changed the World* (1990). This was a summary report of research done under the auspices of the five-year International Motor Vehicle Program (IMVP), based at MIT.[9] (Full disclosure: I was a faculty member in the program at the time and supervised several of the master's theses. I also later served as the program co-director with John Paul MacDuffie.) Again, the IMVP researchers were not allowed to reveal individual company names and focused on regional comparisons. But they clearly documented the striking advantage of Japanese automakers—whether they operated in Japan or in North America—in manufacturing productivity (mainly assembly plant hours) and quality (defects, provided by J. D. Powers), based on the study of sixty assembly plants in Japan, North America, Europe, and developing countries. *The Machine that Changed the World* ended up selling over 600,000 copies in eleven languages.

Both books demonstrate how academics can do research of enormous value to managers—not by leading practice, but by analyzing, measuring, and explaining what the best firms at the time are doing. Both studies also demonstrate the importance for academic researchers to take the time to gain a deep understanding of the phenomena they are analyzing. The quantitative evidence of Japanese superiority during the 1980s and 1990s in efficiency, flexibility, and quality, in both automobile manufacturing and product development, is hard to refute. Some questions remain regarding the effectiveness of these practices in different settings, such as with different union work rules. But carefully collected data and precise qualitative descriptions of *why* the Japanese practices worked so well proved convincing and valuable to managers, and have played a major role in disseminating knowledge about best practices in manufacturing and product development.

All the studies I have mentioned so far, including my own books, encounter challenges when we try to generalize. Are Toyota-style "pull" techniques always associated with higher productivity and manufacturing quality—in any company, industry, country, and time period? How can Toyota and other companies use pull concepts more effectively in design and testing? Are economies of scope and flexibility always important or only under certain conditions? There are many factors to consider before we can confidently state that one underlying concept or specific practice is better than another. Every practice takes place within a certain time period and context, and we need to understand the impact of that context to determine what is a "real" effect and what is specific to a particular firm or industry or due simply to chance.[10] Any college course or textbook dealing with the use of statistics will emphasize the need to identify the proper "control variables," though how to specify these is not always so obvious. The methods of business school researchers are also growing increasingly sophisticated and difficult for non-experts to understand. Nonetheless, I offer some examples below of major contextual variables that researchers in the fields of strategy and innovation have generally considered relevant. More importantly, managers should consider these same issues when deciding which ideas or practices to follow as well as how to implement them in their particular situations.

Imitation versus Implementation

First, we need to acknowledge that, even if a particular practice appears fundamental to competitive advantage, other firms are likely to imitate that practice like lemmings following a leader.[11] Then, at some point, it will no longer be as effective as when the practice was unique or rare. The advantage here may only go to the first mover or pioneers. Successful imitation is perhaps the major reason why industry leaders lose their edge.[12] If most firms adopt a new best practice, then that becomes standard practice and, at least on the surface, may no longer be so useful. At the same time, however, firms can differ greatly in implementation skills. This difference may reflect a deeper, more subtle set of concepts or techniques related to process, people, and technology management.

For example, most firms in the automobile industry followed Toyota and introduced pull systems to reduce parts inventories and assemble vehicles as well as make or receive component deliveries more on a just-in-time basis. But inventory levels as well as productivity and quality levels vary greatly across firms, and even within the same firm, depending on the factory location, product design, worker training, local management commitment, and other factors. Therefore, managers and researchers should not simply think of a firm as following or not following a particular practice. We need to look more deeply into how well an organization implements a practice and why differences across and within firms exist. Then we can better start to understand what really are best practices in particular contexts.

Industry or Technology Life Cycles

Another factor to consider when comparing firms is the stage of the industry or technology *life cycle*. In the 1970s, William Abernathy and James Utterback brought the idea of life cycles into the everyday language of managers as well as scholars concerned with operations and technology strategy. Their studies and follow-on research found that many industries go through periods of "ferment" to maturity and then decline in terms of growth rates, profitability, or number of competitors. They argue that, in industries where maturity sets in, manufacturing companies need to shift their focus from innovation in product design to process efficiency (that is, manufacturing) as designs standardize and companies come to compete on the basis of price. Other scholars such as Stephen Klepper have explored this idea further, with some refinements, but the general observations remain similar. There is now a large literature on this topic, and the research includes both a rich set of industry analyses and large-sample statistical work.[13]

The implications for enduring principles are straightforward. Some strategies or management approaches may be superior in the early stages of an industry, when emphasizing speed to market, applied research, or innovative product designs is most important. But, as competitors move in, industry structure changes, and the technology evolves, specific practices that once led to competitive advantage may no longer be so effective. We know that not all industries follow life cycles, but enough do to make this an important factor for both managers and researchers to consider.

Unfortunately, it is much easier to identify where one is (or whether one is) within a life cycle *after the fact* rather than before or in real time. This is a major problem when trying to use life-cycle or industry stage as a control variable. Nonetheless, it is important for any firm-level or project-level analysis to try to account for differences that may be related to the life-cycle phenomenon, or at least the maturity of an industry.[14]

Nature of the Technology or Innovation

Another variable is the nature or type of technology, or the associated innovation, apart from the maturity of the industry. Academics generally define technology as the knowledge or capabilities used to turn some form of inputs (for example, materials, components, or even raw knowledge itself) into some form of outputs (for example, products and services). But technology, or technological innovation, can take the form of product design, system architecture, materials, manufacturing, or even the distribution process (such as the Web). A product or process technology can also be new to the firm, new to a particular market, or new to the world. Technological innovations can be incremental, radically new, or potentially disruptive to existing players in an industry and their ways of making money. All these variations on the nature of a technology and associated innovations can affect whether or not a particular practice works or does not work in a given context.

For example, practices that work extremely well with a relatively well-understood technology may fail completely with a new technology or one of a different nature. Proven techniques for managing automobile product development may not work so well when applied to very different kinds of tasks, processes, people, and inputs, such as occurs with software, pharmaceuticals, and chemical products. Toyota-style production management may not result in the same benefits outside repetitive assembly operations, although I am not so sure of that. Best practices for managing technological innovation may differ when we are talking about a product versus a process versus a service or a digital good. Or perhaps the nature of the technology does not matter as much as factors such as familiarity with the technology. If so, then a particular practice, such as agile product development, or a pull system, should provide equal benefits in almost any setting.

We do know that the challenges facing managers as well as researchers are not simple. Many products contain a mix of new and old technologies, such as in the form of different components or subsystems and subroutines. Hence, it is often difficult to classify the technology used in many products, including automobiles—which contain some technologies invented a hundred years ago and other technologies invented almost yesterday. Case studies are useful to explore these different issues, but they cannot settle many arguments. Larger sample research using type of technology or type of innovation as a control variable is usually the best way to explore the question of how much does technology matter. But researchers still need to figure out how to measure different technologies in ways that make them comparable and control for meaningful differences in technology or types of innovation.[15]

"Clockspeed" and Other Industry or Business Differences

A related factor is how much does the *industry* matter (apart from technology) when trying to understand differences in company performance. For example, is the fact that a particular firm is a member of the pharmaceutical industry as opposed to the automobile industry more important for predicting its profit and growth rates than other factors? Certainly, industry differences affect other metrics, such as average product development times—which can be a decade for a new drug versus a few years for a new car and perhaps a few days for a new software program. My MIT Sloan colleague Charles Fine has referred to these kinds of variations as "clockspeed"—how fast or slow the cycles of change, or life and death, are in a particular industry.[16] These can vary greatly and impact on what practices work for firms in a particular industry, as well as how long they work. Fine's argument, essentially, is that all competitive advantage is temporary, and this is more obvious in fast-paced industries, such as software and Internet services.

Several economists and economics-trained strategy researchers have also tackled this question of industry differences from somewhat of a different perspective. The studies do not agree completely, though they build upon each other and refine prior results. Point of view also matters here, as it usually does in research. Economists are usually trained to look for macro-effects, such as industry differences, whereas strategy researchers tend to place more importance on firm-level or managerial effects—what the field has come to call differences in "resources" and "capabilities," as discussed in Chapter 3. But we need the tools of the economist here because large-sample econometric analysis is the best methodology for this type of question.

In a 1985 article, the economist Richard Schmalensee (and my former dean at MIT Sloan) reported that industry factors were most important in influencing profitability rates at the business-unit level compared to corporate affiliation (membership in a particular diversified firm) or market share effects, and accounted for about 20 percent of the difference in profitability. This was a landmark study, but Schmalensee used only one year of data. Richard Rumelt, a prominent strategy researcher with an economics bent who had earlier done pioneering work on corporate diversification levels and profitability, used a multi-year dataset and was able to separate out "stable" from "transient" influences. His 1991 article found that industry-level differences did matter somewhat, but that more fine-grained business-unit or segment effects were far more important, accounting for nearly half the variance in profit rates. In other words, business segments within the same industry—for example, luxury cars versus mass-market cars—differed as much or more when compared to each other than when we compare different industries—such as automobiles to steel manufacturing. Another article from a strategy and economics point of view by Anita McGahan and Michael Porter, published in 1997, confirmed results found in both of the earlier studies. They found that industry accounted for about 19 percent of the variation in profitability, and business-segment differences about 32 percent. But, since the business-level differences persisted over time, and these were strongly affected by industry differences, McGahan and Porter concluded that broad industry effects actually did matter more than the other factors, albeit with some sector variations.[17]

For the purposes of this book, the flip side of the basic question about industry impact is: to what extent does *management* matter? At least within an industry, we can say that there are significant differences in how similar organizations perform, and this surely has something to do with the characteristics of those organizations and their managers and employees, rather than just industry-level factors. But, when looking at an entire population of firms, such as the S&P 500, we need to control first for industry differences, direct and indirect, such as at the business-unit level. It is also impossible to generalize convincingly without multiple years of data. It follows that researchers must be careful of market segment differences that may show up more in metrics such as product type, within an industry or business. But the problem remains: it is not always so clear that managerial practices matter more than other factors when it comes to understanding firm performance.

Environmental or Institutional Context: Japan as a Case in Point

A further obstacle that has stood in the way of determining enduring principles and best practices relates to the *environmental or institutional context.* We have to account for the fact that firms in one country may perform differently or do things differently from firms in another country, possibly because of differences in culture or socialization but also because of potential variations in local institutions such as government policy, educational systems, or locally available resources.

Here, my experience as a student of Japanese business history and management practices has heavily shaped my thinking. When I first joined a business school in 1984, a very large number of cases throughout the MBA curriculum—dealing with strategy, organizational behavior, operations management, and industrial policy—touted the excellence of Japanese approaches. Japan was "number one," as Harvard professor Ezra Vogel stated so directly in his best-seller book from 1979.[18] (Full disclosure: I was once Professor Vogel's teaching assistant whilst in graduate school.) Japan in many ways represented overall "best practice" at this point in time, especially when it came to manufacturing and quality control, long-term capabilities development and the management of people, and government industrial policy. A number of best-selling books helped educate the Western public, such as *Kaisha: The Japanese Company, The Art of Japanese Management,* and *MITI and the Japanese Miracle.*[19] Learning from Japan declined in popularity very quickly after its economic bubble burst in 1989; fortunately, most "Japan specialists" in business schools had taken on functional specialties, such as technological innovation, strategy, and international management. But the realization has stayed with me, from observing Japan as well as studies such as by Peters and Waterman, that best practices and top performers in one context and time period may not endure once the environment changes.

The case of Japan is so striking that it deserves more treatment here.[20] Whilst many features of "Japan, Inc." and the country's political, economic, cultural, and social "systems" provided benefits during periods of rapid growth, they proved to be a drag on change and international competitiveness once growth had stagnated.[21] These features as well as a long list of managerial and employment practices once thought to represent the best way to do things also turned out to be inefficient and ineffective in the 1990s (Table I.2). For example, as far back as the 1960s and 1970s, foreign observers of Japan began describing specific aspects of "the Japanese management system" that they believed contributed to the low-cost but high-quality goods produced by major Japanese firms, led by companies such as Toyota and Sony. In particular, management experts in the West praised Japan's preference for lifetime employment in large firms, seniority-based wages, company-based unions, and long-term and consensus-oriented decision making, as well as the just-in-time (lean) production techniques, quality-control practices pushed down to the lowest level of the organization, and extensive use of low-cost but dedicated supplier networks.[22] Ironically, at least some of these Japanese management practices, especially the focus on quality and a broader view of leadership and decision making, were not really Japanese in origin but were initially taught to Japanese managers after the Second World War by American engineers and consultants.[23]

The Japanese economy grew at "miraculous" levels from the mid-1950s through the early 1970s, and still did very well through the 1980s. After 1989, however, the economy went into a tailspin, primarily because of overheated real estate and stock markets, and weak financial regulation—much like the United States and the world experienced in 2008–9. The 1989–90 boom and bust exposed the less attractive side of Japan's supposed strengths. As seen in Europe, too, commitments to lifetime employment can greatly reduce the flexibility of firms to adapt to increasing external competition or changes in technology or knowledge requirements. Seniority wages do not reward merit and can lead to complacency and conservative decisions. Consensus decision making can be good when strategic options are clear, but it can also lead to "lowest-common-denominator" compromises as well as "herd-like" thinking, rather than crisp decisions and strategic focus. The lack of pressure from shareholders may be good to support long-term investments, such as the kind Japan needed to develop a domestic automobile industry or to create innovative consumer electronics products like the home video cassette recorder. But, as global competition has increased, many corporate investments have not brought good returns for Japanese firms and have contributed to their extremely low levels of profitability by international standards.

Table I.2. *Japan: 1980s strengths to 1990s weaknesses?*

1980s strengths	**1990s weaknesses**
Economic system	
• Low wages	• Rising value of yen
• High savings	• Bubbles in stock market and real estate
• High exports	
Financial system	
• Low interest rates	• Inefficient use of capital and poor investment returns
• Lots of capital for investment	• Bankrupt banks
• Protected banks	
• Deficit financing	
Political system	
• Stable, conservative	• Struggles over shrinking pie
• Consensus oriented	• Political "gridlock"
• Sharing of wealth through subsidies	• Slow/negative growth, unemployment
Social and cultural system	
• Centralized and standardized primary education	• Weak universities
	• Too much emphasis on rote learning
• Shared values	• Not enough individualism and creativity
• Hierarchy and authority	
• Group over individual	
Management and employment system	
• Lifetime employment in large firms	• Reduced flexibility
• Seniority-based wages	• Do not reward merit and achievement
• Company-based unions	• Inadequate concern with worker welfare
• Consensus decision making	• Lowest-common denominator decisions
• Long-term view	• Little pressure for efficiency/profits
• Institutional shareholding	• Problem in global competition for some firms
• Just-in-time (lean) production	• Too much focus on manufacturing; traffic jams
• Quality control and *kaizen* (continuous improvement)	• Diminishing returns
	• "Shell game" of transferring costs to suppliers
• Low-cost dedicated supplier networks	

Furthermore, whilst just-in-time practices can be extremely useful in manufacturing, they do not solve design problems or help firms explore new technologies, such as complex software-based systems—which Toyota learned the hard way. A preoccupation with manufacturing also seems to have encouraged Japanese firms and the government to neglect productivity in services—the bulk of the economy in 2010. Japan has also moved too slowly into new information-intensive and science-based industries such as software and biotechnology, which have become new sources of employment and wealth in Western countries. The obsession with "continuous improvement" (*kaizen*) can lead to higher quality and productivity but can also waste time and money when diminishing returns set in or when firms produce much better quality than customers really need. And pressing suppliers to reduce costs or absorb excess labor from large firms can become little more than a "shell game" as demand slows.

It became clear by the mid-1990s that there were limits to Japanese management techniques, such as how far the Japanese could push JIT production, reductions in inventories, cuts in worker headcount, and other efficiency-oriented practices in product development and manufacturing.[24] But another factor that caused Japan's economy to stagnate for two decades was intensifying external competition. Most firms in the USA, Europe, and Asia learned what there was to learn about superior Japanese production, engineering, and quality-control practices, as well as human resource management practices. Japan remains a very wealthy country and, as of early 2010, remained the second largest economy in the world, behind only the United States (albeit with China gaining fast). It still has many firms that are highly competitive in global industries. Nonetheless, the "Japanese management system" that provided a powerful competitive edge in the past no longer seemed so effective in the 1990s or 2000s. Nor did even the best Japanese firms, such as Toyota, seem decisively superior to their foreign competitors in design, production management, or quality control.

"Luck" and Population Ecology

Finally, when trying to determine which factors really matter, we also need to consider the role of luck—good and bad. Some firms or practices may be successful because of random decisions or weak competition at a particular point in time. A large enough sample and multiple years of data help with this problem. But then we must still confront the issue brought up by population ecologists. I take liberties with my paraphrasing, but some scholars of this persuasion seem to argue that "best practice" does not exist in any absolute sense. All firms and organizations, they point out, have trouble adapting to change. Those organizations that seem to thrive just "happen" to fit the requirements of the market and circumstances of the time one is looking at. Success is more like a Darwinian process of natural selection. The actions of managers have little or no impact, precisely because it is so hard to alter the structural characteristics of an organization. Hence, when we think we are looking at an enduring principle, we are simply looking at practices of the remaining population of firms with characteristics matching the market needs of a particular time and place.[25]

I do not fully agree with the view that best practices and corporate success are entirely the artifacts of time and place.[26] But, surely, there is some truth to this argument—otherwise great firms would remain great for much longer periods of time. Did Bill Gates and Microsoft just happen to be in the right place at the right time? Yes, and they were lucky. But Microsoft was also the best PC software company *c.*1980 and Gates thoroughly exploited the opportunities before him—in 1975 for a PC programming language and again in 1980 when IBM came calling for an operating system. Microsoft's power has declined with time, though some of its practices remain essential to superior performance in the PC software business. What about General Motors? Or General Electric, Lehman Brothers, Fannie Mae, Citibank, et al.? Or Toyota? At their peak, these firms embraced practices that clearly dominated the competition and suited a particular time and set of circumstances. Then these practices became less distinctive (or even disastrous) as circumstances changed. Luck clearly cuts both ways—good and bad. But outcomes need not be totally random. As Louis Pasteur once said, "Chance favors the prepared mind."

Notes

1. See the classic references Eisenhardt (1989b) and Yin (2002).
2. See Christiansen and Pascale (1983a, b). See also Pascale (1984).
3. Peters and Waterman (1982).
4. Collins (2001).
5. Collins (2009).
6. Chandler, Jr. (1990).
7. Chandler, Jr. (1962).
8. Clark and Fujimoto (1991).
9. Womack, Jones, and Roos (1990).
10. An excellent discussion of this issue combined with an empirical study of pharmaceutical R&D is Cockburn, Henderson, and Stern (2000).
11. My thanks to David Yoffie for this analogy.
12. A commonly cited strategy article that makes this point is Peteraf (1993).
13. See, e.g., the classic Utterback and Abernathy (1975) and Abernathy and Utterback (1978). See also Utterback (1994) and Klepper (1996, 1997).
14. See, e.g., the analysis in Suarez and Utterback (1995).
15. How to think about the nature of technology and different types of innovation is an enormous subject. For a brief review done because we needed to control for different levels of innovation in automobile projects, see Cusumano and Nobeoka (1998: 101–6).
16. Fine (1998).
17. See Schmalensee (1985), Rumelt (1991), and McGahan and Porter (1997).
18. See Vogel (1979).
19. Johnson (1982); Pascale and Athos (1982); Abegglen and Stalk (1985).
20. See Cusumano (2005b). A version of this is published in Cusumano and Westney (2010).
21. On this topic, see Porter, Takeuchi, and Sakakibara (2000).
22. For more particularly influential books on Japan, see also, as examples from a very long list, Dore (1973), Vogel (1979), and Schonberger (1982).
23. See Hopper and Hopper (2007), esp. pp. 108–24.
24. See, e.g., Cusumano (1994).
25. The most important early works from this point of view are generally considered Hannan and Freeman (1977) and Aldrich (1979). See also Hannan and Freeman (1989). Kay (1993) too is close to this population ecology view, though he also clearly argues that managers can impact on firm performance and survival.
26. I have elaborated a bit more on this in Cusumano (2009a).

APPENDIX II

ADDITIONAL TABLES

Table II.1. *VHS and BETA group alignments, 1983–1984*

Japan	United States	Europe
VHS Group (40 firms)		
JVC	Magnavox (MA)	Blaupunkt (MA)
Matsushita	Sylvania (MA)	Zaba (J)
Hitachi	Curtis Mathes (MA)	Nordmene (J)
Mitsubishi	J. C. Penny (MA)	Telefunken (J)
Sharp	GE (MA)	SEL (J)
Tokyo Sanyo	RCA (H)	Thorn-EMI (J)
Brother (MI)	Zenith (J) (from spring 1984)	Thomson-Brandt (J)
Ricoh (H)		Granada (H)
Tokyo Juki (H)		Hangard (H)
Canon (MA)		Sarolla (H)
Asahi Optical (H)		Fisher (T)
Olympus (MA)		Luxer (MI)
Nikon (MA)		
Akai Trio (J)		
Sansui (J)		
Clarion (J)		
Teac (J)		
Japan Columbia (H)		
Funai		
BETA Group (12 firms)		
Sony	Zenith (S) (until spring 1984)	Kneckerman (SA)
Sanyo	Sears (SA)	Fisher (SA)
Toshiba		Rank (TO)
NEC		
General (TO)		
Aiwa		
Pioneer (S)		

Note: Suppliers indicated by initiations (J=JVC, MA=Matsushita, H=Hitachi, MI=Mitsubishi, T=Tokyo Sanyo, S=Sony, TO=Toshiba, SA=Sanyo).

Source: Cusumano, Mylonadis, and Rosenbloom (1992: 73, table 5).

Table II.2. *VCR production and format shares, 1975–1984 (%)*

Group	1975	1976	1977	1978	1979	1980
BETA Group						
Sony	100	56	51	28	24	22
Others		5	5	12	15	11
Subtotal	100	61	56	40	39	34
VHS Group						
Matsushita		29	27	36	28	29
JVC		9	15	19	22	18
Others		1	2	5	11	19
Subtotal		39	44	60	61	66
Group	**1981**	**1982**	**1983**	**1984**	**...**	**1989**
BETA Group						
Sony	18	14	12	9		
Sanyo	9	10	8	6		
Toshiba	4	4	4	3		
Others	1	1	2	2		
Subtotal	32	28	25	20		0
VHS Group						
Matsushita	28	27	29	25		
JVC	19	20	16	17		
Hitachi	10	10	11	15		
Sharp	7	7	9	9		
Mitsubishi	3	3	3	4		
Sanyo		3	4	5		
Others	2	2	2	5		
Subtotal	68	72	75	80		100

Note: Unit: % of total industry production (both formats).

Source: Cusumano, Mylonadis, and Rosenbloom (1992: 55, table 1).

Table II.3. *How Microsoft made money, 2009*

Total revenue $58.4 bn. and Operating income $20.4 bn. (35% of revenue)
Products (90%) + Services (5%) + Online Services (5%)

Client (Windows desktop)
$14.7 bn. revenue (25% of total) and $10.9 bn. operating income (38% of profit before corporate charges & adjustments)
Gross margin = 74%
80% of revenues = OEM (PC manufacturers) sales, 20% = consumer sales

Servers and Tools
$14.1 bn. revenue (24% of total) and $5.3 bn. operating income (19% of profit before corporate charges & adjustments)
Gross margin = 38%
50% of revenues = multi-year license contracts
20% = volume licensing deals, 10% = OEM sales
20% = services (consulting, technical support =$2.8 bn. or 5% of total revenues)

Online Services (MSN, Bing, etc.)
$3.1 bn. revenue (5% of total) and –$2.3 bn. operating loss
Consumer sales = 100%

Microsoft Business Division (Office)
$18.9 bn. revenue (32% of total) and $12.1 bn. operating income (42% of profit before corporate charges & adjustments)
Gross margin = 64%
Office sales = 90%, Other applications = 10%
Enterprise sales = 80%, Consumer sales = 20%

Entertainment and Devices (Xbox, Zune, Windows Mobile OS)
$7.8 bn revenue (13% of total) and $0.2 bn. operating income (1% of profit before corporate charges & adjustments)
Gross margin = 3%

Additional analysis

- Windows client (desktop) and Office = 57% of sales and 80% of sector profits before corporate adjustments
- Consumer sales = about 30% of total revenues (20% of Windows client, 100% of Online Services and Entertainment and Devices, 20% of Business/Office)

Source: Calculated from Microsoft Corporation, *Form 10-K* (fiscal year ended June 2009).

Table II.4. *Financial comparison of GM, Ford, and Toyota, 2001–2009*

	2001	2002	2003	2004	2005	2006	2007	2008	2001–2008	2009
General Motors										
Total revenues ($ m.)	177,260	186,763	185,837	193,517	193,050	204,467	179,984	148,979	*1,469,857*	104,589
Financial services ($ m.)	25,769	27,026	30,006	31,972	34,427	33,816	2,390	1,247	*186,653*	n.a.
Total rev./services (%)	15	16	16	17	18	17	1	1	*13*	n.a.
Total net income ($ m.)	601	1,736	3,822	2,805	(10,308)	(1,978)	(38,732)	(30,860)	*(72,914)*	n.a.
Services income ($ m.)	2,675	1,882	2,827	2,894	3,614	4,628	181	(45)	*18,656*	n.a.
Services income (%)	445	108	74	103	(loss)	(loss)	(loss)	(loss)	(loss)	n.a.
Ford Motor										
Total revenues ($ m.)	160,504	162,256	166,040	172,255	176,835	160,065	172,455	146,277	*1,316,687*	118,308
Financial services ($ m.)	29,768	27,983	25,754	25,197	23,422	16,816	18,076	17,111	*184,127*	12,415
Total rev./services (%)	19	17	16	15	13	11	10	12	*14*	10
Total net income ($ m.)	(5,453)	(980)	239	3,038	1,440	(12,613)	(2,723)	(14,672)	*(31,724)*	2,717
Services income ($ m.)	1,438	2,104	3,327	4,827	4,953	1,966	1,224	(2,581)	*17,258*	1,814
Services income (%)	(loss)	1,392	672	159	344	(loss)	(loss)	(loss)	*(loss)*	67
Toyota Motor										
Total revenues ($ m.)	107,960	118,252	128,965	163,637	172,749	179,083	202,864	262,394	*1,335,904*	208,995
Financial services ($ m.)	4,759	5,816	6,031	6,791	7,275	8,486	11,017	14,955	*65,130*	14,024
Total rev./services (%)	4	5	5	4	4	5	5	6	*5*	7
Total net income ($ m.)	5,624	4,638	6,247	10,995	10,907	11,681	20,182	17,146	*115,240*	4,448
Services income (%)	320	376	252	1,381	1,870	1,326	1,343	863	*7,731*	732
Services income (%)	6	8	4	13	11	11	7	5	*9*	(loss)

Notes: Units: fiscal year, $ million, %.
GM and Ford have restated results in several years because of sale of assets.
n.a. = not available or not applicable because of bankruptcy process.
Exchange rate of 120 yen per dollar used for 2001–2; otherwise annual dollars, as stated in *Form 20-F*.

Sources: Form 10-K (GM and Ford) and *Form 20-F* (Toyota) annual reports.

Table II.5. *Nissan and Toyota motor vehicle production, 1934–1945*

Year	Nissan				Toyota			
	Units	Trucks and buses (%)	Year-end employees	Units per employee	Units	Trucks and buses (%)	Year-end employees	Units per employee
1934	940	31	1,218	—	—	—	—	—
1935	3,800	31	1,914	2.4	20	100	n.a.	—
1936	6,163	58	2,874	2.6	1,142	91	n.a.	—
1937	10,227	60	5,804	2.4	4,013	86	3,000	—
1938	16,591	75	9,057	2.2	4,615	88	4,000	1.3
1939	17,781	92	9,173	2.0	11,981	91	5,200	2.6
1940	15,925	93	7,706	1.9	14,787	98	5,200	2.8
1941	19,688	92	7,570	2.6	14,611	99	5,200	2.8
1942	17,434	95	7,744	2.3	16,302	100	6,500	2.8
1943	10,753	95	n.a.	—	9,827	99	7,500	1.4
1944	7,083	100	n.a.	—	12,720	100	6,000	1.9
1945	2,001	100	n.a.	—	3,275	100	4,000	0.7
Totals	128,386	85	—	—	93,293	98	—	—

Notes: Units per employee reflect unit production in the current year divided by average employees, which is calculated by averaging the number of employees at the end of the prior year plus the current year.
n.a. = not available.

Source: Compiled from Cusumano (1985: 46, 61).

Table II.6. *Cluster analysis of strategy types*

	Group 1	Group 2	Group 3	Group 4
	New design	Concurrent technology transfer	Sequential technology transfer	Design modification
Total sample	26	15	11	13
By region				
Japan	6	9	0	5
USA	4	1	3	4
Europe	16	5	8	4
Multi-project strategy usage (%)				
New design	*92.0*	46.0	22.6	25.2
Concurrent	0.6	*41.0*	0.0	1.0
Sequential	4.4	5.8	*72.1*	19.1
Modification	3.0	7.3	5.3	*54.7*
New-product introduction rate (%)	45	69	51	47
Average platform age (years)***	0.37	0.91	3.75	5.54
Market-share change (%)***	3.4	23.4	9.1	-15.6

Notes: *** Differences among the strategy types are significant at the 1% level (ANOVA).
Market-share change measures a ratio and therefore does not total zero.

Source: Cusumano and Nobeoka (1998: 150).

REFERENCES

Aaker, D., and B. Mascarenhas (1984). "The Need for Strategic Flexibility," *Journal of Business Strategy*, 5 (2): 74–82.

Abegglen, J., and G. Stalk (1985). *Kaisha: The Japanese Corporation* (New York: Basic Books).

Abernathy, W. J. (1978). *The Productivity Dilemma: Roadblock to Innovation in the Automobile Industry* (Baltimore: Johns Hopkins University Press).

Abernathy, W. J., and K. B. Clark (1985). "Innovation: Mapping the Winds of Creative Destruction," *Research Policy*, 14 (1): 3–22.

Abernathy, W. J., and J. M. Utterback (1978). "Patterns of Innovation in Industry," *Technology Review*, 80 (7): 40–7.

Abernathy, W. J., and K. Wayne (1974). "Limits of the Learning Curve," *Harvard Business Review*, 52 (5): 109–18.

Ackman, D. (2001). "Tire Trouble: The Ford-Firestone Blowout," Forbes.com, June 20 (accessed Feb. 6, 2010).

Adler, P. S. (2005). "The Evolving Object of Software Development," *Organization*, 12 (3): 401–35.

Adler, P. S., B. Goldoftas, and D. Levine (1999). "Flexibility versus Efficiency? A Case Study of Model Changeovers in the Toyota Production System," *Organization Science*, 10 (1): 43–68.

Adner, R. (2006). "Match your Innovation Strategy to your Innovation Ecosystem," *Harvard Business Review*, 84 (4): 98–107.

Adobe Systems Incorporated, *Form 10-K*, annual.

Aldrich, H. (1979). *Organizations and Environments* (Palo Alto: Stanford University Press; classic edn., 2008).

Allen, T. J. (1984). *Managing the Flow of Technology: Technology Transfer and the Flow of Technological Information within the R&D Organization* (Cambridge, MA: MIT Press).

Allen, T. J., and G. Henn (2006). *The Organization and Architecture of Innovation: Managing the Flow of Technology* (Oxford: Butterworth-Heinemann).

Allison, K. (2007). "Facebook Plans to Open Site to Outside Widgets," *Financial Times*, May 26, p. 19.

Andrews, K. (1980). *The Concept of Corporate Strategy* (New York: Richard D. Irwin).

Aoshima, Y., M. Cusumano, and A. Takeishi (2010) (eds.). *Made-in-Japan wa owarunoka? Kiseki to shuen no saki ni arumono* [Is Made-in-Japan over? After the Miracle and End of Japan] (Tokyo: Toyo Keizai Shimbunsha).

Aoyama, M. (1993). "Concurrent-Development Process Model," *IEEE Software*, 12 (3): 20–4.

Aoyama, Y., and H. Izushi (2003). "Hardware Gimmick or Cultural Innovation: Technological, Cultural, and Social Foundations of the Japanese Video Game Industry," *Research Policy*, 32 (3): 423–44.

Arango, T., and A. Vance (2009). "New Corp. Weighs Exclusive Alliance with Bing," *New York Times*, Nov. 24, p. B1.

Arthur, W. B. (1989). "Competing Technologies, Increasing Returns, and Lock-in by Historical Events," *Economic Journal*, 99 (Mar.): 116–31.

Athreye, S. S. (2005). "The Indian Software Industry," in A. Arora and A. Gambardella (eds.), *From Underdogs to Tigers: The Rise and Growth of the Software Industry in Brazil, China, India, Ireland, and Israel* (Oxford: Oxford University Press), 7–40.

Auchard, E. (2007). "Facebook Becomes Hub for Outside Software Makers," *Boston Globe*, May 24.

Bailey, D. (2010). "Ford Offers Fix for Fusion Hybrid Brake Glitch," www.reuters.com, Feb. 4 (accessed Feb. 8, 2010).

Bakos, Y., and E. Brynjolfsson (1999). "Bundling Information Goods: Pricing, Profits and Efficiency," *Management Science*, 45 (12): 1613–30.

Baldwin, C. Y., and K. B. Clark (1999). *Design Rules: The Power of Modularity, Volume 1* (Cambridge, MA: MIT Press).

Barnard, C. (1938). *The Functions of the Executive* (Cambridge, MA: Harvard University Press).

Barney, J. (1991). "Firm Resources and Sustained Competitive Advantage," *Journal of Management*, 17 (1): 99–120.

Barney, J. B., and D. N. Clark (2007). *Resource-Based Theory: Creating and Sustaining Competitive Advantage* (New York: Oxford University Press).

Barras, R. (1986). "Towards a Theory of Innovation in Services," *Research Policy*, 15 (3): 161–73.

Basili, V. R., and A. J. Turner (1975). "Iterative Enhancement: A Practical Technique for Software Development," *IEEE Transactions on Software Engineering*, 1 (4): 390–6.

Baum, C. (1981). *The System Builders: The Story of SDC* (Santa Monica, CA: System Development Corporation).

Bell, D. (1973). *The Coming of Postindustrial Society: A Venture in Social Forecasting* (New York: Basic Books).

Bemer, B. (undated). "The Software Factory Principle" (www.trailing-edge.com/bobbemer/FACTORY.HTM (accessed Mar. 28, 2009)).

Bemer, R. W. (1969). "Position Papers for Panel Discussion: The Economics of Program Production," in *Information Processing 68*, Amsterdam: North-Holland, 1626–7.

Bettis, R. (1981). "Performance Differences in Related and Unrelated Diversified Firms," *Strategic Management Journal*, 2 (4): 379–93.

Bhattacharjee, R. (2009). "An Analysis of the Cloud Computing Platforms," unpublished Master's thesis in System Design and Management, Massachusetts Institute of Technology, June.

Birkinshaw, J., and P. Hagström (2002) (eds.). *The Flexible Firm: Capability Management in Network Organizations* (Oxford and New York: Oxford University Press).

Boehm, B. W. (1988). "A Spiral Model of Software Development and Enhancement," *IEEE Computer*, 21 (5): 61–72.

Boston Globe (2007). "Wii Fans Give Nintendo a Boost," May 29, p. C4.

Boudreau, K. (2006). "Too Many Complementors? Evidence on Software Firms," unpublished working paper, HEC-Paris School of Management, Nov.

Bowen, D. E., C. Siehl, and B. Schneider (1991). "Developing Service-Oriented Manufacturing," in I. Kilmann (ed.), *Making Organizations Competitive* (San Francisco: Jossey-Bass), 397–418.

Brass, D. (2010). "Microsoft's Creative Destruction," *New York Times*, Feb. 4, p. A27.

Bratman, H., and T. Court (1975). "The Software Factory," *IEEE Computer*, 8 (5): 28–37.

Bratman, H., and T. Court (1977). "Elements of the Software Factory: Standards, Procedures, and Tools," in Infotech International Ltd., *Software Engineering Techniques* (Infotech International Ltd., Berkshire, England), 117–43.

Bresnahan, T., and S. Greenstein (1999). "Technological Competition and the Structure of the Computer Industry," *Journal of Industrial Economics*, 47 (1): 1–40.

Brown, S. L., and K. M. Eisenhardt (1995). "Product Development—Past Research, Present Findings, and Future Directions," *Academy of Management Review*, 20 (2): 343–78.

Brown, S. L., and K. M. Eisenhardt (1997). "The Art of Continuous Change: Linking Complexity Theory and Time-Paced Evolution in Relentlessly Shifting Organizations," *Administrative Science Quarterly*, 42 (1): 1–34.

Brown, S. L., and K. M. Eisenhardt (1998). *Competing on the Edge: Strategy as Structured Chaos* (Boston: Harvard Business School Press).

Campbell-Kelly, M. (2004). *A History of the Software Industry: From Airline Reservations to Sonic the Hedgehog* (Cambridge, MA: MIT Press).

Campbell-Kelly, M., and D. Garcia-Schwartz (2007). "From Products to Services: The Software Industry in the Internet Era," *Business History Review*, 81 (4): 735–64.

Carnoy, D. (2009). "Sony PlayStation 3 Slim," CNET Reviews, Nov. 20 (http://reviews.cnet.com/ consoles/ sony-playstation-3-slim/4505-10109_7-33765068.html#cnetReview).

Carpenter, G. S., and K. Nakamoto (1989). "Consumer Preference Formation and Pioneering Advantage," *Journal of Marketing Research*, 26 (3): 285–98.

CBC News (2008). "Google Global Search Share down in February," Mar. 20 (www.cbc.ca/technology/ story/2008/03/20/tech-google-search.html).

Chandler, Jr., A. D. (1962). *Strategy and Structure: Chapters in the History of the American Industrial Enterprise* (Cambridge, MA: MIT Press).

Chandler, Jr., A. D. (1990). *Scale and Scope: The Dynamics of Industrial Capitalism* (Cambridge, MA: Harvard University Press).

Change, V., and J. Pfeffer (2004). "Gary Loveman and Harrah's Entertainment" (Boston: Harvard Business School, Case #OB45-PDF-ENG).

Chesbrough, H. (2003). *Open Innovation: The New Imperative for Creating and Profiting from Technology* (Boston: Harvard Business School Press).

Christensen, C. M. (1997). *The Innovator's Dilemma: When New Technologies Cause Great Firms to Fail* (Boston: Harvard Business School Press).

Christensen, C. M., J. H. Grossman, and J. Hwang (2008). *The Innovator's Prescription: A Disruptive Solution for Healthcare* (New York: McGraw Hill).

Christiansen, E. T., and R. T. Pascale (1983a). "Honda (A)" (Boston: Harvard Business School, Case #384049).

Christiansen, E. T., and R. T. Pascale (1983b). "Honda (B)" (Boston: Harvard Business School, Case #384050).

Clark, K. B., and T. Fujimoto (1991). *Product Development Performance: Strategy, Organization, and Management in the World Auto Industry* (Boston: Harvard Business School Press).

Coase, R. H. (1937). "The Nature of the Firm," *Economica*, 4 (16): 386–405.

Cockburn, I. M., R. M. Henderson, and S. Stern (2000). "Untangling the Origins of Competitive Advantage," *Strategic Management Journal*, 21 (10/11): 1123–45.

Cohen, W. M., and D. A. Levinthal (1990). "Absorptive Capacity: A New Perspective on Learning and Innovation," *Administrative Science Quarterly*, 35 (1): 128–52.

Collins, J. (2001). *From Good to Great: Why Some Companies Make the Leap... and Others Don't* (New York: Harper Business).

Collins, J. (2009). *How The Mighty Fall: And Why Some Companies Never Give in* (New York: HarperCollins).

Cusumano, M. A. (1985). *The Japanese Automobile Industry: Technology and Management at Nissan and Toyota* (Cambridge, MA: Harvard University Press).

Cusumano, M. A. (1988). "Manufacturing Innovation: Lessons from the Japanese Auto Industry," *MIT Sloan Management Review*, 30 (1): 34–5.

Cusumano, M. A. (1989a). " 'Scientific Industry': Strategy, Technology, and Entrepreneurship in Prewar Japan," in William Wray (ed.), *Managing Industrial Enterprise: Cases from Japan's Prewar Experience* (Cambridge, MA: Council on East Asian Studies/Harvard University Press).

Cusumano, M. A. (1989b). "The Software Factory: A Historical Interpretation," *IEEE Software*, 6 (2): 23–30.

Cusumano, M. A. (1991a). "Factory Concepts and Practices in Software Development," *IEEE Annals of the History of Computing*, 13 (1): 3–32.

Cusumano, M. A. (1991b). *Japan's Software Factories: A Challenge to US Management* (New York: Oxford University Press).

Cusumano, M. A. (1992). "Shifting Economies: From Craft Production to Flexible Systems and Software Factories," *Research Policy*, 21 (5): 453–80.

Cusumano, M. A. (1994). "The Limits of Lean," *MIT Sloan Management Review*, 35 (4): 27–32.

Cusumano, M. A. (1997). "How Microsoft Makes Large Teams Work like Small Teams," *MIT Sloan Management Review*, 39 (1): 9–20.

Cusumano, M. A. (2000). "That's Some Fine Mess You've Made, Mr Gates," *Wall Street Journal*, Apr. 5, p. A26.

Cusumano, M. A. (2004). *The Business of Software: What Every Manager, Programmer, and Entrepreneur Must Know to Thrive and Survive in Good Times and Bad* (New York: Free Press).

Cusumano, M. A. (2005a). "Google: What it is and What it is Not," *Communications of the ACM*, 48 (2): 15–17.

Cusumano, M. A. (2005b). "The Japan Problem as Paradox: Views from Abroad, in Good Times and Bad," unpublished working paper, "End of Japan? Conference," Honolulu, Hawaii, Jan.

Cusumano, M. A. (2006a). "Microsoft to Toyota jidosha – komodetei-ka no teisetsu o kutsugae-shita saikyo ni-sha no kyotsuten" [Microsoft and Toyota Motor – Common Points between Two of the Strongest Companies in their Approach to Overcoming Commoditization], *Daimond Weekly*, June 10, p. 23.

Cusumano, M. A. (2006b). "What Road Ahead for Microsoft and Windows," *Communications of the ACM*, 49 (7): 21–3.

Cusumano, M. A. (2007a). "The Changing Labyrinth of Software Pricing," *Communications of the ACM*, 50 (7): 19–22.

Cusumano, M. A. (2007b). "Extreme Programming Compared with Microsoft-Style Iterative Development," *Communications of the ACM*, 50 (10): 15–18.

Cusumano, M. A. (2008a). "The Changing Software Business: Moving from Products to Services," *IEEE Computer*, 41 (1): 78–85.

Cusumano, M. A. (2008b). "The Puzzle of Apple," *Communications of the ACM*, 51 (9): 22–4.

Cusumano, M. A. (2009a). "Strategies for Difficult (and Darwinian) Economic Times," *Communications of the ACM*, 52 (4): 27–8.

Cusumano, M. A. (2009b). "The Legacy of Bill Gates," *Communications of the ACM*, 52 (1): 25–6.

Cusumano, M. A., and A. Gawer (2002). "The Elements of Platform Leadership," *MIT Sloan Management Review*, 43 (3): 51–8.

Cusumano, M., S. Kahl, and F. Suarez (2008). "A Theory of Services in Product Firms," Dec. 30, Center for Digital Business, MIT.

Cusumano, M. A., and C. F. Kemerer (1990). "A Quantitative Analysis of US and Japanese Practice and Performance in Software Development," *Management Science*, 36 (11): 1384–1406.

Cusumano, M. A., Y. Mylonadis, and R. S. Rosenbloom (1992). "Strategic Maneuvering and Mass-Market Dynamics: The Triumph of VHS over Beta," *Business History Review*, 66 (1): 51–94.

Cusumano, M. A., and K. Nobeoka (1998). *Thinking beyond Lean: How Multi-Project Management is Transforming Product Development at Toyota and Other Companies* (New York: Free Press).

Cusumano, M. A., and R. W. Selby (1995). *Microsoft Secrets: How the World's Most Powerful Software Company Creates Technology, Shapes Markets, and Manages People* (New York: Free Press).

Cusumano, M. A., and R. W. Selby (1997). "How Microsoft Builds Software," *Communications of the ACM*, 40 (6): 53–62.

Cusumano, M. A., and S. A. Smith (1997). "Beyond the Waterfall: Software Development at Microsoft," in David B. Yoffie (ed.), *Competing in the Age of Digital Convergence* (Boston: Harvard Business School Press), 371–412.

Cusumano, M. A., and A. Takeishi (1991). "Supplier Relations and Management: A Survey of Japanese, Japanese-Transplant, and US Auto Plants," *Strategic Management Journal*, 12 (8): 563–88.

Cusumano, M. A., and D. E. Westney (2010). "Nihon no kyosoryoku ni taisuru obei roncho no henkan" [The Change in Western Commentary on Japan's Competitiveness] (Tokyo: Toyo Keizai Shimbunsha), in Y. Aoshima, M. Cusumano, and A. Takeishi (eds.), *Made-in-Japan wa owarunoka? Kiseki to shuen no saki ni arumono* [Is Made-in-Japan over? After the Miracle and End of Japan] (Tokyo: Toyo Keizai Shimbunsha).

Cusumano, M. A., and D. B. Yoffie (1998). *Competing on Internet Time: Lessons from Netscape and its Battle with Microsoft* (New York: Free Press).

Cusumano, M. A., and D. B. Yoffie (1999). "What Netscape Learned from Cross-Platform Software Development," *Communications of the ACM*, 42 (10): 72–8.

Cusumano, M. A., A. MacCormack, C. F. Kemerer, and W. Crandall (2003). "Software Development Worldwide: The State of the Practice," *IEEE Software*, 20 (6): 28–34.

Cusumano, M. A., A. MacCormack, C. F. Kemerer, and W. Crandall (2009). "Critical Decisions in Software Development: Updating the State of the Practice," *IEEE Software*, 25th Anniversary Issue, 26 (5): 66–9.

Cyert, R., and J. March (1963). *A Behavioral Theory of the Firm* (Oxford: Wiley-Blackwell).

D'Aveni, R. (1994). *Hypercompetition: Managing the Dynamics of Strategic Maneuvering* (New York: Free Press).

David, P. (1985). "Clio and the Economics of QWERTY," *American Economic Review*, 75 (2): 332–7.

Davis, J. P., K. M. Eisenhardt, and C. B. Bingham (2009). "Optimal Structure, Market Dynamism, and the Strategy of Simple Rules," *Administrative Science Quarterly*, 54 (Sept.): 413–52.

De Toni, A., M. Caputo, and A. Vinelli (1988). "Production Management Techniques: Push–Pull Classification and Application Conditions," *International Journal of Operations and Production Management*, 8 (2): 35–51.

Dodgson, M., D. Gann, and A. Salter (2005). *Think, Play, Do: Technology, Innovation, and Organization* (Oxford: Oxford University Press).

Dore, R. (1973). *British Factory–Japanese Factory: The Origins of National Diversity in Industrial Relations* (Berkeley and Los Angeles: University of California Press).

The Economist (2007). "Console Wars—Video Games," Mar. 24, p. 6.

The Economist (2009). "Briefing: Japan's Technology Champions—Invisible but Indispensable," Nov. 7, pp. 64–6.

Eisenhardt, K. M. (1989a). "Agency Theory: An Assessment and Review," *Academy of Management Review*, 14 (1): 57–74.

Eisenhardt, K. M. (1989b). "Building Theories from Case Study Research," *Academy of Management Review*, 14 (4): 532–50.

Eisenhardt, K. M., and S. L. Brown (1998). "Time Pacing: Competing in Markets that Won't Stand Still," *Harvard Business Review*, 76 (2): 59–69.

Eisenhardt, K. M., and J. Martin (2000). "Dynamic Capabilities: What are They?" *Strategic Management Journal*, 21 (10): 1105–21.

Eisenhardt, K. M., and B. N. Tabrizi (1995). "Accelerating Adaptive Processes: Product Innovation in the Global Computer Industry," *Administrative Science Quarterly*, 42 (3): 501–29.

Eisenmann, T. (2006). "Internet Companies Growth Strategies: Determinants of Investment Intensity and Long-Term Performance," *Strategic Management Journal*, 27 (12): 1183–1204.

Eisenmann, T., G. Parker, and M. W. Van Alstyne (2006). "Strategies for Two-Sided Markets," *Harvard Business Review*, 84 (10): 92–101.

Eisenmann, T., G. Parker, and M. W. Van Alstyne (2007). "Platform Envelopment," unpublished working paper, Mar.

Elenkov, D., and M. A. Cusumano (1994). "Linking International Technology Transfer with Strategy and Management: A Literature Commentary," *Research Policy*, 23 (2): 195–215.

Farrell, J., and G. Saloner (1986). "Installed Base and Compatibility: Innovation, Product Preannouncements and Predation," *American Economic Review*, 76 (5): 940–55.

Farrell, J., and C. Shapiro (1988). "Dynamic Competition with Switching Costs," *RAND Journal of Economics*, 29 (1): 123–37.

Fine, C. F. (1998). *Clockspeed: Winning Industry Control in the Age of Temporary Advantage* (Reading, MA: Perseus).

Friedman, T. L. (2005). *The World is Flat: A Brief History of the Twenty-First Century* (New York: Farrar, Straus and Giroux).

Fujimoto, T. (1999). *The Evolution of a Manufacturing System at Toyota* (New York: Oxford University Press).

Gadiesh, O., and J. L. Gilbert (1998). "Profit Pools: A Fresh Look at Strategy," *Harvard Business Review*, 76 (3): 139–47.

Garvin, D. A. (1988). *Managing Quality: The Strategic and Competitive Edge* (New York: Free Press).

Garvin, D. A. (1993). "Building a Learning Organization," *Harvard Business Review*, 71 (4): 78–91.

Gates, B. (1995). "The Internet Tidal Wave," Internal Microsoft company memo, May 26 (www.usdoj.gov/atr/cases/exhibits/20.pdf).

Gawer, A. (2000). "The Organization of Platform Leadership: An Empirical Investigation of Intel's Management Processes Aimed at Fostering Complementary Innovation by Third Parties," unpublished Ph.D. dissertation, MIT Sloan School of Management.

Gawer, A. (2009) (ed.). *Platforms, Markets and Innovation* (Cheltenham: Edward Elgar).

Gawer, A., and M. A. Cusumano (2002). *Platform Leadership: How Intel, Microsoft, and Cisco Drive Industry Innovation* (Boston: Harvard Business School Press).

Gawer, A., and M. A. Cusumano (2008). "How Companies Become Platform Leaders," *MIT Sloan Management Review*, 49 (2): 29–30.

Gelman, J. R., and S. C. Salop (1983). "Judo Economics: Capacity Limitation and Coupon Competition," *Rand Journal of Economics*, 14 (2): 315–25.

Gemawat, P., and T. Lenk (1998). "De Beers Consolidated Mines, Ltd. (A)," Harvard Business School, Case #9-391-076.

General Electric Company (2009). *Form 10-K*, fiscal year ended Dec. 31.

Gerstner, Jr., L. V. (2002). *Who Says Elephants Can't Dance? Inside IBM's Historic Turnaround* (New York: HarperBusiness).

Gerwin, D. (1993). "Manufacturing Flexibility: A Strategic Perspective," *Management Science*, 39 (4): 395–410.

Ghemawat, P., and P. Del Sol (1998). "Commitment versus Flexibility," *California Management Review*, 40 (4): 26–42.

Gill, G. (1990). "Microsoft Corporation," Boston, Harvard Business School, Case #9-691-033.

Goeldi, A. (2007). "The Emerging Market for Web-Based Enterprise Software," unpublished Master's thesis in Management of Technology, Sloan Fellows Program, MIT Sloan School of Management.

Guillen, M. F. (1994). *Models of Management: Work, Authority, and Organization in a Comparative Perspective* (Chicago: University of Chicago Press).

Hall, R. W. (1983). *Zero Inventories* (New York: McGraw Hill).

Halligan, B., and D. Shah (2010). *Inbound Marketing: Get Found Using Google, Social Media, and Blogs* (Hoboken, NJ: John Wiley & Sons).

Hallowell, R. (1999). "Exploratory Research: Consolidation and Economies of Scope," *International Journal of Service Industry Management*, 10 (4): 359–68.

Hamel, G., and C. K. Pralahad (1989). "Strategic Intent," *Harvard Business Review*, 67 (3): 63–76.

Hamm, S. (2007). *Bangalore Tiger: How Indian Tech Upstart Wipro is Rewriting the Rules of Global Competition* (New Delhi: Tata-McGraw Hill).

Hannan, M., and J. Freeman (1977). "The Population Ecology of Organizations," *American Journal of Sociology*, 82 (5): 929–64.

Hannan, M., and J. Freeman (1989). *Organizational Ecology* (Cambridge, MA: Harvard University Press).

Hansen, M., N. Nohria, and T. Tierney (1999). "What's Your Strategy for Managing Knowledge?" *Harvard Business Review*, 77 (2): 106–16.

Harrigan, K. R. (1984). *Strategic Flexibility: A Management Guide for Changing Times* (Lexington, MA: Lexington Books).

Hax, A., and N. Majluf (1984). *Strategic Management: An Integrative Perspective* (Englewood Cliffs, NJ: Prentice-Hall).

Hayes, B. (2008). "Cloud Computing," *Communications of the ACM*, 51 (7): 9–11.

Hayes, R. H., and S. C. Wheelright (1984). *Restoring our Competitive Edge: Competing through Manufacturing* (New York: John Wiley & Sons).

Hein, B. (2007). "0+0 = 1: The Appliance Model of Selling Software Bundled with Hardware," unpublished Master's thesis in Management of Technology, Sloan Fellows Program, MIT Sloan School of Management.

Helfat, C. E., and K. M. Eisenhardt (2004). "Inter-Temporal Economies of Scope, Organizational Modularity, and the Dynamics of Diversification," *Strategic Management Journal*, 25 (13): 1217–32.

Henderson, R. H., and K. B. Clark (1990). "Architectural Innovation: The Reconfiguration of Existing Product Technologies and the Failure of Established Firms," *Administrative Science Quarterly*, 35 (1): 9–30 (Special Issue: Technology, Organizations, and Innovation).

Henderson, R., and I. Cockburn (1996). "Scale, Scope, and Spillovers: The Determinants of Research Productivity in Drug Discovery," *RAND Journal of Economics*, 27 (1): 32–59.

Highsmith, J. (2009). *Agile Project Management: Creating Innovative Products* (2nd edn.; Reading, MA: Addison-Wesley).

Holweg, M. (2007). "The Genealogy of Lean Production," *Journal of Operations Management*, 25 (2): 420–37.

Holweg, M., and F. K. Pil (2004). *The Second Century: Reconnecting Customer and Value Chain through Build-to-Order* (Cambridge, MA: MIT Press).

Hopper, K., and W. Hopper (2007). *The Puritan Gift: Reclaiming the American Dream amidst Global Financial Chaos* (London and New York: I. B. Taurus).

Horwitch, M. (1987). "The Emergence of Post-Modern Strategic Management," unpublished working paper, MIT Sloan School of Management.

Hounschell, D. (1985). *From the American System to Mass Production, 1800–1932: The development of manufacturing technology in the United States* (Baltimore: Johns Hopkins University Press).

Huang, Y. (2008). *Capitalism with Chinese Characteristics: Entrepreneurship and the State* (Cambridge: Cambridge University Press).

Iansiti, M., and R. Levien (2004). *The Keystone Advantage: What the New Dynamics of Business Ecosystems Mean for Strategy, Innovation, and Sustainability* (Boston: Harvard Business School Press).

IBM Systems Journal (2008). Service Science, Management, and Engineering, 47 (1).

Infosys Technologies, Ltd. (2003). "Infosys Technologies Announces Results for the Quarter Ended December 31, 2002," US GAAP Press Release, Jan. 10, http://infosys.com/investor/reports/quarterly/2002-2003/Q3/US_gaap_pr.pdf.

Infosys Technologies, Ltd. (2008). *2008 Annual Report*, 52.

Jacobides, M. G., and S. Billinger (2006). "Designing the Boundaries of the Firm: From 'Make, Buy, or Ally' to the Dynamic Benefits of Vertical Architecture," *Organization Science*, 17 (2): 249–61.

Jaikumar, R. (1986). "Postindustrial Manufacturing," *Harvard Business Review*, 64 (6): 69–76.

Johnson, C. (1982). *MITI and the Japanese Miracle: The Growth of Industrial Policy, 1925–1975* (Palo Alto, CA: Stanford University Press).

Judd, R. C. (1964). "The Case for Redefining Services," *Journal of Marketing*, 28 (1): 58–9.

Kane, S., Liberman, E., DiVesti, T., and Click, F. (2010). *Toyota Sudden Unintended Acceleration* (Rehoboth, MA: Safety Research & Strategies, Inc., Feb. 4); www.safetyresearch.net (accessed Feb. 6, 2010).

Kane, Y. (2009). "Breaking Apple's Grip on the iPhone: Firms Launch Sites Selling Unauthorized Software for Device, Posing Challenge to Official Online Store," *Wall Street Journal*, Mar. 6, p. B1.

Karki, R. (2008). *Competing with the Best: Strategic Management of Indian Companies in a Globalizing Arena* (New Delhi: Penguin Books India).

Katz, M., and C. Shapiro (1992). "Product Introduction with Network Externalities," *Journal of Industrial Economics*, 40 (1): 55–83.

Kay, J. (1993). *Foundations of Corporate Success* (Oxford: Oxford University Press).

Kekre, S., and K. Srinivasan (1990). "Broader Product Line: A Necessity to Achieve Success?" *Management Science*, 36 (10): 1216–31.

Kelley, T. (2004). *The Art of Innovation: Lessons in Creativity from IDEO* (New York: Broadway Business).

Kim, K. H. (1983). "A Look at Japan's Development of Software Engineering Technology," *IEEE Computer*, 16 (5): 26–37.

Kirzner, I. M. (1937). *Competition and Entrepreneurship* (Chicago: University of Chicago Press).

Kirzner, I. M. (1985). *Perception, Opportunity and Profit Studies in the Theory of Entrepreneurship* (Chicago: University of Chicago Press).

Kishida, K., M. Toramoto, K. Torii, and Y. Urano (1987). "Quality Assurance in Japan," *IEEE Software*, 4 (5): 11–18.

Klepper, S. (1996). "Entry, Exit, Growth, and Innovation over the Product Lifecycle," *American Economic Review*, 86: 562–83.

Klepper, S. (1997). "Industry Lifecycles," *Industrial and Corporate Change*, 6: 145–81.

Knecht, T., R. Leszinski, and F. A. Weber (1993). "Memo to a CEO: Making Profits after the Sale," *McKinsey Quarterly*, Nov. (4): 79–86.

Kotter, J. P. (1996). *Leading Change* (Boston: Harvard Business School Press).

Krafcik, J. F. (1988a). "Comparative Analysis of Performance Indicators at World Auto Assembly Plants," unpublished Master of Science thesis, MIT Sloan School of Management.

Krafcik, J. F. (1988b). "Triumph of the Lean Production System," *MIT Sloan Management Review*, 30 (1): 41–52.

Krishnan, V., and K. T. Ulrich (2001). "Product Development Decisions: A Review of the Literature," *Management Science*, 47 (1): 1–21.

Lall, S. (1987). *Learning to Industrialize: The Acquisition of Technological Capability by India* (London: Macmillan).

Langlois, R. (1992). "External Economies and Economic Progress: The Case of the Microcomputer Industry," *Business History Review*, 66 (1): 1–50.

Larus, J., T. Ball, M. Das, R. DeLine, M. Fahndrich, J. Pincus, S. Rajamani, and R. Venkatapathy (2004). "Righting Software," *IEEE Software*, 21 (3): 92–100.

Lawrence, P. R., and J. W. Lorsch (1967). *Organization and Environment: Managing Differentiation and Integration* (Boston: Harvard Business School Press).

Lee, L. C. (1989). "A Comparative Study of the Push and Pull Production Systems," *International Journal of Operations and Production Management*, 9 (4): 5–18.

Leggette, J. A., and B. Killingsworth (1983). "An Empirical Study of Economies of Scope: The Case of Air Carriers," *Studies in Economics and Finance*, 7 (2): 27–33.

Leonard-Barton, D. (1992). "Core Capabilities and Core Rigidities: A Paradox in Managing New Product Development," *Strategic Management Journal*, 13 (1): 111–25.

Levitt, T. (1972). "Production-Line Approach to Service," *Harvard Business Review*, 50 (5): 41–52.

Levitt, T. (1976). "The Industrialization of Service," *Harvard Business Review*, 54 (5): 63–74.

Liker, J. (2003). *The Toyota Way: 14 Management Principles from the World's Greatest Manufacturer* (New York: McGraw-Hill).

Liker, J., and M. Hoseus (2007). *Toyota Culture: The Heart and Soul of the Toyota Way* (New York: McGraw-Hill).

Lohr, S. (2002). "He Loves to Win: At I.B.M, He Did," *New York Times*, Mar. 10, Sec. 3, pp. 1, 11.

Lohr, S. (2009). "Microsoft and Yahoo are Linked up: Now what?" *New York Times*, July 29, p. B1.

MacCormack, A., R. Verganti, and M. Iansiti (2001). "Developing Products on Internet Time: The Anatomy of a Flexible Development Process," *Management Science*, 47 (1): 133–50.

MacCormack, A., C. F. Kemere, M. A. Cusumano, and W. Crandall (2003). "Trade-offs between Productivity and Quality in Selecting Software Development Practices," *IEEE Software*, 20 (5): 78–85.

MacDuffie, J. P. (1995). "Human Resource Bundles and Manufacturing Performance: Organizational Logic and Flexible Production Systems in the World Auto Industry," *Industrial and Labor Relations Review*, 48 (2): 197–221.

MacDuffie, J. P., K. Sethuraman, and M. L. Fisher (1996). "Product Variety and Manufacturing Performance: Evidence from the International Automotive Assembly Plant Study," *Management Science*, 42(3): 350–69.

McGahan, A. M., and M. E. Porter (1997). "How Much Does Industry Matter, Really?" *Strategic Management Journal*, 18 (Summer Special Issue): 15–30.

McIlroy, M. D. (1969). "Mass Produced Software Components," in Peter Naur and Brian Randell (eds.), *Software Engineering: Report on a Conference Sponsored by the NATO Science Committee* (Brussels: Scientific Affairs Division, NATO, January), 151–5.

Malone, T. (2004). *The Future of Work* (Boston: Harvard Business School Press).

Manes, S., and P. Andrews (1993). *Gates: How Microsoft's Mogul Reinvented an Industry—and Made himself the Richest Man in America* (New York: Doubleday).

Mansharamani, V. (2007). "Scale and Differentiation in Services: Using IT to Manage Customer Experiences at Harrah's Entertainment and Other Companies," unpublished Ph.D. dissertation, MIT Sloan School of Management, Feb.

Mason, C. (1989). "Zero-Defect Code," Microsoft memo, June 20.

Matsumoto, Y. (1984). "Management of Industrial Software Production," *IEEE Computer*, 17 (2): 59–72.

Matsumoto, Y. (1987). "A Software Factory: An Overall Approach to Software Production," in Peter Freeman (ed.), *Software Reusability* (Washington: IEEE Computer Society).

Maynard, M. (2010). "With Eye on its Reputation, Toyota Issues Repair for Pedal," *New York Times*, Feb. 2, p. A1

Meyer, M. H., and D. Dalal (2002). "Managing Platform Architectures and Manufacturing Processes for Nonassembled Products," *Journal of Production Innovation Management*, 19 (4): 277–93.

Meyer, M. H., and A. DeTore (2001). "Perspective: Creating a Platform-Based Approach for Developing New Services," *Journal of Production Innovation Management*, 18 (3): 188–204.

Meyer, M. H., and A. P. Lehnerd (1997). *The Power of Product Platforms* (New York: Free Press).

Meyer, M. H., and J. M. Utterback (1993). "The Product Family and the Dynamics of Core Capability," *MIT Sloan Management Review*, 34 (3): 29–47.

Microsoft Corporation. *Form 10-K*, annual.

Microsoft Corporation (2009). "Microsoft Promotes Steven Sinofsky to President, Windows Division," Press Release, July 8.

Mintzberg, H. (1979). *The Structuring of Organizations* (Englewood Cliffs, NJ: Prentice-Hall).

Mintzberg, H. (1987). "Crafting Strategy," *Harvard Business Review*, 65 (4): 66–74.

Mintzberg, H. (1994). *The Rise and Fall of Strategic Planning* (New York and London: Prentice-Hall).

MIT News (2008). "3 Questions: Michael Cusumano on Letting US Automakers Fail," Nov. 21 (http://web.mit.edu/newsoffice/2008/3q-cusumano-1121.html).

Mizuno, Y. (1983). "Software Quality Improvement," *IEEE Computer*, 16 (3): 66–72.

Monden, Y. (1981a). "What Makes the Toyota Production System Really Tick?" *Industrial Engineering*, 13 (1): 36–46.

Monden, Y. (1981b). "Adaptable Kanban System Helps Toyota Maintain Just-in-Time Production," *Industrial Engineering*, 13 (5): 29–46.

Monden, Y. (1982). *The Toyota Production System* (Portland, OR: Productivity Press).

Monteverde, K., and D. J. Teece (1982). "Supplier Switching Costs and Vertical Integration in the Automobile Industry," *Bell Journal of Economics*, 13 (1): 206–12.

Morgenson, G. (2004). "Market Place: IBM Shrugs Off Loss of a Contract it Once Flaunted," *New York Times*, Sept. 16, p. C1.

Mossberg, W. S. (2009). "A Windows to Help You Forget: Microsoft's New Operating System is Good Enough to Erase Bad Memory of Vista," *Wall Street Journal*, Oct. 8 (www.wsj.com).

Nader, R. (1965). *Unsafe at Any Speed: The Designed-in Dangers of the American Automobile* (New York: Grossman Publishers).

Nalebuff, B. (2004). "Bundling as an Entry Deterrent," *Quarterly Journal of Economics*, 119 (1): 159–87.

Nambisan, S. (2001). "Why Services Businesses are not Product Businesses," *MIT Sloan Management Review*, 42 (4): 72–80.

Needleman, R. (2009). "Microsoft Bing: Much Better than Expected," *CNET News*, May 28 (http://news.cnet.com/8301-17939_109-10251432-2.html).

Neely, A. (2009). "Exploring the Financial Consequences of the Servitization of Manufacturing," *Operations Management Research*, 2 (1).

Nishiguchi, T., and A. Beaudet (1997). "The Toyota Group and the Aisin Fire," *MIT Sloan Management Review*, 40 (1): 49–59.

Nobeoka, K. (1993). "Multi-Project Management: Strategy and Organization in Automobile Product Development," unpublished Ph.D. dissertation, MIT Sloan School of Management.

Nobeoka, K. (1996). *Maruchi-purojiekuto senryaku: posuto-riin no seihin kaihatsu manejimento* [Multi-Project Strategies: Post-Lean Product Development Management] (Tokyo: Yuhikaku).

Nobeoka, K., and M. A. Cusumano (1995). "Multi-Project Strategy, Design Transfer, and Project Performance: A Survey of Automobile Development Projects in the US and Japan," *IEEE Transactions on Engineering Management*, 42 (4): 397–409.

Nobeoka, K., and M. A. Cusumano (1997). "Multi-Project Strategy and Sales Growth: The Benefits of Rapid Design Transfer in New Product Development," *Strategic Management Journal*, 18 (3): 169–86.

Nussbaum, B. (2004). "The Power of Design," *Business Week*, May 17, pp. 86–94.

O'Brien, K. (2009). "Europe Drops Microsoft Antitrust Case," *New York Times*, Dec. 17, p. B2.

Ohno, T. (1978). *Toyota seisan hoshiki* [The Toyota Production System] (Tokyo: Daiyamondo).

Ohno, T. (1988). *The Toyota Production System: Beyond Large-Scale Production* (Portland, OR: Productivity Press).

Oliva, R., and R. Kallenberg (2003). "Managing the Transition from Products to Services," *International Journal of Service Industry Management*, 14 (2): 160–72.

Oracle Corporation. *Form 10-K*, annual.

O'Reilly III, C. A., and M. L. Tushman (2004). "The Ambidextrous Organization," *Harvard Business Review*, 82 (4): 74–81.

Osono, E., N. Shimizu, and H. Takeuchi (2008). *Extreme Toyota: Radical Contradictions that Drive Success at the World's Best Manufacturer* (New York: John Wiley & Sons).

Oster, S. (1999). *Modern Competitive Analysis* (3rd edn.; New York and Oxford: Oxford University Press).

Pagell, M., and D. R. Krause (2004). "Re-Exploring the Relationship between Flexibility and the External Environment," *Journal of Operations Management*, 21 (6): 629–49.

Panzar, J. C., and R. D. Willig (1981). "Economies of Scope," *American Economic Review*, 71 (2): 268–72.

Parker, G., and M. W. Van Alstyne (2005). "Two-Sided Network Effects: A Theory of Information Product Design," *Management Science*, 51 (10): 1494–1504

Pascale, R. T. (1984). "Perspectives on Strategy: The Real Story behind Honda's Success," *California Management Review*, 26 (3): 47–72.

Pascale, R. T., and A. G. Athos (1982). *The Art of Japanese Management: Applications for American Executives* (New York: Warner Books).

Patni Computer Systems, Ltd. (2004). "Patni Assessed at SEI-CMMI Level 5," company press release, Mar. 11.

Peers, M. (2009). "Microsoft's Dropped Call," *Wall Street Journal*, Dec. 30, p. 32.

Penrose, E. T. (1959). *The Theory of the Growth of the Firm* (New York: John Wiley; republished by Oxford University Press, 1995).

Perez, J. (2007). "Update: Facebook Picks Microsoft over Google for minority stake," *Computerworld*, Oct. 24, (www.computerworld.com).

Perrow, C. (1967). "A Framework for the Comparative Analysis of Organizations," *American Sociology Review*, 32 (2): 194–208.

Peteraf, M. A. (1993). "The Cornerstones of Competitive Advantage: A Resource-Based View," *Strategic Management Journal*, 14 (3): 179–91.

Peters, T. J., and R. H. Waterman (1982). *In Search of Excellence: Lessons from America's Best Run Companies* (New York: HarperCollins).

Pine, J. (1992). *Mass Customization: The New Frontier in Business Competition* (Boston: Harvard Business School Press).

Pogue, D. (2009). "Windows 7 Keeps the Good and Tries to Fix Flaws," *New York Times*, Oct. 21 (www.nytimes.com).

Porter, M. E. (1980). *Competitive Strategy: Techniques for Analyzing Industries and Competitors* (New York: Free Press).

Porter, M. E. (1985). *Competitive Advantage: Creating and Sustaining Superior Performance* (New York: Free Press).

Porter, M. E. (1987). "From Competitive Advantage to Corporate Strategy," *Harvard Business Review*, 65 (3): 43–59.

Porter, M. E. (1996). "What is Strategy?" *Harvard Business Review*, 74 (6): 61–80.

Porter, M. E., H. Takeuchi, and M. Sakakibara (2000). *Can Japan Compete?* (Cambridge, MA: Perseus).

Potts, G. W. (1988). "Exploiting your Product's Service Lifecycle," *Harvard Business Review*, 68 (5): 58–67.

Pralahad, C. K., and G. Hamel (1990). "The Core Competence of the Corporation," *Harvard Business Review*, 68 (3): 79–91.

Quinn, J. B. (1978). "Strategic Change: Logical Incrementalism," *MIT Sloan Management Review*, 20 (1): 7–19.

Quinn, J. B. (1992). *Intelligent Enterprise: A Knowledge and Service Based Paradigm for Industry* (New York: Free Press).

Ramanujam, V., and P. Varadarjan (1989). "Research on Corporate Diversification: A Synthesis," *Strategic Management Journal*, 10 (6): 523–51.

Rathmell, J. M. (1966). "What Is Meant by Services?" *Journal of Marketing*, 30 (4): 32.

Ravenscroft, D. J., and F. M. Sherer (1987). *Mergers, Sell-offs, and Economic Efficiency* (Washington: Brookings).

Roberts, E. B. (1991a). *Entrepreneurs in High Technology: Lessons from MIT and Beyond* (New York: Oxford University Press).

Roberts, E. B. (1991b). "Technological Entrepreneurship: Birth, Growth, and Success," *MIT Management* (Winter), 21–5.

Rochet, J. C., and J. Tirole (2003). "Platform Competition in Two-Sided Markets," *Journal of the European Economic Association*, 1 (4): 990–1029.

Rochet, J. C., and J. Tirole (2006). "Two-Sided Markets: A Progress Report," *RAND Journal of Economics*, 37 (3): 645–67.

Rosenberg, N., and C. Frischtak (1985) (eds.). *International Technology Transfer: Concepts, Measures, and Comparisons* (New York: Praeger).

Rosenbloom, R. S., and M. A. Cusumano (1987). "Technological Pioneering and Competitive Advantage: The Birth of the VCR Industry," *California Management Review*, 29 (4): 51–76.

Rumelt, R. (1974). *Strategy, Structure, and Economic Performance* (Boston: Harvard Business School Press).

Rumelt, R. P. (1991). "How Much Does Industry Matter?" *Strategic Management Journal*, 12 (3): 167–85.

Saghbini, J. J. (2005). "Standards in the Data Storage Industry: Emergence, Sustainability, and the Battle for Platform Leadership," unpublished M.Sc. thesis, System Design and Management Program, Massachusetts Institute of Technology.

Sanchez, R. (1993). "Strategic Flexibility, Firm Organization, and Managerial Work in Dynamic Markets: A Strategic Options Perspective," in P. Shrivastava, A. Huff, and J. Dutton (eds.), *Advances in Strategic Management, Vol. 9* (Greenwich, CT: JAI Press), 251–91.

Sanchez, R. (1995). "Strategic Flexibility in Product Competition," *Strategic Management Journal*, 16 (Special Summer Issue): 63–76.

Sanchez, R., and J. T. Mahoney (1996). "Modularity, Flexibility, and Knowledge Management in Product Organization Design," *Strategic Management Journal*, 17 (Special Winter Issue): 63–76.

Sanderson, S. W., and M. Uzumeri (1996). *Managing Product Families* (New York: Irwin).

Scanlon, J. (2009). "Health Robocalls People Want to Receive," *Business Week*, July 13 (www.businessweek.com/innovate/content/jul2009/id20090713_231581.htm?chan=innovation_innovation+%2B+design_top+stories).

Schmalensee, R. (1985). "Do Markets Differ Much?" *American Economic Review*, 75 (3): 341–51.

Schmalensee, R., D. Evans, and A. Hagiu (2006). *Invisible Engines: How Software Platforms Drive Innovation and Transform Industries* (Cambridge, MA: MIT Press).

Schmenner, R. (1990). *Production/Operations Management: Concepts and Situations* (New York: Macmillan).

Schonberger, R. J. (1982). *Japanese Manufacturing Techniques: Nine Lessons in Simplicity* (New York: Free Press/MacMillan).

Schroeder, S. (2009). "Google Voice Thoroughly Banned from the iPhone; So Much for an Open Platform," Mashable: The Social Media Guide, July 28 (http://mashable.com/2009/07/28/google-voice-banned (accessed Sept. 20, 2009)).

Schumpeter, J. A. (1934). *The Theory of Economic Development* (Cambridge, MA: Harvard University Press; repr. Oxford University Press, 1961, and first published in German, 1912).

Senge, P. M. (1990). *The Fifth Discipline: The Art and Practice of the Learning Organization* (New York: Doubleday).

Sethi, A. K., and P. S. Sethi (1990). "Flexibility in Manufacturing: A Survey," *International Journal of Flexible Manufacturing Systems*, 2: 289–328.

Shankland, S. (2006). "IBM to Give Birth to 'Second Life' Business Group," CNET news.com, Dec. 12 (http://news.com.com/2100-1014_3-6143175.html).

Shapiro, C., and H. Varian (1998). *Information Rules: A Strategic Guide to the Network Economy* (Boston: Harvard Business School Press).

Sharma, D. (2009). *The Long Revolution: The Birth and Growth of India's IT Industry* (New Delhi: HarperCollins Publishers India).

Sheff, D. (1993). *Game Over: How Nintendo Zapped an American Industry, Captured your Dollars, and Enslaved your Children* (New York: Random House).

Shimokawa, K., and T. Fujimoto (2009) (eds.). *The Birth of Lean: Conversations with Taiichi Ohno, Eiji Toyoda, and Other Figures who Shaped Toyota Management* (Cambridge, MA: Lean Enterprise Institute).

Singel, R. (2009). "Hands on with Microsoft's New Search Engine: Bing, but no Boom," *Wired*, May 28 (www.wired.com/epicenter/2009/05/microsofts-bing-hides-its-best-features).

Skinner, W. (1974). "The Focused Factory," *Harvard Business Review*, 52 (3): 113–21.

Slack, N. (2005). "Patterns of Servitization: Beyond Products and Service," unpublished working paper, University of Cambridge, Institute for Manufacturing.

Slack, N., and H. Correa (1992). "The Flexibilities of Push and Pull," *International Journal of Operations and Production Management*, 12 (4): 82–92.

Smith, S. A. (1993). "Software Development in Established and New Entrant Companies: Case Studies of Leading Software Producers," unpublished Master's thesis, Management of Technology Program, Massachusetts Institute of Technology.

Sobek, D., J. Liker, and A. Ward (1998). "Another Look at How Toyota Integrates Product Development," *Harvard Business Review*, 76 (4): 36–50.

Spear, S. (2008). *Chasing the Rabbit: How Market Leaders Outdistance the Competition and How Great Companies Can Catch Up and Win* (New York: McGraw Hill).

Spear, S., and H. K. Bowen (1999). "Decoding the DNA of the Toyota Production System," *Harvard Business Review*, 77 (5): 97–106.

Staudenmayer, N. (1997). "Managing Interdependencies in Large-Scale Software Development," unpublished Ph.D. thesis, MIT Sloan School of Management.

Staudenmayer, N., and M. A. Cusumano (1998). "Alternative Designs for Product Component Integration," MIT Sloan Working Paper, #4016 (Apr.).

Stigler, G. J. (1951). "The Division of Labor is Limited by the Extent of the Market," *Journal of Political Economy*, 59 (3): 185–93.

Stokes, D. (1997). *Pasteur's Quadrant: Basic Science and Technological Innovation* (Washington: Brookings Institution).

Stone, B., and C. C. Miller (2009). "Music Forecast: The Cloud," *New York Times*, Dec. 16, p. B1.

Suarez, F. (1992). "Strategy and Manufacturing Flexibility: A Case Study on the Assembly of Printed Circuit Boards," unpublished Ph.D. dissertation, MIT Sloan School of Management.

Suarez, F. F., and M. A. Cusumano (2009). "The Role of Services in Platform Markets," in A. Gawer (ed.), *Platforms, Markets and Innovation* (Cheltenham and Northampton, MA: Edward Elgar).

Suarez, F. F., M. A. Cusumano, and C. H. Fine (1995). "An Empirical Study of Flexibility in Manufacturing," *MIT Sloan Management Review*, 37 (1): 25–32.

Suarez, F. F., M. A. Cusumano, and C. H. Fine (1996). "An Empirical Study of Manufacturing Flexibility in Printed-Circuit Board Assembly," *Operations Research*, 44 (1): 223–40.

Suarez, F. F., M. A. Cusumano, and S. Kahl (2008). "Services and the Business Models of Product Firms: An Empirical Analysis of the Software Products Business," unpublished working paper, Center for Digital Business, MIT Sloan School of Management.

Suarez, F. F., and J. M. Utterback (1995). "Dominant Designs and the Survival of Firms," *Strategic Management Journal*, 16 (6): 415–30.

Sugimori, Y. K., K. Kusunoki, F. Cho, and S. Uchikawa (1977). "Toyota Production System and Kanban System: Materialization of Just-in-Time and Respect-for-Human System," *International Journal of Production Research*, 15 (6): 553–64.

Sullivan, D. (2009). "Search Market Share 2008: Google Grew, Yahoo and Microsoft Dropped and Stabilized," *Search Engine Land*, Jan. 26 (http://searchengineland.com/search-market-share-2008-google-grew-yahoo-microsoft-dropped-stabilized-16310).

Tajima, D., and T. Matsubara (1984). "Inside the Japanese Software Industry," *IEEE Computer*, 17 (3): 34–43.

Taleb, N. N. (2007). *The Black Swan: The Impact of the Highly Improbable* (New York: Random House).

Teece, D. (1980). "Economies of Scope and the Scope of the Enterprise," *Journal of Economic Behavior & Organization*, 1 (3): 223–47.

Teece, D. (1986). "Profiting from Technological Innovation: Implications for Integration, Collaboration, Licensing and Public Policy," *Research Policy*, 15 (6): 285–305.

Teece, D. (2007). "Explicating Dynamic Capabilities: The Nature and Microfoundations of (Sustainable) Enterprise Performance," *Strategic Management Journal*, 28 (13): 1319–50.

Teece, D. (2009). *Dynamic Capabilities and Strategic Management* (Oxford and New York: Oxford University Press).

Teece, D., G. Pisano, and A. Shuen (1997). "Dynamic Capabilities and Strategic Management," *Strategic Management Journal*, 18 (7): 509–33.

Thomke, S. (2003a). "R&D Comes to Services: Bank of America's Pathbreaking Experiments," *Harvard Business Review*, 81 (4): 77–9.

Thomke, S. (2003b). *Experimentation Matters: Unlocking the Potential of New Technologies for Innovation* (Boston: Harvard Business School Press).

Thurrott, P. (2001). "Internet Home Alliance Integrates Its Way into US Homes," *Connected Home Magazine*, Nov. 7 (www.connectedhomemag.com/Articles/Print.cfm?ArticleID=23133&Path=Networking).

Tushman, M. L., and C. A. O'Reilly III (1996). "Ambidextrous Organizations: Managing Evolutionary and Revolutionary Change," *California Management Review*, 38 (4): 8–30.

Ulrich, K. (1995). "The Role of Product Architecture in the Manufacturing Firm," *Research Policy*, 24 (3): 419–40.

Ulrich, K. T., and S. D. Eppinger (2004). *Product Design and Development* (New York: McGraw-Hill).

Upton, D. M. (1994). "The Management of Manufacturing Flexibility," *California Management Review*, 36 (2): 88–104.

Utterback, J. M. (1994). *Mastering the Dynamics of Innovation* (Boston: Harvard Business School Press).

Utterback, J. M., and W. J. Abernathy (1975). "A Dynamic Model of Process and Product Innovation," *Omega*, 3 (6): 639–56.

Vascellaro, J., and Y. Kane (2009). "Schmidt Resigns his Seat on Apple's Board," *Wall Street Journal*, Aug.

Vascellaro, J. E. and S. Morrison (2009). "Google Outlines Chrome OS Plans," *Wall Street Journal*, Nov. 20 (http://online.wsj.com/article/SB10001424052748704204304574545874128475870.html).

Vascellaro, J. E., and N. Sheth (2009). "Google Set to Market Own Phone Next Year," *Wall Street Journal*, Dec. 13 (http://online.wsj.com).

Vogel, E. F. (1979). *Japan as Number One: Lessons for America* (Cambridge, MA: Harvard University Press).

Volberda, H. W. (1996). "Toward the Flexible Firm: How to Remain Vital in Hypercompetitive Environments," *Organization Science*, 7 (4): 359–74.

Volberda, H. W. (1999). *Building the Flexible Firm: How to Remain Competitive* (New York: Oxford University Press).

von Hippel, E. (2005). *Democratizing Innovation* (Cambridge, MA: MIT Press).

Walters, C. (2008). "Toyota Announces Tacoma Buyback Program for Severe Rust Corrosion," *Consumerist*, Apr. 5 (http://consumerist.com/379734/toyota-announces-tacoma-buyback-program-for-severe-rust-corrosion).

Walton, M. (1999). "Strategies for Lean Product Development," Lean Aerospace Initiative, Center for Technology, Policy, and Industrial Development, Massachusetts Institute of Technology, unpublished working paper, #WP99-01-911, Aug.

Weisman, R. (2004). "At the Center of a Cultural Shift," *Boston Globe*, May 25 (www.boston.com/business/articles/2004/05/25/at_the_center_of_a_culture_shift).

Wernerfeldt, B. (1984). "A Resource-Based View of the Firm," *Strategic Management Journal*, 5 (2): 171–80.

Wheelright, S. C., and K. B. Clark (1992). *Revolutionizing Product Development* (New York: Free Press).

Wikipedia (2010). "Firestone and Ford Tire controversy," Wikipedia.com (accessed Feb. 6, 2010).

Williamson, O. E. (1975). *Markets and Hierarchies: Analysis and Antitrust Implications* (New York: Free Press).

Williamson, O. E. (1985). *The Economic Institutions of Capitalism* (New York: Free Press).

Wise, R., and P. Baumgartner (1999). "Go Downstream: The New Imperative in Manufacturing," *Harvard Business Review*, 77 (5): 133–41.

Womack, J. P., D. T. Jones, and D. Roos (1990). *The Machine that Changed the World* (New York: Rawson Associates).

Woodward, J. (1965). *Industrial Organization: Theory and Practice* (Oxford: Oxford University Press).

Wortham, J. (2009). "Unofficial Software Incurs Apple's Wrath," *New York Times*, May 13, p. B1.

Yin, R. K. (2002). *Case Study Research: Design and Methods* (Thousand Oaks, CA: Sage Publications).

Yoffie, D. B., and M. A. Cusumano (1999). "Judo Strategy: The Competitive Dynamics of Internet Time," *Harvard Business Review*, 77 (1): 70–82.

Yoffie, D. B., and M. Kwak (2001). *Judo Strategy: Turning Your Competitors' Strength to Your Advantage* (Boston: Harvard Business School Press).

Yoffie, D. B., and M. Kwak (2006). "With Friends Like These: The Art of Managing Complementors," *Harvard Business Review*, 84 (9): 89–98.

Yoffie, D. B., and M. Slind (2008). "Apple, Inc., 2008," Harvard Business School, Case #9-708-480, p. 6.

Yoffie, D. B., P. Yin, and L. Kind (2005). "Qualcomm, Inc. 2004," Harvard Business School, Case #N1-705-401.

Zachary, G. P. (1994). *Show-Stopper* (New York: Free Press).

Zahra, S. A., and G. George (2002). "Absorptive Capacity: A Review, Reconceptualization, and Extension," *Academy of Management Review*, 27 (2): 185–203.

INDEX

Made in the USA
Lexington, KY
06 February 2013